SpringerWienNewYork

CISM COURSES AND LECTURES

The series presents lecture notes, monographs, edited works and proceedings in the field of Mechanics, Engineering, Computer Science and Applied Mathematics.
Purpose of the series is to make known in the international scientific and technical community results obtained in some of the activities organized by CISM, the International Centre for Mechanical Sciences.

INTERNATIONAL CENTRE FOR MECHANICAL SCIENCES

COURSES AND LECTURES - No. 502

MODERN TESTING TECHNIQUES FOR STRUCTURAL SYSTEMS
DYNAMICS AND CONTROL

EDITED BY

ORESTE S. BURSI
UNIVERSITY OF TRENTO, ITALY

DAVID WAGG
UNIVERSITY OF BRISTOL, UK

SpringerWienNewYork

This volume contains 127 illustrations

Printed in Italy
SPIN 12325672

All contributions have been typeset by the authors.

ISBN 978-3-211-09444-0 SpringerWienNewYork

PREFACE

Experimental testing techniques for structural systems have been undergoing a significant evolution over the last ten years or so. The advent of powerful and easily available computing power has led to rapid development of new hybrid experimental-computational techniques. This in turn has led to a new level of understanding of more established techniques, and spawned new areas of research in related topics such as numerical integration methods and real-time control.

This comes at a time when the complexity of machines, structures and systems are escalating considerably leading to higher demands on the analysis, design and testing techniques. In such an atmosphere there is increasing pressure on the design, development and deployment process of engineering structures to be as quick as possible with first time quality and high profit margin. Although significant savings can be achieved by using increasingly powerful and scalable numerical simulations, experimental testing cannot be eliminated entirely. Instead, modern testing techniques need to be more targeted, faster, and more closely integrated with numerical techniques.

The principal aim of this book is to describe the new developments in the area of structural testing, particularly those based upon the principle of fusing numerical and experimental methods such as real-time dynamic substructuring and hardware-in-the-loop testing. These methods have significant potential for component testing which will enable engineering designers to significantly reduce design uncertainty and lead times. In addition to the hybrid methods, chapters on the latest developments in more established techniques, such as shaking table testing, give a completely up-to-date survey of structural testing methods. The chapter contents reflect the multidisciplinary nature of the work by integrating cutting-edge research from the fields of non-linear dynamics, automatic control, numerical analysis, system modelling and mechatronics. The result is a detailed treatment of state-of-the-art techniques and tools required for experimental testing of structures.

In Chapter 1 by Bursi the concepts upon which ad hoc *numerical methods for the prediction of the transient dynamic response of monolithic structural systems as well as of heterogeneous (numerical/physical) subsystems decomposed by computational or physical considera-*

tions are presented in a common framework. In Chapter 2 by Bernal the formulation of state-space reduced order models from the analysis of input-output data is presented. In Chapter 3 by Iervolino and Manfredi the current best practice in record selection is reviewed for the case of probabilistic seismic risk analysis and for code-based seismic assessment and design with special attention to the prescriptions of Eurocode 8 for both buildings and bridges. Chapter 4 by Bairrão describes the state-of-the-art laboratory methodologies for seismic engineering experimentation, with a more clear focus on the shaking table tests.

In Chapter 5 by Pegon the difference between conventional and continuous pseudo-dynamic testing and its impact on substructuring is discussed. Issues, algorithms, implementation & results relating to continuous pseudo-dynamic testing are covered in detail . Chapter 6 by Shing presents an overview of real-time hybrid test methods from a general perspective, and discusses schemes of structural partitioning and issues related to numerical computation, system dynamics, and system performance. Finally Chapter 7 by Wagg, Neild and Gawthrop deals with control and implementation issues for numerical-experimental substructuring in real time.

This volume is a direct result of a course on "Modern Testing Techniques for Structural Systems" held at the International Centre for Mechanical Sciences (CISM) Udine, Italy 18-22 June 2007. We would like to thank all those at CISM who helped to make the course and the production of this volume such a enjoyable experience, in particular Mrs P. Agnola who dealt with the administration of the meeting. In addition we would like to thank the Rectors of CISM: G. Maier, J. Salençon, W. Schneider and the Secretary General, B.A. Schrefler. We would also like to the MTS Systems Corporation, particularly Scott Jiran, Carlo Maria Ornati and Brad Thoen, for their input to the industrial aspects of the meeting. Finally, we would like to thank Prof. Serafini, Simon Neild, Chuanguo Jia, and Pierre Pegon for their valuable assistance with the preparation of this book.

Oreste S. Bursi (University of Trento, Trento, Italy)
David J. Wagg (University of Bristol, Bristol, UK)
March 2008

CONTENTS

Computational Techniques for Simulation of Monolithic and Heterogeneous Structural Dynamic Systems

Oreste S. Bursi *

* Department of Mechanical and Structural Engineering, University of Trento, Trento, Italy

Abstract The prediction of the transient dynamic response of monolithic structural systems, as well as of heterogeneous (numerical/ physical) subsystems, decomposed by computational or physical considerations typical of hardware-in-the-loop and pseudo-dynamic tests using numerical integration, has become an accepted practice almost to the extent that such solutions in non-linear problems often are considered to be *exact solutions*. It is for this reason that this chapter is placed immediately at the beginning of the book.

In light of the large body of literature on computational methods developed for both testing and control techniques applied to linear and non-linear systems, no attempt is made to cover this subject in greater depth. Rather the concepts upon which *ad hoc* computational methods rely are presented in a common framework along with a few applications.

1 Introduction

Nowadays, there is a significant development of models and scalable computational techniques that enable very complex simulations in various engineering and scientific fields essentially in multi-physics, mechatronics and design and control analysis. Distinguishing features are the use of partitioned techniques (Felippa et al., 2001), modular and distributed simulations of multibody systems (Kübler and Schiehlen, 2004), real-time simulations (Cuadrado et al., 1997), time-parallel frameworks for parallel simulations (Farhat et al., 2006) and reduced models (Biondi and Muscolino, 2000). Several techniques rely on a top-bottom approach which begins with the full global equations of motion and then partitions them into substructures by taking their interactions into account. Other ones employ a bottom-up approach, that consists in approximating each substructure and then in assembling the approximated substructural equations in order to form the global equations of motion. These techniques do integrate cutting-edge research on the fields of numerical analysis, linear/non-linear dynamics, multibody dynamics, digital control, system modelling and also experimental techniques. In such a scenario, the role of the transient analysis of multi-physics dynamical systems is still at the forefront of computational methods for solving problems of non-linear dynamics. Moreover, because simulations, *i.e.* experimentations with models, are necessarily limited, techniques such as the Hardware-in-the-loop (*HiL*) and Pseudo-Dynamic (*PsD*) testing based on the principle of Dynamic Substructuring (*DS*) (Pinto et al., 2004; Misselhorn et al., 2006; Bursi et al., 2008) are becoming more and more important in testing, design and control. Thus, computational techniques for the simulation of monolithic and heterogeneous (numerical/physical) dynamic systems presented here reflected a shift in features from discipline-oriented dynamics to system-oriented dynamics, from sequential to parallel computations, from accuracy concerns to system-model refinements. In addition, *HiL* is actually employed to test more complex mechanical systems (Wagg et al., 2008) than just a piece of hardware, *e.g.* a controller; whilst *PsD* is used nowadays to perform tests on civil-engineering components in fast or real time (Pegon, 2008; Shing, 2008). Hence, these developments are necessarily not uniform and difficult to fit into established categories and such unified approaches, as for instance, in mechatronics (Samin et al., 2007).

Since this chapter is the first, in this section we present a brief state-of-the-art on real-time testing with *DS* in the context of *HiL* and *PsD* testing. So this section introduces and interleaves particular cases of the Generalized-α methods both in *HiL* and *PsD* testing. Up to now, these structural integrators represent the most common time integration methods used in finite element *FE* simulations and testing. After the introduction, different forms of equations of structural dynamics are presented in Section 2 with correction and compensation techniques specific to *PsD* testing. Thus, Section 1 and 2 present also the notation used in this chapter and the two sets of equations employed by different time integration methods, and linearly implicit Rosenbrock methods are interleaved. A major classification of time integration algorithms is then described in Section 3, followed by analysis techniques and metrics useful for structural dynamics problems. A subsection devoted to error propagation in time integration algorithms is also included, being particularly useful for experimental techniques. In Section 4 we introduce problems and analysis techniques that arise and are faced when coupling of numerical and physical substructures occurs. Because *HiL* and *PsD* techniques require transfer systems, i.e. actuators and controllers for their actuation, numerical techniques used to discretize and linearize adaptive controllers are introduced. Moreover, filter design tools like bilinear transformations, are employed to characterize time integrators. Successively, Section 5 and 6 present variants of general algorithms specialized in the domain of real time *HiL* and *PsD* testing. In detail, Section 5 reports time integration methods for the solution of coupled-field real-time problems in *HiL*, *i.e.* primarily methods for structure-structure and structure-control interaction problems. In fact the analysis and design of algorithms for the operation of structure-control systems, multibody and mechatronic systems operating in real time still remain one of the liveliest research activities in recent years (Arnold et al., 2007). Conversely, Section 6 presents innovative parallelized partitioned methods based on Lagrange multipliers able to guarantee fast pseudo-dynamic tests of complex substructures. Conclusions are summarized in Section 7 with future perspectives of numerical techniques for *HiL* and *PsD* testing with *DS*. This chapter is not meant to be an exhaustive state-of-the-art of the research, but is intended to assist researchers and practitioners in selecting time integration algorithms in a numerical/physical context.

1.1 Algorithms for *HiL* and Real-time Testing with *DS*

The aim of this section is to review some numerical methods that were used for Simulation of Heterogeneous Systems with Dynamic substructuring (*SHSDS*) and underline some problems derived from their application. Several properties of algorithms mentioned herein are discussed at length in Subsection 3.2.

SHSDS is a novel form of hybrid numerical-experimental testing, which can be used to test structural or mechanical systems under dynamic loading (Williams and Blakeborough, 2001). The technique involves splitting the system being tested into two parts: the Physical Subsystem (*PS*), that contains a key region of interest and is experimentally tested; the Numerical Subsystem (*NS*), that contains the remainder of the system and is numerically simulated as illustrated in Figure 1. By imposing compatibility and equilibrium conditions at the interface between the *NS* and the *PS*, respectively, the subsystems are made to interact in real time such that they emulate the dynamic behaviour of the full system. In detail, a *real-time simulation* of the emulated system requires a computational time per time interval, *i.e.* an integration time Δt plus the time for display, smaller than the physical time taken by the actual motion of the emulated system. In order to interconnect the *PS* to the *NS*, a transfer system that acts on the *PS* is typically controlled to follow *NS* interface kinematic quantities or other outputs. This transfer system normally consists of an electric or electro-hydraulic actuator and a controller as shown in Figure 1.

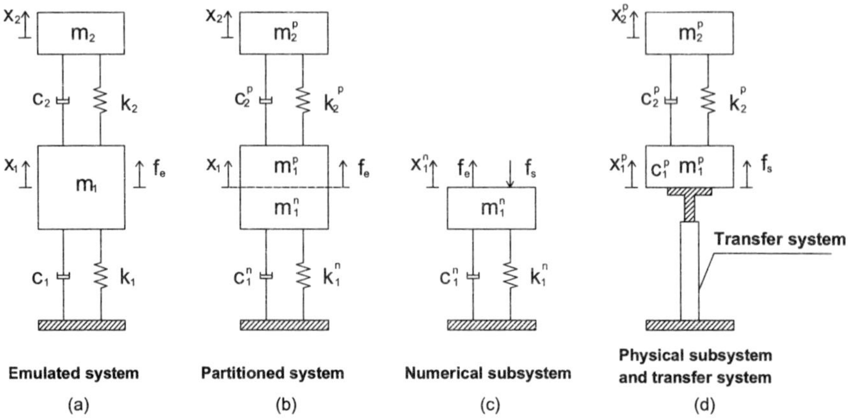

Figure 1. Schematic representation of a 2-DoF system with dynamic substructuring.

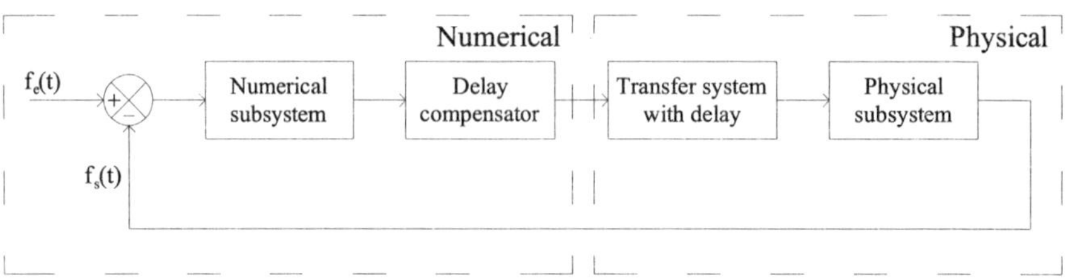

Figure 2. Block diagram representation of a heterogeneous simulation with dynamic substructuring.

The relevant block diagram representation is depicted in Figure 2 in which both the transfer system with delay and a delay compensator are shown. Note that coupling can be applied in the opposite sense by operating the transfer system in force control: however, operating a transfer system in force control is much more involved owing to the mechanical characteristics of the actuator itself, the load cell measurement errors as well as the relevant control (Reinhorn et al., 2004).

The technique exploited herein to partition a system is called differential partitioning (Felippa et al., 2001) and is based on the governing differential equations describing the *NS* and *PS* involved, respectively. Historically, differential partitioning was the original approach used in the derivation of staggered solution procedures for Fluid-Structure Interaction (*FSI*), and remains a natural way to handle physically diverse systems. Starting from the governing differential equations, the system of equations of motion of a discretised structure reads

$$\mathbf{M}\ddot{\mathbf{x}}(t) + \mathbf{C}\dot{\mathbf{x}}(t) + \mathbf{K}\mathbf{x}(t) = \mathbf{f}_{\mathrm{e}}(t), \tag{1.1}$$

where $\mathbf{M}$, $\mathbf{C}$ and $\mathbf{K}$ are the global mass, damping and stiffness matrices and $\mathbf{f}_{\mathrm{e}}(t)$ corresponds to the externally applied time-varying forces vector. $\mathbf{x}$, $\dot{\mathbf{x}}$ and $\ddot{\mathbf{x}}$ are respectively displacement, velocity and acceleration vectors of the global structure. By splitting the global matrices such that $\mathbf{M} = \mathbf{M}^{\mathrm{n}} + \mathbf{M}^{\mathrm{p}}$, $\mathbf{C} = \mathbf{C}^{\mathrm{n}} + \mathbf{C}^{\mathrm{p}}$ and $\mathbf{K} = \mathbf{K}^{\mathrm{n}} + \mathbf{K}^{\mathrm{p}}$, where the superscripts "n" and "p" refer to the *NS* and the *PS*, respectively, one gets

$$\mathbf{M}^{n}\ddot{\mathbf{x}}(t) + \mathbf{C}^{n}\dot{\mathbf{x}}(t) + \mathbf{K}^{\mathbf{n}}\mathbf{x}(t) = \mathbf{f}_{e}(t) + \mathbf{f}_{s}(t), \tag{1.2}$$

where the coupling force vector $\mathbf{f}_{\mathrm{s}}(t)$ reads:

$$\mathbf{f}_{s}(t) = -\mathbf{M}^{p}\ddot{\mathbf{x}}(t) - \mathbf{C}^{p}\dot{\mathbf{x}}(t) - \mathbf{K}^{\mathbf{p}}\mathbf{x}(t). \tag{1.3}$$

Thus, by means of differential partitioning, the global system can be uncoupled using a set of two differential equations, where (1.2) defines the *NS* and (1.3) describes the *PS*, respectively. Since the modes of the structure partition are not neccssarily modes of the coupled system, the modal decomposition does not hold and a typical model problem cannot be defined.

A standard approach for discrete time integration is to formulate *SHSDS* in a continuous-time environment, see for instance Wallace et al. (2004), and then use Runge-Kutta integrators, like the embedded Dormand-Prince method ODE45 (Dormand and Prince, 1980), which is implemented in Simulink (Mat, 2005). Typically, these algorithms are only conditionally stable and computationally expensive since they need many stages to achieve high order accuracy. For instance, the ODE45 uses 7 stages and exhibits a fifth-order accuracy entailing a high computational effort. Other numerical approaches based on *structural integrators* were also suggested by Darby et al. (2002). They pointed out that the Central Difference *(CD)* method (Hughes, 1987) introduces inaccuracies on both Single- and Multiple-Degrees-of-Freedom (MDoF) systems. Note that the *CD* algorithm is second-order accurate with no substructure coupling and does not provide any explicit target velocity (Wu et al., 2005).

Bonnet (2006) analysed different *structural integrators* for *SHSDS*. At the end, only three of them were considered suitable for most applications. The Newmark explicit scheme (Hughes, 1987), which is derived from the

Newmark scheme (Newmark, 1959) by setting $\beta = 0$ and $\gamma = 0.5$:

$$\begin{aligned} \mathbf{M}^n \mathbf{a}_{k+1} + \mathbf{C}^n \mathbf{v}_{k+1} + \mathbf{K}^\mathbf{n} \mathbf{d}_{k+1} &= \mathbf{f}_{e,k+1} + \mathbf{f}_{s,k+1} \\ \mathbf{d}_{k+1} &= \mathbf{d}_k + \Delta t \mathbf{v}_k + \tfrac{\Delta t^2}{2} \mathbf{a}_k \\ \mathbf{v}_{k+1} &= \mathbf{v}_k + \tfrac{\Delta t}{2} (\mathbf{a}_k + \mathbf{a}_{k+1}) \end{aligned} \tag{1.4}$$

In detail, the first difference equation is used to derive explicitly the displacement command signal to be imposed on the specimen. The force vector $\mathbf{f}_{s,k+1}$ is measured and the combination of the equation of motion and of the second difference equation is exploited to derive $\mathbf{a}_{k+1}$ and $\mathbf{v}_{k+1}$, necessary to carry on the scheme for $k+2$. In its standard form this scheme is second-order accurate and requires very little computation. When stability issues become important, the modified Newmark explicit algorithm proposed by Chang (2002) was shown to be very efficient. It reads,

$$\begin{aligned} \mathbf{M}^n \mathbf{a}_{k+1} + \mathbf{C}^n \mathbf{v}_{k+1} + \mathbf{K}^\mathbf{n} \mathbf{d}_{k+1} &= \mathbf{f}_{e,k+1} + \mathbf{f}_{s,k+1} \\ \mathbf{d}_{k+1} &= \mathbf{d}_k + \beta_1 \Delta t \mathbf{v}_k + \beta_2 \Delta t^2 \mathbf{a}_k \\ \mathbf{v}_{k+1} &= \mathbf{v}_k + \frac{\Delta t}{2} (\mathbf{a}_k + \mathbf{a}_{k+1}) \end{aligned} \tag{1.5}$$

where

$$\begin{aligned} \beta_1 &= [\mathbf{I} + \frac{\mathbf{\Delta t}}{\mathbf{2}} \mathbf{M}^{-1} \mathbf{C} + \frac{\mathbf{\Delta t^2}}{\mathbf{4}} \mathbf{M}^{-1} \mathbf{K}_0]^{-1} [\mathbf{I} + \frac{\mathbf{\Delta t}}{\mathbf{2}} \mathbf{M}^{-1} \mathbf{C}] \\ \beta_2 &= \frac{1}{2} [\mathbf{I} + \frac{\mathbf{\Delta t}}{\mathbf{2}} \mathbf{M}^{-1} \mathbf{C} + \frac{\mathbf{\Delta t^2}}{\mathbf{4}} \mathbf{M}^{-1} \mathbf{K}_0]^{-1} \end{aligned} \tag{1.6}$$

The matrices $\boldsymbol{\beta}_1$ and $\boldsymbol{\beta}_2$ which are determined from the initial stiffness matrix $\mathbf{K_0}$, render the method unconditionally stable, with no numerical dissipation and no overshooting effects like the trapezium rule for linear structures with $\mathbf{K_0}$ (Hughes, 1987). The limits of these methods lie on the facts that the velocity vector expression is implicit and for their efficient use, the damping matrix $\mathbf{C}$ must be constant. Moreover, these schemes do not provide an explicit target velocity, thus limiting the control of the transfer system (Wu et al., 2005).

If numerical damping is needed, Bonnet (2006) suggested the use of the

$\boldsymbol{\alpha}$-Operator-Splitting (OS-$\boldsymbol{\alpha}$) method, which can be expressed by

$$\mathbf{M}^n\mathbf{a}_{k+1} + (1+\alpha)\mathbf{C}^n\mathbf{v}_{k+1} - \alpha\mathbf{C}^n\mathbf{v}_k + (1+\alpha)\mathbf{K}^n\mathbf{d}_{k+1} - \alpha\mathbf{K}^n\mathbf{d}_k$$
$$= (1+\alpha)(\mathbf{f}_{e,k+1} + \mathbf{f}_{s,k+1}) - \alpha(\mathbf{f}_{e,k} + \mathbf{f}_{s,k}) \quad (1.7)$$
$$\mathbf{d}_{k+1} = \tilde{\mathbf{d}}_{k+1} + \beta\Delta t^2\mathbf{a}_{k+1} \quad (1.8)$$
$$\tilde{\mathbf{d}}_{k+1} = \mathbf{d}_k + \Delta t\mathbf{v}_k + \Delta t^2(\frac{1}{2} - \beta)\mathbf{a}_k \quad (1.9)$$
$$\mathbf{v}_{k+1} = \mathbf{v}_k + (1-\gamma)\Delta t\mathbf{a}_k + \gamma\Delta t\mathbf{a}_{k+1} \quad (1.10)$$
$$\gamma = \frac{1}{2} - \alpha \quad \text{and} \quad \beta = \frac{(1-\alpha)^2}{4} \quad (1.11)$$

Also this scheme suffers of the limits underlined for the two previous ones and Wu et al. (2006) proposed an Operator-splitting method able to deal explicitly with systems endowed with non-linear damping. Other integration algorithms, originated from the control theory, were developed: see, among others, Lopez-Almansa et al. (1998); Zhang et al. (2005); Chu et al. (2005). Nonetheless, the first two methods were only conditionally stable, while the third method implied the calculation of the exponential of the system matrix (Moler and Loan, 1978) which is of limited use for large *FE* models.

In summary, we recall that all unconditionally, *i.e.* A-stable methods, for uncoupled problems suggested for *SHSDS* and derived from *structural integrators* obey to the second Dahlquist barrier (Lambert (1991), p. 243), *viz.* : (i) an explicit linear multistep method cannot be A-stable; (ii) *the order of an A-stable linear multistep method cannot exceed two*; (iii) the second-order A-stable linear multistep method with smallest error constant is the *trapezium rule*. As a result, the algorithms used so far are utmost second-order accurate for uncoupled problems. To perform real-time computation, it appears more convenient to employ unconditionally single step integration formula with fixed time step and order, to achieve the same computational cost in each integration step without iterations (Arnold et al., 2007). Accuracy is the feature that must be sacrificed in conflict with other properties and mainly stability; in detail, the time integration method must be computationally inexpensive, with few evaluations in each step and without introducing excessive loss of accuracy. Along these line, L-stable real-time compatible algorithms based on Rosenbrock methods have been developed (Bursi et al., 2008). They are introduced in Subsection 3.1 and analysed at length in Section 5 for both monolithic and heterogeneous coupled systems.

1.2 Algorithms for *PsD* Testing

After decades of development, the *PsD* method is recognised to be as a reliable technique for testing full scale structures. A conceptual flow of the method is shown in Figure 3. By its very numerical nature, the *PsD* method allows substructuring and requires only a part of the structure to be tested in the laboratory, the non-linear experimental part, while the linearly responding part of the structure, the analytical part, can be treated numerically (Shing et al., 1994). The time integration of multiple-degrees-of-freedom (MDoF) systems is most frequently performed using implicit or linearly inflict time-stepping schemes. Examples of such schemes are: (i) Thewalt and Mahin's method (Thewalt and Mahin, 1987) based on the *HHT*-α method of Hilber et al. (1977) and on iterations carried out at a hybrid digital-analog level; (ii) the method of Nakashima et al. (1990), based on the mixed implicit-explicit *OS* (Hughes et al., 1979) which does not require any numerical iteration; (iii) the scheme suggested by Shing et al. (1991), in short the *C*-α algorithm, based on the *HHT*-α method which relies on a modified Newton-Raphson type iterative procedure; (iv) the scheme suggested by Bonelli and Bursi (2004), in short the *IPC*-ρ_∞ algorithm, based on the *CH*-α method (Chung and Hulbert, 1993) and a predictor-single corrector procedure to avoids iterations.

Implicit and Explicit *CH*-α Algorithms Unconditional stability may not be sufficient to ensure a robust time integration; some high-frequency components of the numerical solution may damp out very slowly and show up as oscillations altering the solution. The low effectiveness of numerical dissipation and the overshoot consequences on the response of *HHT*-α method applied to *stiff* dynamic systems were highlighted by Erlicher et al. (2002). Some of these issues motivated the development of the *CH*-α method. Let us consider a structure to be tested, whose dynamic behaviour can be condensed on a reduced number of DoF, each of which is controlled by an actuator. According to the *CH*-α method, its motion can be analysed by means of a system of weighted dynamic equilibrium equations written in matrix notation

$$\mathbf{M}\mathbf{a}_{i+1-\alpha_m} + \mathbf{C}\mathbf{v}_{i+1-\alpha_f} + \mathbf{r}_{i+1-\alpha_f} = \mathbf{f}_{e,i+1-\alpha_f} \tag{1.12}$$

which arises from the temporal discretization of semi-discrete equations. The solution is computed inside the time step size $\Delta t_i = t_{i+1} - t_i$ with given initial conditions. $\mathbf{r}_{i+1-\alpha_f}$ is the non-linear nodal restoring force vector, $\mathbf{f}_{e,i+1-\alpha_f}$ denotes the external excitation vector applied to the system.

Let $0 = t_0 < t_1 < \ldots < t_n = t_f$, where n is the number of time steps, define a partition of the time domain $[0, t_f]$ and assume a constant time step

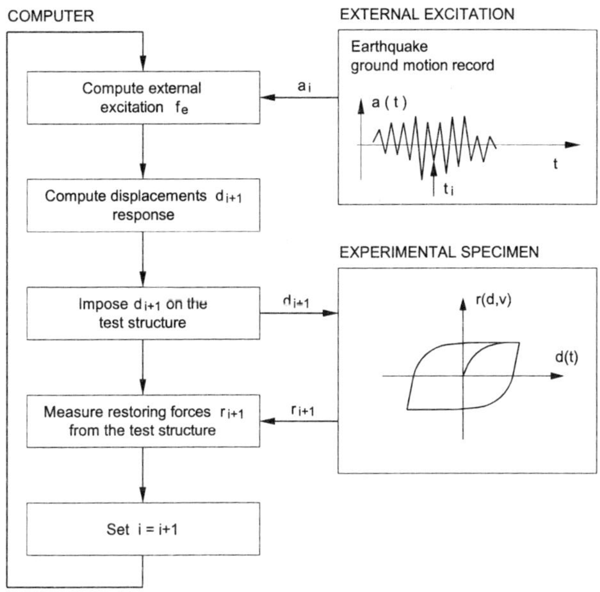

Figure 3. Flow of the *PsD* method

size $\Delta t = \Delta t_i$. The CH-α method relies on the temporal approximation of $\mathbf{v}$ and $\mathbf{a}$ proposed by Newmark (1959)

$$\mathbf{a}_{i+1} = \frac{1}{\Delta t^2 \beta} \left(\mathbf{d}_{i+1} - \tilde{\mathbf{d}}_{i+1} \right) \tag{1.13}$$

$$\mathbf{v}_{i+1} = \mathbf{v}_i + \Delta t \left[(1-\gamma)\, \mathbf{a}_i + \gamma \frac{1}{\Delta t^2 \beta} \left(\mathbf{d}_{i+1} - \tilde{\mathbf{d}}_{i+1} \right) \right] \tag{1.14}$$

where $\mathbf{a}_i$ is the numerical acceleration at time t_i and

$$\tilde{\mathbf{d}}_{i+1} = \mathbf{d}_i + \Delta t \mathbf{v}_i + \Delta t^2 \left(\frac{1}{2} - \beta \right) \mathbf{a}_i \tag{1.15}$$

with the initial conditions $\mathbf{d}_0 = \mathbf{d}$, $\mathbf{v}_0 = \mathbf{v}$ and $\mathbf{a}_0 = \mathbf{M}^{-1}(\mathbf{f}_0(0) - \mathbf{C}\mathbf{v}_0 - \mathbf{r}(\mathbf{d}_0))$. The time discrete combinations of displacements, velocities and accelerations read

$$\begin{aligned} \mathbf{d}_{i+1-\alpha_f} &= (1-\alpha_f)\, \mathbf{d}_{i+1} + \alpha_f \mathbf{d}_i \\ \mathbf{v}_{i+1-\alpha_f} &= (1-\alpha_f)\, \mathbf{v}_{i+1} + \alpha_f \mathbf{v}_i \\ \mathbf{a}_{i+1-\alpha_m} &= (1-\alpha_m)\, \mathbf{a}_{i+1} + \alpha_m \mathbf{a}_i \end{aligned} \tag{1.16}$$

where $\boldsymbol{i} \in \{\mathbf{0}, \mathbf{1}, ..., \boldsymbol{n} - \mathbf{1}\}$. With regard to the evaluation of the non-linear internal forces $\mathbf{r}_{i+1-\alpha_f} = \mathbf{r}(\alpha_f, \mathbf{d}_i, \mathbf{d}_{i+1})$ of (1.12), various quadrature rules can be employed (Erlicher et al., 2002). Nonetheless, only the generalized trapezoidal rule, the most suitable for the *PsD* method is considered here:

$$\mathbf{r}_{i+1-\alpha_f} = (\mathbf{1} - \alpha_f)\,\mathbf{r}(\mathbf{d}_{i+1}) + \alpha_f \mathbf{r}(\mathbf{d}_i) \tag{1.17}$$

$$\mathbf{f}_{i+1-\alpha_f} = (\mathbf{1} - \alpha_f)\,\mathbf{f}(t_{i+1}) + \alpha_f \mathbf{f}(t_i)\,. \tag{1.18}$$

The *PsD* method is mainly exploited in the non-linear case, where algorithms regarded as unconditionally stable in linear dynamics can exhibit severe numerical instabilities in the non-linear case (Erlicher et al., 2002). When applied to the non-linear regime, the algorithm defined in (1.12-1.16) can be implemented through a matrix $\overline{\mathbf{K}}$, which considers inertial and viscous damping effects and an effective stiffness matrix $\mathbf{K}^{*(k)}_{i+1}$. They read

$$\overline{\mathbf{K}} = \frac{\mathbf{1} - \alpha_m}{(\mathbf{1}-\alpha_f)\,\beta\Delta t^2}\mathbf{M} + \gamma\frac{\mathbf{1} - \alpha_m}{\beta\Delta t}\mathbf{C} \text{ and} \tag{1.19}$$

$$\mathbf{K}^{*(k)}_{i+1} = \overline{\mathbf{K}} + \mathbf{K}^{(k)}_{a,i+1}, \tag{1.20}$$

respectively, where $\mathbf{K}^{(k)}_{a,i+1}$ is the tangent stiffness matrix of the structure. Defining the residual as

$$\begin{aligned}\mathbf{R}^{(k)}_{i+1} = \mathbf{f}_{i+1} + \frac{\alpha_f}{\mathbf{1} - \alpha_f}\mathbf{f}_i - \frac{\mathbf{1} - \alpha_m}{\mathbf{1} - \alpha_f}\mathbf{M}\mathbf{a}^{(k)}_{i+1} - \frac{\alpha_m}{\mathbf{1} - \alpha_f}\mathbf{M}\mathbf{a}_i - \mathbf{C}\mathbf{v}^{(k)}_{i+1} \\ - \frac{\alpha_f}{\mathbf{1} - \alpha_f}\mathbf{C}\mathbf{v}_i - \mathbf{r}^{(k)}_{i+1} - \frac{\alpha_f}{\mathbf{1} - \alpha_f}\mathbf{r}_i,\end{aligned} \tag{1.21}$$

the displacement increment $\Delta\mathbf{d}^{(k)}_{i+1}$ can be evaluated by means of a Newton-Raphson method

$$\Delta\mathbf{d}^{(k)}_{i+1}\left(\mathbf{r}^{(k)}_{i+1}, \mathbf{v}^{(k)}_{i+1}, \mathbf{a}^{(k)}_{i+1}\right) = \left(\mathbf{K}^{*(k)}_{i+1}\right)^{-1}\mathbf{R}^{(k)}_{i+1} \tag{1.22}$$

and therefore

$$\mathbf{d}^{(k+1)}_{i+1} = \mathbf{d}^{(k)}_{i+1} + \Delta\mathbf{d}^{(k)}_{i+1} \tag{1.23}$$

$$\mathbf{a}^{(k+1)}_{i+1}\left(\mathbf{d}^{(k+1)}_{i+1}\right) = \frac{\mathbf{1}}{\Delta t^2\beta}\left(\mathbf{d}^{(k+1)}_{i+1} - \tilde{\mathbf{d}}_{i+1}\right) \tag{1.24}$$

$$\mathbf{v}^{(k+1)}_{i+1} = \mathbf{v}_i + \Delta t\,(\mathbf{1} - \gamma)\,\mathbf{a}_i + \Delta t\gamma\mathbf{a}^{(k+1)}_{i+1}\,. \tag{1.25}$$

In the linear case, $\mathbf{r}_{i+1-\alpha_f} = (1-\alpha_f)\,\mathbf{K}\mathbf{d}_{i+1} + \alpha_f \mathbf{K}\mathbf{d}_i$ and the $\boldsymbol{CH-\alpha}$ scheme is endowed with: (i) second-order accuracy; (ii) optimized dissipation characteristics of higher modes, *i.e.* the high-frequency components of the response are filtered with minimum influence on the low-frequency region. The amount of dissipation can be user-specified by means of the high-frequency spectral radius ρ_∞ ($= \rho(\Omega)$ for $\Omega \to \infty$ where Ω represents the non-dimensional frequency); in detail, $\rho_\infty \in [0,1]$ and the choice $\rho_\infty = 0$ corresponds to asymptotic annihilation of the high-frequency response, *i.e.* L-stability, while $\rho_\infty = 1$ corresponds to no algorithmic dissipation. The relations between the user-specified spectral radius at the high-frequency limit ρ_∞ and the free parameters β, γ, α_m and α_f, obtained from optimizations in the linear case (Chung and Hulbert, 1993) read

$$\begin{array}{ll} \beta = \dfrac{1}{(1+\rho_\infty)^2} \in [\dfrac{1}{4}, 1]; & \gamma = \dfrac{1}{2}\dfrac{3-\rho_\infty}{1+\rho_\infty} \in [\dfrac{1}{2}, \dfrac{3}{2}]; \\ \alpha_m = \dfrac{2\rho_\infty - 1}{1+\rho_\infty} \in [-1, \dfrac{1}{2}]; & \alpha_f = \dfrac{\rho_\infty}{1+\rho_\infty} \in [0, \dfrac{1}{2}] \end{array}. \tag{1.26}$$

As far as unforced systems are concerned, a detailed convergence analysis of the *CH-α* method in the non-linear case can be found in Erlicher et al. (2002). Summing up, it has been proved that the assumptions (1.26) entail a consistent second order scheme, viz.

$$\mathbf{d}_i = \mathbf{u}(t_i) + \mathbf{O}\left(\Delta t^2\right), \tag{1.27}$$

where $\mathbf{u}(t)$ is the exact solution of the parent system of differential equations. Accelerations evaluated by means of (1.24) are just first order accurate, while a second order approximation can be found using the following combination

$$[(1-\alpha_m)\mathbf{a}_{i+1} + \alpha_m \mathbf{a}_i] - [(1-\alpha_f)\ddot{\mathbf{d}}(t_{i+1}) + \alpha_f \ddot{\mathbf{d}}(t_i)] = \mathbf{O}(\Delta t^2) \tag{1.28}$$

This relationship was used by Lacoma and Romero (2007) and Brüls and Golinval (2008), among others, in their applications.

Moreover, it was demonstrated that the dissipative properties of the algorithm can be still controlled in the non-linear case by the parameter ρ_∞. In detail, assume $\mathbf{a}_i = r_{a,i}\mathbf{a}_0$, $\mathbf{v}_i = r_{v,i}\mathbf{v}_1$ and $\mathbf{r}_{i+1-\alpha_f} = r_{r,i}\mathbf{r}_{1-\alpha_f}$; if $\Delta\tau = \frac{\Delta t}{T} \gg 1$, where T represents the period of the highest frequency component, the following relations hold

$$\lim_{i\to\infty} \frac{r_{a,i+1}}{r_{a,i}} = \frac{r_{v,i+1}}{r_{v,i}} = \frac{r_{r,i+1}}{r_{r,i}} = -\rho_\infty. \tag{1.29}$$

If $\mathbf{C} = \mathbf{0}$, denote with $\hat{E}_i = E_i + \Delta t^2 \left(\beta - \frac{\gamma}{2}\right) \frac{1}{2}\mathbf{a}_i^T \mathbf{M} \mathbf{a}_i$ the generalized energy, where $E_i = T_i + U_i = \frac{1}{2}\mathbf{v}_i^T \mathbf{M} \mathbf{v}_i + U\left(\mathbf{d}_i\right)$ is the total mechanical energy of the system at the beginning of the *i-th* time step, the following relations hold

$$\frac{\hat{E}_{i+1}}{\hat{E}_i} = \rho_\infty^2 \frac{(i+1)^2 \frac{(1-\rho_\infty)^4}{16} + \frac{(1-\rho_\infty)^2}{4(1+\rho_\infty)^2}\left(-\rho_\infty + \frac{i+1}{2}\left(1-\rho_\infty^2\right)\right)^2}{i^2 \frac{(1-\rho_\infty)^4}{16} + \frac{(1-\rho_\infty)^2}{4(1+\rho_\infty)^2}\left(-\rho_\infty + \frac{i}{2}\left(1-\rho_\infty^2\right)\right)^2} \tag{1.30}$$

$$\lim_{i\to\infty} \frac{\hat{E}_{i+1}}{\hat{E}_i} = \rho_\infty^2. \tag{1.31}$$

for $\mathbf{a}_0 \neq \mathbf{0},\ i \geqslant 0$. (1.31) confirms the effectiveness of the dissipative characteristics of the $CH-\alpha$ scheme in the high frequency range.

With regard to non-linear forced systems, the *CH*-α method is capable to limit the error near resonance both in terms of frequency location and intensity of the resonant peak (Baldo et al., 2006).

When the time step Δt is very small, as in the continuous *PsD* method introduced in Section 6, explicit schemes can be used, owing to their reduced numerical effort and accuracy. For instance, the *ECH*-α method still employs the Newmark approximations of velocity and acceleration (1.13-1.14) as well as (1.12). First, let us consider the undamped case and two parameter ρ_p and ρ_s which represent the modulus of the principal roots and of the spurious root at the bifurcation frequency Ω_b, respectively. To maximize the high-frequency numerical dissipation, Hulbert and Chung (1996) impose $\rho_s \leq \rho_p$. By denoting the value of the spectral radius at Ω_b by ρ_b, the condition $\rho_s = \rho_p = \rho_b$ entails the optimal balance of dissipation at low and high frequencies. The relations among the algorithmic parameters then become

$$\alpha_m = \frac{2\rho_b - 1}{\rho_b + 1} \tag{1.32}$$

$$\beta = \frac{\rho_b^3 - 2\alpha_f\rho_b^3 + \alpha_f^2\rho_b^3 - 3\rho_b^2 + 3\alpha_f\rho_b^2 + 3\alpha_f\rho_b - 3\alpha_f^2\rho_b + 3\rho_b - 2\alpha_f^2 - 1 - 2\alpha_f}{(-2+\rho_b)(\rho_b+1)^2\alpha_f} \tag{1.33}$$

$$\gamma = \frac{1}{2} - \alpha_m + \alpha_f. \tag{1.34}$$

where (1.34) entails second order accuracy and α_f is a free parameter;

moreover, $\boldsymbol{\Omega}_b$ and $\boldsymbol{\Omega}_{cr}$ read

$$\begin{aligned} \boldsymbol{\Omega}_b &= (\rho_b + 1)\sqrt{2 - \rho_b} \\ \boldsymbol{\Omega}_{cr} &= \sqrt{\frac{12(\rho_b+1)^3(\rho_b-2)}{\left(\rho_b^4-\rho_b^3+\rho_b^2-15\rho_b-10\right)}} \end{aligned} \tag{1.35}$$

respectively, where $\boldsymbol{\Omega}_{cr}$ represents the critical frequency corresponding to the stability limit.

For damped systems, Hulbert and Chung (1996) proposed both an explicit and an implicit treatment of physical damping. Therefore two implementations are possible. If $\boldsymbol{\alpha}_f = \mathbf{1}$, the *ECH-$\alpha$* method can be formulated with an explicit treatment of damping to attain second-order accuracy even in the non-linear case, and implemented in a predictor-corrector form:
Predictor phase:

$$\tilde{\mathbf{v}}_{i+1} = \mathbf{v}_i + (1-\gamma)\,\Delta t \mathbf{a}_i \tag{1.36}$$

$$\tilde{\mathbf{d}}_{i+1} = \mathbf{d}_i + \Delta t \mathbf{v}_i + \left(\frac{1}{2} - \beta\right)\Delta t^2 \mathbf{a}_i \tag{1.37}$$

Corrector phase:

$$(1-\alpha_m)\,\mathbf{M}\,\mathbf{a}_{i+1} = \mathbf{f}_{ex}(t_i) - \alpha_m \mathbf{M}\,\mathbf{a}_i - \mathbf{C}\mathbf{v}_i - \mathbf{r}\left(\tilde{\mathbf{d}}_i\right) \tag{1.38}$$

$$\mathbf{v}_{i+1} = \tilde{\mathbf{v}}_{i+1} + \gamma \Delta t \mathbf{a}_{i+1} \tag{1.39}$$

$$\mathbf{d}_{i+1} = \tilde{\mathbf{d}}_{i+1} + \beta \Delta t^2 \mathbf{a}_{i+1}. \tag{1.40}$$

With an implicit damping treatment and $\boldsymbol{\alpha}_f \neq \mathbf{1}$, Hulbert and Chung (1996) in the correction phase substitute the equilibrium equation (1.38) with the following relationship

$$\begin{aligned} &((1-\alpha_m)\,\mathbf{M}+\gamma\Delta t\,(1-\alpha_f)\,\mathbf{C})\;\mathbf{a}_{i+1} = \mathbf{f}_{ex}\left(t_{i+1-\alpha_f}\right) \\ &-\alpha_m \mathbf{M}\,\mathbf{a}_i - (1-\alpha_f)\,\mathbf{C}\tilde{\mathbf{v}}_{i+1} - \alpha_f\,\mathbf{C}\mathbf{v}_i - \mathbf{r}\left(\tilde{\mathbf{d}}_{i+1-\alpha_f}\right) \end{aligned} \tag{1.41}$$

***CH-α* Algorithms Applied to *PsD* Testing** The application of an implicit scheme to the *PsD* method is not trivial as a consistent linearization via the full Newton-Raphson method is required to solve the non-linear system (1.12). Moreover, $\mathbf{r}_{i+1}^{(k)}$ in (1.21) is measured during a *PsD* test. As a result, every time step needs generally more than one movement of each actuator. The *C-α* method (Shing and Vannan, 1990) adopts a system of modified equilibrium equations (1.12) with $\boldsymbol{\alpha}_m = \mathbf{0}$ and $\boldsymbol{\alpha}_f = -\boldsymbol{\alpha}$ and the Newmark approximations (1.13-1.14). It is essentially endowed with

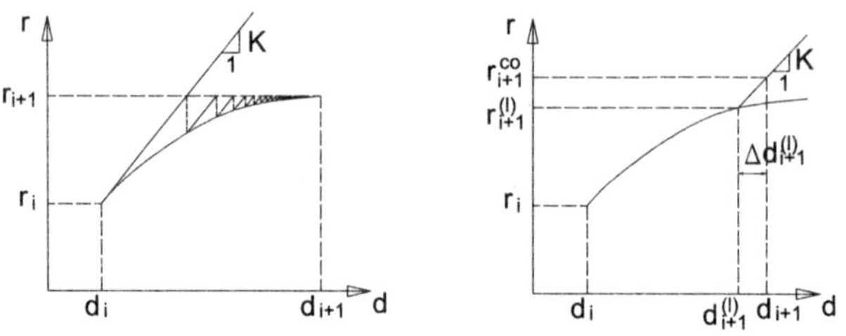

Figure 4. Non-linear solution method for the C-$\boldsymbol{\alpha}$ algorithm: a) modified Newton iterations; b) correction residual errors.

modified Newton-Raphson iterations with linear convergence and the error-correction procedure depicted in Figure (4)a and described in (1.21-1.25).

If more than one movement of each actuator is needed inside one time step like in the C-$\boldsymbol{\alpha}$-method, there is a double negative effect: i) the overall test time increases; ii) the time scaling factor between earthquake time and test time varies in each step. To avoid this problem the OS-$\boldsymbol{\alpha}$ algorithm was proposed (Nakashima et al., 1990). This scheme is the synthesis of the OS technique and the C-$\boldsymbol{\alpha}$ method (Bursi and Shing, 1996). It results in an explicit predictor-implicit corrector or linearly implicit algorithm characterized by the following equations:

$$\begin{aligned}\mathbf{Ma}_{i+1} + (\mathbf{1}+\boldsymbol{\alpha})\mathbf{Cv}_{i+1} - \boldsymbol{\alpha}\mathbf{Cv}_i + (\mathbf{1}+\boldsymbol{\alpha})(\mathbf{Kd}_{i+1}+\tilde{\mathbf{r}}_{i+1}-\mathbf{K}\tilde{\mathbf{d}}_{i+1})\\ -\boldsymbol{\alpha}(\mathbf{Kd}_i+\tilde{\mathbf{r}}_i-\mathbf{K}\tilde{\mathbf{d}}_i) = (\mathbf{1}+\boldsymbol{\alpha})\mathbf{f}_{i+1} - \boldsymbol{\alpha}\mathbf{f}_i \qquad (1.42)\end{aligned}$$

The deviation of the structural response from linearity, as depicted in Figure 5(a), is approximated as

$$\mathbf{Kd}_{i+1} - \mathbf{r}_{i+1} \approx \mathbf{K}\tilde{\mathbf{d}}_{i+1} - \tilde{\mathbf{r}}_{i+1} \qquad (1.43)$$

in which $\mathbf{K}$ represents a predictor stiffness matrix, and $\tilde{\mathbf{r}}_{i+1}$ is the restoring force vector developed by the structure when the predictor displacements $\tilde{\mathbf{d}}_{i+1}$ have been imposed.

The predictor-one pass corrector scheme $\boldsymbol{IPC}-\rho_\infty$ suggested by Bonelli and Bursi (2004) requires, like the OS-$\boldsymbol{\alpha}$ algorithm, only one single movement of the actuator per time step. But differently from previous schemes it evaluates the movement to be imposed by the actuator by means of the implicit formula (1.23) instead of the explicit one (1.37) employed by the OS-$\boldsymbol{\alpha}$ method.

Let's superscript $\boldsymbol{m}$ denote experimental measures and, for brevity, consider only the main steps. The solution procedure of the $\boldsymbol{IPC}-\rho_\infty$ scheme is summarized in an algorithmic form in what follows.

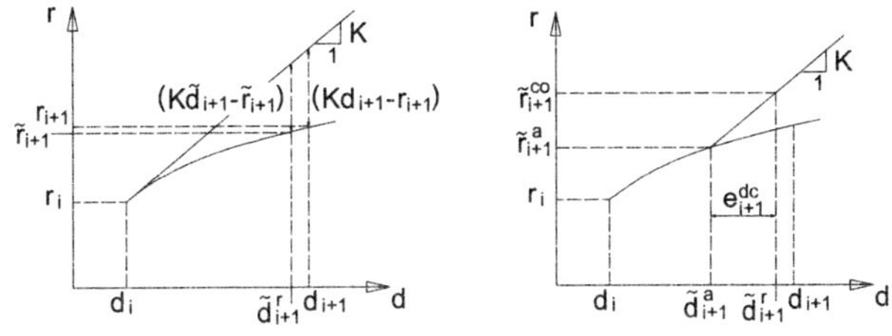

Figure 5. Solution methods for the *OS* algorithms: a) corrections for non-linearity; b) I-Modification for displacement control error.

1. Predictor phase:
   a) Compute $\mathbf{a}_{i+1}(\mathbf{d}_i)$ and $\mathbf{v}_{i+1}(\mathbf{d}_i)$ ((1.24) and (1.25))
   b) Compute $\boldsymbol{\Delta}\mathbf{d}_{i+1}^{(0)}\left(\mathbf{r}_i, \mathbf{v}_{i+1}, \mathbf{a}_{i+1}\right)$ (1.22) and $\mathbf{d}_{i+1}^{(1)}$ (1.23)
   c) Impose $\mathbf{d}_{i+1}^{(1)}$ to the structure
   d) Measure $\mathbf{d}_{i+1}^{m}$ and $\mathbf{r}_{i+1}^{m}$ developed by the structure
   e) Compute $\mathbf{a}_{i+1}(\mathbf{d}_{i+1}^{m})$ and $\mathbf{v}_{i+1}(\mathbf{d}_{i+1}^{m})$ ((1.24) and (1.25))
2. Corrector phase:
   a) Compute $\boldsymbol{\Delta}\mathbf{d}_{i+1}^{(1)}(\mathbf{r}_{i+1}^{m}, \mathbf{v}_{i+1}, \mathbf{a}_{i+1})$ ((1.22)) and $\mathbf{d}_{i+1} = \mathbf{d}_{i+1}^{m} + \boldsymbol{\Delta}\mathbf{d}_{i+1}^{(1)}$
   b) Compute $\mathbf{r}_{i+1} = \mathbf{r}_{i+1}^{m} + \mathbf{K}\boldsymbol{\Delta}\mathbf{d}_{i+1}^{(1)}$
   c) Compute $\mathbf{a}_{i+1}(\mathbf{d}_{i+1})$ and $\mathbf{v}_{i+1}(\mathbf{d}_{i+1})$ ((1.24) and (1.25))
   d) Set $\boldsymbol{i = i + 1}$, $\boldsymbol{t_{i+1} = t_i + \Delta t}$ and go to Step 1b

where Step 1b and 2a are the predictor and the corrector steps, respectively. Each predictor-corrector step is equivalent to the first iteration of a modified Newton-Raphson procedure.

For a SDoF system the predictor-one pass corrector process is depicted in Figure 6a, where the evolution of the secant stiffness $\boldsymbol{K}_i^s$ from step $\boldsymbol{i}$ to step $\boldsymbol{i+1}$ is clearly illustrated. The error corrections performed in Step 2a and 2b define a solution which satisfies (1.12)-(1.14), i.e. the space of solutions at step $\boldsymbol{i+1}$; but they trace only approximately, see the hatched line in Figure 6a, the actual non-linear response of the structure, *viz.* the continuous line. In detail, the solution space corresponds to a straight line with negative slope. In fact for a SDoF system, (1.12) by virtue of (1.13) and (1.14) can be recasted as follows

$$\boldsymbol{r_{i+1} = -\bar{K} d_{i+1} + \bar{F}_{i+1}} \tag{1.44}$$

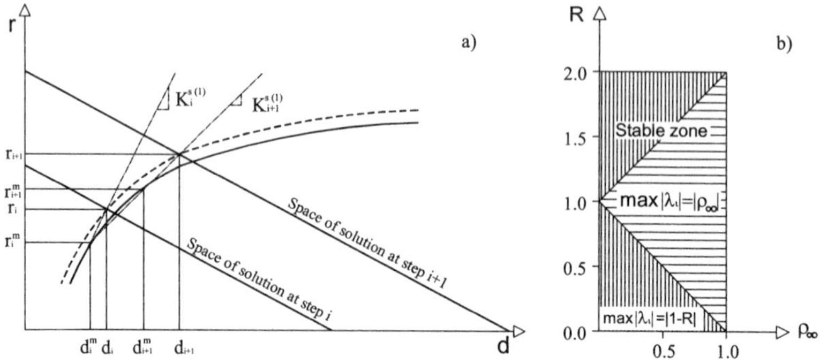

Figure 6. $\boldsymbol{IPC-\rho_\infty}$ scheme: a) non-linear solution method and correction for residual errors; b) zone of unconditional stability in $\boldsymbol{\rho_\infty - R}$ space.

where

$$\overline{K} = \frac{1-\alpha_m}{(1-\alpha_f)\,\beta\Delta t^2}m + \gamma\frac{1-\alpha_m}{\beta\Delta t}c \tag{1.45}$$

$$\begin{aligned}\bar{F}_{i+1} &= \bar{K}\tilde{d}_{i+1} + f_{i+1} + \frac{\alpha_f}{1-\alpha_f}f_i - \frac{\alpha_f}{1-\alpha_f}cv_i - \\ &\left[(1-\gamma)\,\Delta tc + \frac{\alpha_m}{1-\alpha_f}m\right]a_i - \frac{\alpha_f}{1-\alpha_f}r_i\end{aligned} \tag{1.46}$$

With regard to consistency, it can be proved that the $\boldsymbol{IPC-\rho_\infty}$ method is second order accurate. As far as stability properties are concerned, the following relationships

$$\max_i |\lambda_i| = \begin{cases} |\rho_\infty| & \text{if } \rho_\infty \geq |1-R| \\ |1-R| & \text{if } \rho_\infty < |1-R| \end{cases} \tag{1.47}$$

can be proved, where $\boldsymbol{R = \frac{K_a}{K}}$ is the stiffness ratio and $\boldsymbol{K_a}$ the actual stiffness of the SDoF system. They lead to an unconditional stable scheme if and only if

$$0 < R < 2. \tag{1.48}$$

Figure 6(b) depicts the zone of unconditional stability in the $\boldsymbol{\rho_\infty - R}$ space. In detail, one may observe that the condition $\boldsymbol{\rho = \max_i |\lambda_i| = 0}$, i.e. the L-stability property of the $\boldsymbol{IPC - \rho_\infty}$ scheme is attained in *PsD* tests only if $\boldsymbol{R = 1}$ and $\boldsymbol{\rho_\infty = 0}$. The $\boldsymbol{IPC - \rho_\infty}$ method exhibits dissipative favourable properties; moreover differently from the *OS*-$\boldsymbol{\alpha}$ implicit implementation (Combescure and Pegon, 1997), the scheme maintains unconditional stability properties even with $\boldsymbol{R > 1}$.

In general, the application of explicit schemes to the *PsD* method is easier than the one of implicit schemes as movements to be imposed to actuators are explicitly evaluated without approximations. Along the line of the predictor-corrector *ECH-α* algorithm summarized in (1.36-1.40), Bonelli and Bursi (2004) proposed the *EPC-ρ_b* method for *PsD* tests.

2 Forms of Equations of Structural Dynamics and Solution Techniques

In this section two forms of the equations of structural dynamics are presented, from which popular integration algorithms employed in *PsD* and *SHSDS* techniques stemmed from. Moreover, some corrections and compensation techniques employed in testing are commented upon.

2.1 Euler-Lagrange form

The semi-discrete dynamic system of equations of motion reads

$$\mathbf{M}\ddot{\mathbf{x}}(t) + \mathbf{C}\dot{\mathbf{x}}(t) + \mathbf{K}\mathbf{x}(t) = \mathbf{f}_e(t), \tag{2.1}$$

where $\mathbf{M}$, $\mathbf{C}$ and $\mathbf{K}$ are the global mass, damping and stiffness matrices, respectively, $\mathbf{f}_e(t)$ is the vector of applied loads, $\mathbf{x}(t)$ is the displacement vector and superimposed dots indicate time differentiation. The associated initial value problem consists in determining the function $\mathbf{x} = \mathbf{x}(t)$ fulfilling (2.1) for all $t \in [0, t_f]$, $t_f > 0$, for given initial conditions $\mathbf{x}(0) = \mathbf{x}_0$ and $\dot{\mathbf{x}}(0) = \mathbf{v}_0$.

2.2 Hamilton form

Instead of expressing the inertial effects in terms of acceleration, the Hamilton form of (2.1) is given by

$$\begin{aligned} \dot{\mathbf{x}}(t) &= \mathbf{y}_2(t) \\ \mathbf{M}\dot{\mathbf{y}}_2(t) + \mathbf{C}\mathbf{y}_2(t) + \mathbf{K}\mathbf{x}(t) &= \mathbf{f}_e(t) \end{aligned} \tag{2.2}$$

where $\mathbf{y}_2(t)$ is the vector of velocity unknowns. The initial value problem consists in determining the function $\mathbf{x} = \mathbf{x}(t)$ and $\mathbf{y}_2(t)$ which satisfies (2.2) for all $t \in [0, t_f]$, $t_f > 0$, for given initial conditions $\mathbf{x}(0) = \mathbf{x}_0$ and $\mathbf{y}_2(t) = \mathbf{v}_0$. Some integration methods used in control and multibody dynamics are designed for first-order differential equations. For these methods the Hamilton form can be applied, by introducing the state vectors, $\mathbf{y} = \{\mathbf{x}(t) \quad \dot{\mathbf{x}}(t)\}^T = \{\mathbf{y}_1(t) \quad \mathbf{y}_2(t)\}^T$. The first-order momentum form is then written as

$$\dot{\mathbf{y}} = \mathbf{f}(t, \mathbf{y}) \tag{2.3}$$

with $\mathbf{y}(0) = \{\mathbf{x}_0 \quad \mathbf{v}_0\}^T$ where

$$\mathbf{f} = \begin{bmatrix} \mathbf{0} & \mathbf{I} \\ -\mathbf{M}^{-1}\mathbf{K} & -\mathbf{M}^{-1}\mathbf{C} \end{bmatrix} \mathbf{y} + \begin{Bmatrix} \mathbf{0} \\ \mathbf{M}^{-1}\mathbf{f}_e(t) \end{Bmatrix} \tag{2.4}$$

Other forms exist like, for instance, the time-discontinuous space-time *FE* formulation developed by Hughes and Hulbert (1988). In these schemes, the space is discretized using conventional *FE*, the time interval is partitioned in a number of subintervals, while the response is approximated by means of trial functions in the time variable. The use of discontinuous displacement and momentum fields leads to the class of implicit schemes, named time discontinuous Galerkin (*TDG*) methods. A comprehensive review of these methods can be found in Hulbert (2004). These unconditionally stable schemes can damp out any undesirable high-frequency mode, without introducing excessive algorithmic damping in the low-frequency response. Nonetheless, the factorization of a matrix larger than the one exploited in standard implicit schemes is required owing to unknown extra displacement and momentum fields (Bursi and Mancuso, 2002). To limit the computational effort, Li and Wiberg (1996) implemented a predictor-multicorrector solution algorithm in the *TDG* method. Based on similar strategies, a new class of explicit *TDG* methods, with a user-defined dissipation, was developed by Bonelli et al. (2001). The new proposed schemes were implemented in a predictor-multicorrector form, exhibited third-order accuracy and accrued a stability limit higher than the one of the *CD* scheme. This scheme was also extended to non-linear elastodynamics (Bonelli et al., 2002) showing that both third-order accuracy and energy-decaying properties can be retained. These techniques could be explored in *SHSDS* and *PsD* techniques.

2.3 Corrections and Compensation Techniques for Testing

The efficiency of a time-stepping scheme for *PsD* is connected to its reduced sensitivity to errors, as in a heterogeneous environment several errors arise. While in Subsection 3.2 an error propagation analysis of some algorithms is presented, herein we show modified solution procedures to reduce unfavourable error effects.

Correction techniques for PsD testing The multicorrector phase (1.21) -(1.25) used in the implicit *CH*-α scheme results to be very sensitive to experimental errors (Shing and Vannan, 1990). To overcome these difficulties, Shing et al. (1991) proposed a numerical correction to be evaluated after convergence:

$$\begin{aligned} \mathbf{d}_{i+1} &= \mathbf{d}_{i+1}^{(l)} + \mathbf{\Delta d}_{i+1}^{(l)} \\ \mathbf{r}_{i+1} &= \mathbf{r}_{i+1}^{(l)} + \mathbf{K}\mathbf{\Delta d}_{i+1}^{(l)} \end{aligned} \tag{2.5}$$

in which $\mathbf{\Delta d}_{i+1}^{(l)}$ represents a residual displacement vector which satisfies the convergence criteria $\left|\mathbf{\Delta d}_{i+1}^{(l)}\right| \leqslant \mathbf{\Delta d}^n$, where $\mathbf{\Delta d}^n$ is a preset displacement tolerance vector. This error correction, depicted in Figure 4(b) reduces the effect of stepwise residual errors on numerical result and eliminates the spurious higher-mode response that could otherwise be excited by residual errors.

To mitigate the effects of displacement control errors in *PsD* tests similar modifications were proposed for the *OS*-$\boldsymbol{\alpha}$ algorithm introduced in Subsection 1.2. In detail, the *I-Modification* was proposed. It simply corrects the restoring force vector as follows:

$$\tilde{\mathbf{r}}_{i+1}^{co} = \tilde{\mathbf{r}}_{i+1}^{a} - \mathbf{K}(\tilde{\mathbf{d}}_{i+1}^{a} - \tilde{\mathbf{d}}_{i+1}^{r}) \tag{2.6}$$

in which $\tilde{\mathbf{d}}_{i+1}^{r}$ denotes the exact numerical solution, $\tilde{\mathbf{d}}_{i+1}^{a}$ and $\tilde{\mathbf{r}}_{i+1}^{a}$ represent experimental feedback quantities, and $\tilde{\mathbf{r}}_{i+1}^{co}$ denotes the corrected restoring force vector as sketched in Figure 5(b). The relevant *OSM*-$\boldsymbol{\alpha}$ algorithm offers the main benefit of simplicity in implementation and no numerical iteration. However, in such a scheme, each predictor-corrector step is equivalent to the first iteration of a modified Newton procedure. As a result, errors increase when the predictor stiffness is very different from the actual stiffness of a structure, or when the structural specimen exhibits a severe non-linearity (Bursi and Shing, 1996). To reduce those effects, small integration time intervals as well as accurate displacement measurements and servo-controls have to be used. Moreover, to maintain the stability properties of the original *OS*-$\boldsymbol{\alpha}$ scheme, the predictor stiffness has to be greater than the actual stiffness in (1.42).

The $\boldsymbol{IPC} - \boldsymbol{\rho_\infty}$ method introduced in Subsection 1.2 is endowed with an error corrections in Step 2a and 2b along the line of (2.5) to reduce the effects of stepwise residual errors on numerical results.

Explicit schemes are very sensitive to errors near their stability limit. Typically, they are used with small time step to avoid special correction formulae (Bonelli and Bursi, 2004).

Substepping and Compensation in SHSDS Standard or modified Newton-Raphson linearizations were used in Subsection 1.2 to solve non-linear problems in structural dynamics. Other techniques to solve *SHSDS* use a digital feedback technique, an implicit integration scheme based on the

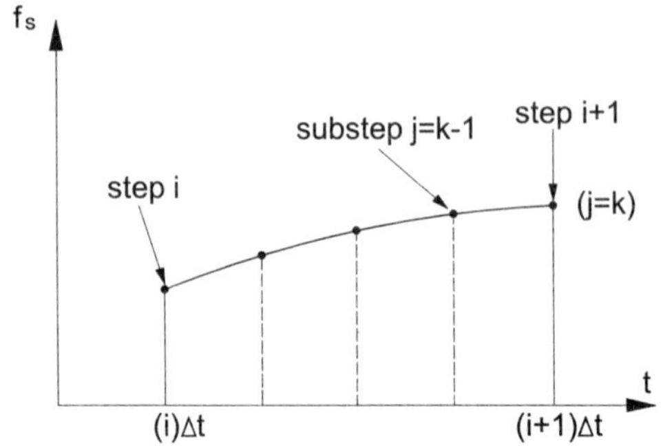

Figure 7. The history of the coupling force f_s in a substepping strategy

Newmark method and a substepping technique (Bayer et al., 2005). Since the coupling force $\mathbf{f_c}$ is unknown at time step $(\mathbf{i}+\mathbf{1})\boldsymbol{\Delta t}$, the response $\boldsymbol{d}_{i+1}$ and its derivatives cannot be obtained directly. Through substepping, the coupled system of equations of motion reads,

$$\mathbf{Ma}_{i+1} + \mathbf{Cv}_{i+1} + \mathbf{Kd}_{i+1} = \mathbf{f}_{l,i+1}, + \mathbf{f}_{s,i+1,k-1} \tag{2.7}$$

where the index $\boldsymbol{i+1,k-1}$ denotes the substep $\boldsymbol{k-1}$ in order to reach $\boldsymbol{i+1}$ and $\boldsymbol{k}$ is the number of substeps.

In order to satisfy equilibrium, an error force compensation method was applied, and a compensating force $\boldsymbol{\Delta}\mathbf{f}$ was added into the right side of (2.7); thus, the equilibrium equation becomes

$$\mathbf{Ma}_{i+1} + \mathbf{Cv}_{\mathrm{i+1}} + \mathbf{Kd}_{i+1} = \mathbf{f}_{l,i+1}, + \mathbf{f}_{s,i+1,k-1} + \boldsymbol{\Delta}\mathbf{f}_{i+1} \tag{2.8}$$

To approach the correct solution, the compensated force $\boldsymbol{\Delta}\mathbf{f}$ should approach the true force $\boldsymbol{\Delta}\mathbf{f}^*_{i+1}$, such that,

$$\boldsymbol{\Delta}\mathbf{f}_{i+1} \rightarrow \boldsymbol{\Delta}\mathbf{f}^*_{i+1} = \mathbf{f}_{s,i+1} - \mathbf{f}_{s,i+1,k-1} \tag{2.9}$$

The compensation force $\boldsymbol{\Delta}\mathbf{f}_{i+1}$ is strongly depending on the response $\boldsymbol{d}$ and can be treated as

$$\boldsymbol{\Delta}\mathbf{f}_{i+1} = \boldsymbol{P}[\mathbf{f}_{u,i} + \boldsymbol{I\Delta t}\sum_{i}^{i} \mathbf{f}_{u,i} + \frac{\boldsymbol{D}}{\boldsymbol{\Delta t}}(\mathbf{f}_{u,i} - \mathbf{f}_{u,i-1})] \tag{2.10}$$

where P, I and D are the proportional, integral and differential gains of a controller,respectively. Numerical studies on these parameters need to be performed before actual experiments to gather experience for optimal parameter settings. $\mathbf{f}_u$ is the so-called unbalanced force in substep control.

$$\mathbf{f}_u = (\mathbf{f}_{l,i} + \mathbf{f}_{s,i} - (\mathbf{M}\mathbf{a}_i + \mathbf{C}\mathbf{v}_i + \mathbf{K}\mathbf{d}_i)] \tag{2.11}$$

The PID error force compensation algorithm can reduce of about 50% $\mathbf{f}_u$ when $P \approx 1,\ I = 0,\ D = 0.$

3 Classification, Analysis and Properties of Time Integration Algorithms

We provide in Subsection 3.1 an essential classification of time integration algorithms for solving structural dynamic problems; while we define in Subsection 3.2 notions of accuracy, stability, numerical dissipation, dispersion and error propagation in order to understand the attributes of some integration schemes.

3.1 Classification of time integration algorithms

For convenience, we deal with the Hamilton form (2.3) of the equations of structural dynamics. If we consider the problem (2.3) in the interval $[0, t_f]$, it admits one and only one solution $\mathbf{y}(t)$, $t \in [0, t_f]$, if the function $\mathbf{f}(t, \mathbf{y})$ is continuous in t and Lispschitz continuous in $\mathbf{y}$, that is,

$$\|\mathbf{f}(t, \mathbf{y}) - \mathbf{f}(t, \hat{\mathbf{y}})\| < M \|\mathbf{y} - \hat{\mathbf{y}}\| \tag{3.1}$$

for some positive constant M (Lambert, 1991). As we anticipated, there are many numerical methods capable of solving (2.3). In a greater detail, methods may be classified as *explicit*, when the direct solution of one or more of the dynamical states does not require factorizing a combination of the damping and stiffness matrices; or *implicit*, when the direct solution of the dynamical states requires a factorization of some form of a matrix that comprises a combination of the mass, damping and stiffness matrices; *one-step* if they rely solely on the initial conditions provided at the beginning of a time step t_k and on the equations of motion and the dynamical states at time points within the time interval $[t_k, t_{k+1}]$; or *multistep* if they employ a sequence of earlier dynamical states and equation of motion evaluations; *single* or *multistage* according to the number of times the equations of motion must be evaluated to advance the solution by one time step. We recall two such methods to orient the reader.

The simplest explicit method is the one-step forward *Euler* method:

$$\mathbf{y}_{k+1} = \mathbf{y}_k + \Delta t \mathbf{f}(t_k, \mathbf{y}_k) \tag{3.2}$$

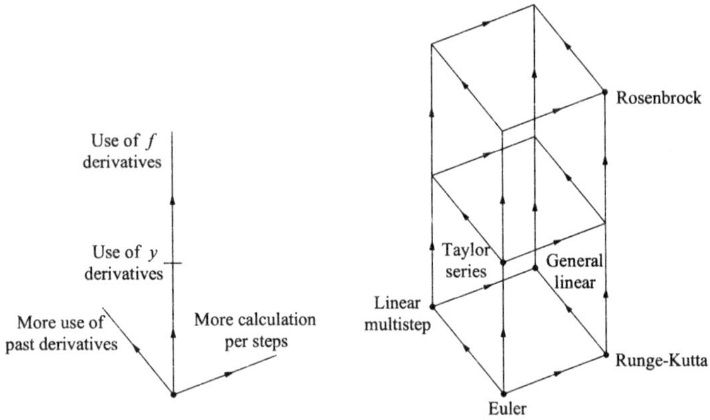

Figure 8. Classification of integration methods.

where Δt is the step size and $\mathbf{y}_{k+1}$ is the numerical solution. The simplest implicit method is the one-step backward *Euler* method:

$$\mathbf{y}_{k+1} = \mathbf{y}_k + \Delta t \mathbf{f}\left(t_{k+1}, \mathbf{y}_{k+1}\right) \tag{3.3}$$

This equation is usually solved for $\mathbf{y}_{k+1}$ by Newton-Raphson linearization, see Subsection 1.2, and involves the Jacobian matrix $\mathbf{J} = \partial_{\mathbf{y}} \mathbf{f}\left(t, \mathbf{y}\right)$. Both the *Euler* and backward *Euler* methods are simple but not very accurate and may require very small step sizes. More accurate finite difference methods were developed from *Euler's* method in two streams:

1. Linear multistep methods (LMS) which combine values $\mathbf{y}_{k+1}$, $\mathbf{y}_k$, $\mathbf{y}_{k-1}$, . . ., and $\mathbf{f}_{k+1}$, $\mathbf{f}_k$, $\mathbf{f}_{k-1}$, . . ., in a linear way to achieve higher accuracy, but sacrifice the one-step format. Linearity permits simple local error estimates, but makes it complicate to change step size.
2. *Runge–Kutta* methods which achieve higher accuracy by retaining the one-step form but sacrificing linearity. One-step form makes it easy to change step size but renders it difficult the estimation of the local error.

Other methods use the computation of the derivatives of the function $\mathbf{y}$, *viz.* the Taylor series techniques; or of the function $\mathbf{f}$, *e.g.* the Rosenbrock's methods which will be introduced hereafter. All these methods can be summarized in Figure 8 using the notation proposed by Butcher (2003); the methods based on the derivative of the $\mathbf{f}$ function are preferred, as they allow more flexibility.

Linear multistep methods LMS methods are typically applied to the Hamilton form (2.3); see, for instance, Hairer and Wanner (1996) for a treatment of these methods. For brevity, an example of LMS method is derived from the single step *CH-α* method (Erlicher et al., 2002) introduced in Subsection 1.2.

Eqs. (1.23)-(1.25) allow $\mathbf{d}_{i+1}, \mathbf{v}_{i+1}$ and $\mathbf{a}_{i+1}$ to be determined starting from known values of $\mathbf{d}_i, \mathbf{v}_i$ and $\mathbf{a}_i$. If the same relations are applied to the time intervals $[t_{i+1}, t_{i+2}]$ and $[t_{i+2}, t_{i+3}]$, nine equations are obtained, in which four velocities and four accelerations can be eliminated. Thereby, a recurrence relation is obtained among displacements $\mathbf{d}_{i+j}$, $j = 0, 1, 2, 3$, that leads to the LMS formulation of the *CH-α* method. The resulting three-step algorithm reads

$$\sum_{j=0}^{3}[\mathbf{M}\alpha_j\mathbf{d}_{i+j} + \Delta t\mathbf{C}\gamma_j\mathbf{d}_{i+j}] + \Delta t^2 \sum_{j=1}^{3} \delta_j(\mathbf{r}_{i+j-\alpha_f} - \mathbf{f}_{i+j-\alpha_f}) = 0 \tag{3.4}$$

where $\mathbf{f}_{i+1-\alpha_f} = \mathbf{f}[(1-\alpha_f)t_{i+1}+\alpha_f t_i]$ and $\mathbf{r}_{i+1-\alpha_f} = \mathbf{r}[(1-\alpha_f)\mathbf{d}_{i+1}+ \alpha_f\mathbf{d}_i]$ are approximated by means of the mid-point rule. Moreover, the parameters in (3.4) read

$$\begin{array}{lll} \alpha_0 = \alpha_m & \gamma_0 = \alpha_f(-1+\gamma) & \delta_1 = \frac{1}{2} + \beta - \gamma \\ \alpha_1 = 1 - 3\alpha_m & \gamma_1 = -1 + 2\alpha_f + \gamma - 3\gamma\alpha_f & \delta_2 = \frac{1}{2} - 2\beta + \gamma \\ \alpha_2 = -2 + 3\alpha_m & \gamma_2 = 1 - \alpha_f - 2\gamma + 3\gamma\alpha_f & \delta_3 = \beta \\ \alpha_3 = 1 - \alpha_m & \gamma_3 = (1-\alpha_f)\gamma. & \end{array} \tag{3.5}$$

Runge-Kutta methods In order to introduce the Runge-Kutta methods in structural dynamics, we employ the form (**2.3**). Generally, a Runge-Kutta s-stage method applied to (**2.3**) may be expressed by

$$\mathbf{y}_{k+1} = \mathbf{y}_k + \sum_{i=1}^{s} b_i\mathbf{k}_i \tag{3.6}$$

where b_i are the weights of the function and $\mathbf{k}_i$ are the estimates of the right-hand side of the differential equation, function of s intermediate substep inside the time step (Hairer et al., 1987). Their values are expressed as

$$\mathbf{k}_i = \mathbf{f}(t_k + c_i\Delta t, \mathbf{y}_k + \sum_{j=1}^{s} a_{ij}\mathbf{k}_j)\Delta t \quad (i = 1, \ldots, s) \tag{3.7}$$

In practice, the method sends feelers into the solution space, to gather samples of the derivatives and the solution is obtained as a linear combination of these feelers. The method can be summarized in the *Butcher tableau* (Butcher, 2003)

$$\begin{array}{c|c} \mathbf{c} & \mathbf{A} \\ \hline & \mathbf{b}^T \end{array}$$

where $\mathbf{b}$ is the vector of the weights of the estimates $\mathbf{k}_i$ in (3.6), $\mathbf{c}$ is the vector of the time estimate of the function $\mathbf{f}$, while $\mathbf{A}$ is the matrix of the coefficients a_{ij} that define the weights of the estimates $\mathbf{k}_j$ in the evaluation of the estimate $\mathbf{k}_i$ in (3.7). The matrix $\mathbf{A}$, the vectors $\mathbf{b}$ and $\mathbf{c}$ are made of real numbers. For instance, the explicit three-stage Runge-Kutta method can be summarized in the *Butcher tableau*

$$\begin{array}{c|ccc} 0 & 0 & 0 & 0 \\ 1/3 & 1/3 & 0 & 0 \\ 2/3 & 0 & 2/3 & 0 \\ \hline & 1/4 & 0 & 3/4 \end{array}$$

and it relies on the following solution:

$$\mathbf{y}_{k+1} = \mathbf{y}_k + \frac{1}{4}\mathbf{k}_1 + \frac{3}{4}\mathbf{k}_3, \tag{3.8}$$

where

$$\begin{aligned} \mathbf{k}_1 &= \mathbf{f}(\mathbf{f}_{\mathrm{e,k}}, \mathbf{y}_k)\Delta t, \\ \mathbf{k}_2 &= \mathbf{f}(\mathbf{f}_{e,k+\frac{1}{3}}, \mathbf{y}_k + \frac{1}{3}\mathbf{k}_1)\Delta t, \\ \mathbf{k}_3 &= \mathbf{f}(\mathbf{f}_{e,k+\frac{2}{3}}, \mathbf{y}_k + \frac{2}{3}\mathbf{k}_2)\Delta t, \end{aligned} \tag{3.9}$$

in which the external forces $\mathbf{f}_{e,k+\frac{1}{3}}$ and $\mathbf{f}_{e,k+\frac{2}{3}}$ are defined at time $t_{k+\frac{1}{3}} = t_k + \frac{1}{3}\Delta t$ and $t_{k+\frac{2}{3}} = t_k + \frac{2}{3}\Delta t$, respectively. All these intermediate evaluations are sketched in Figure 9. Note that both the forward *Euler* and the explicit three-stage Runge-Kutta algorithm are *real-time compatible*, as they do not require the knowledge of the value of the function f or its derivatives at the end of Δt, to obtain the solution of (2.3). This is an important property for *SHSDS* applications.

To study the properties of time integration algorithms in Subsection 3.2, we introduce shortly the real-time compatible Rosenbrock method with one stage (Rosenbrock, 1963). Let's consider an autonomous problem, characterized by

$$\mathbf{y}' = \mathbf{f}(\mathbf{y}). \tag{3.10}$$

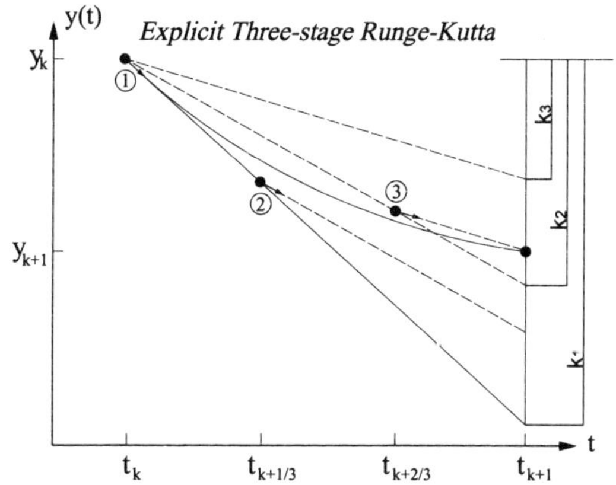

Figure 9. Representation of the Explicit three-stage Runge-Kutta method.

The solution of the problem by means of an s-stage Rosenbrock method is given by the formulas:

$$\mathbf{y}_{k+1} = \mathbf{y}_k + \sum_{i=1}^{s} b_i \mathbf{k}_i \tag{3.11}$$

$$\mathbf{k}_i = \Delta t(\mathbf{f}(\mathbf{y}_k + \sum_{j=1}^{i-1} \alpha_{ij}\mathbf{k}_j) + \mathbf{J}\sum_{j=1}^{i} \gamma_{ij}\mathbf{k}_j), \quad (i = 1, \ldots, s)$$

where α_{ij}, γ_{ij} and b_j are the algorithm coefficients and $\mathbf{J}$ is evaluated at the initial step solution $\mathbf{y}_k$. Each stage of this method consists of a system of linear equations with unknowns $\mathbf{k}_i$ and with the inversion of the matrix $[\mathbf{I} - \Delta t\gamma_{ii}\mathbf{J}]$. We assume that $\gamma_{11} = \gamma_{22} = \ldots = \gamma_{ss} = \gamma$, so that we need only one LU-decomposition per step. Non-autonomous problems, characterized by the differential equation

$$\mathbf{y}' = \mathbf{f}\,(t, \mathbf{y}) \tag{3.12}$$

can be converted to autonomous form by adding $t' = 1$. If the method is applied to the augmented system, the components corresponding to the t

variable can be computed explicitly, thus arriving to the equations

$$\mathbf{k}_i = \Delta t(\mathbf{f}(t_k + \alpha_i \Delta t, \mathbf{y}_k + \sum_{j=1}^{i-1} \alpha_{ij}\mathbf{k}_j) + \gamma_i \Delta t \left.\frac{\partial \mathbf{f}}{\partial t}\right|_{t_k, \mathbf{y}_k} + \mathbf{J}\sum_{j=1}^{i} \gamma_{ij}\mathbf{k}_j) \tag{3.13}$$

$$\mathbf{k}_i = \Delta t(\mathbf{f}(t_k + \alpha_i \Delta t, \mathbf{y}_k + \sum_{j=1}^{i-1} \alpha_{ij}\mathbf{k}_j) + \mathbf{J}\sum_{j=1}^{i} \gamma_{ij}\mathbf{k}_j) \tag{3.14}$$

$$\mathbf{y}_{k+1} = \mathbf{y}_k + \sum_{i=1}^{s} b_i \mathbf{k}_i \tag{3.15}$$

where (3.14) is a simplified form of (3.13), and the additional coefficients are given by

$$\alpha_i = \sum_{j=1}^{i-1} \alpha_{ij} \quad \text{and} \quad \gamma_i = \sum_{j=1}^{i} \gamma_{ij} \tag{3.16}$$

The method with one stage reads:

$$y_{k+1} = y_k + k_1 b_1 \tag{3.17}$$

with

$$\begin{aligned} k_1 &= f(t_k, y_k)\Delta t + \Delta t \frac{\partial f}{\partial y}(t_k, y_k)\gamma_{11} k_1 \\ k_1 &= \frac{f(t_k, y_k)\Delta t}{(1 - \Delta t J \gamma_{11})} \end{aligned} \tag{3.18}$$

The parameters values

$$b_1 = 1 \quad \text{and} \quad \gamma_{11} = 1 \tag{3.19}$$

entail L-stability to this method, labelled *LSRT1*. Additional details are provided hereafter.

3.2 Analysis of time integration algorithms

In order to understand the pros and cons of the plethora of time integration algorithms employed both for *SHSDS* and *PsD* testing, accuracy, stability, numerical dissipation and dispersion and error propagation are discussed herein.

Accuracy A time integration algorithm can be considered accurate if the computed numerical response approaches as much as desired to the exact response. Usually, the load vector $\mathbf{f}_e(t)$ is not included in the analysis assuming that the power of the leading error term of its approximation is greater than the order of accuracy of the method. With reference to the Hamilton form (2.3) of the equations of structural dynamics and along the lines of Hulbert and Hughes (1987) we consider the error both owing to the initialization process and to the local truncation error. Accordingly, the global error $\mathbf{e}_k$ at t_k is affected by the initial error $\mathbf{e}_0$ and by the error at each integration step Δt, i.e.

$$\mathbf{e}_N = \mathbf{A}^N \mathbf{e}_0 + \sum_{k=0}^{N-1} \mathbf{A}^{N-(k+1)} \boldsymbol{\tau}_k \tag{3.20}$$

where $\mathbf{e}_k$ is defined as

$$\mathbf{e}_k = \mathbf{e}\left(t_k\right) = \mathbf{y}_k - \mathbf{y}\left(t_k\right) \tag{3.21}$$

being $\mathbf{y}(t_k)$ the corresponding exact solution of the state vector $\mathbf{y}_k$ at t_k and $\boldsymbol{\tau}_k = \boldsymbol{\tau}(t_k)$ the local truncation error defined as

$$\boldsymbol{\tau}_k = \boldsymbol{\tau}\left(t_k\right) = \mathbf{A}\mathbf{y}\left(t_k\right) + \mathbf{L}_k - \mathbf{y}\left(t_{k+1}\right). \tag{3.22}$$

If $\tau = O\left(\Delta t^{k+1}\right)$ the algorithm is said to be of the k-th order (Hairer et al., 1987). A treatment of $\mathbf{e}_0$ will be performed in Subsection 6.1 with reference to *PsD* tests; whilst the conditions to assure a specific accuracy order of $\boldsymbol{\tau}_k$ for the Rosenbrock method (3.13)-(3.15) are performed here comparing the derivative terms of the Taylor series of the exact solution with those of the numerical solution. Along this line, we expand the Taylor series of the SDoF differential equation $y\prime = f\left(t, y\left(t\right)\right)$ up to the third order:

$$\begin{aligned} y_{k+1} &= y_k + f_k \Delta t + \frac{\Delta t^2}{2}\left(f_k \left.\frac{\partial f}{\partial y}\right|_k + \left.\frac{\partial f}{\partial t}\right|_k\right) + \frac{\Delta t^3}{6} f_k \left(\left.\frac{\partial f}{\partial y}\right|_k\right)^2 \\ &+ \frac{\Delta t^3}{6}\left(f_k^2 \left.\frac{\partial^2 f}{\partial y^2}\right|_k + \left.\frac{\partial f}{\partial y}\right|_k \left.\frac{\partial f}{\partial t}\right|_k + 2 f_k \left.\frac{\partial^2 f}{\partial t \partial y}\right|_k + \left.\frac{\partial^2 f}{\partial t^2}\right|_k\right) + O(\Delta t^4) \end{aligned} \tag{3.23}$$

where $f_k = f\left(t_k, y_k\right)$ and $k = \left(t_k, y_k\right)$. In a similar way the Taylor series expansion of the problem integrated by a three stages Rosenbrock method introduced in Subsection 5.1 is obtained:

$$
\begin{aligned}
y_{k+1} &= y_k + f_k \Delta t (b_1 + b_2 + b_3) + \Delta t^2 (b_1 \alpha_1 + b_2 \alpha_2 + b_3 \alpha_3) \left.\frac{\partial f}{\partial t}\right|_k \\
&+ f_k \Delta t^2 (b_2 \beta_{21} + b_3 (\beta_{31} + \beta_{32}) + \gamma) \left.\frac{\partial f}{\partial y}\right|_k + \\
&+ f_k \Delta t^3 (\gamma - \gamma^2 + (\alpha_{32} + \gamma_{32})(\alpha_{21} + \gamma_{21}) b_3) (\left.\frac{\partial f}{\partial y}\right|_k)^2 + \\
&+ \frac{f_k^2 \Delta t^3}{2} (\alpha_{21}^2 b_2 + (\alpha_{31} + \alpha_{32})^2 b_3) \left.\frac{\partial^2 f}{\partial y^2}\right|_k + \\
&+ \Delta t^3 (\beta_{21} b_2 \alpha_1 + (\alpha_1 \beta_{31} + \alpha_2 \beta_{32}) b_3 + \frac{\gamma}{2}) \left.\frac{\partial f}{\partial y}\right|_k \left.\frac{\partial f}{\partial t}\right|_k + \\
&+ f_k \Delta t^3 (\alpha_{21} b_2 \alpha_2 + (\alpha_{31} + \alpha_{32}) b_3 \alpha_3) \left.\frac{\partial^2 f}{\partial t \partial y}\right|_k \\
&+ \frac{\Delta t^3}{2} (b_1 \alpha_1^2 + b_2 \alpha_2^2 + b_3 \alpha_3^2) \left.\frac{\partial^2 f}{\partial t^2}\right|_k + O(\Delta t^4)
\end{aligned}
\tag{3.24}
$$

where $\beta_{ij} = \alpha_{ij} + \gamma_{ij}$. Comparing each term of (3.23) with those of (3.24) the accuracy conditions for a third-order system can be obtained. For non-autonomous systems, by substituting the values of (3.16), the accuracy conditions read

$$
\begin{array}{ll}
1^{st} \text{ order} & \{ b_1 + b_2 + b_3 = 1 \\
2^{nd} \text{ order} & \left\{ \begin{array}{l} b_2 \alpha_2 + b_3 \alpha_3 = 1/2 \\ b_2 \beta_{21} + b_3 (\beta_{31} + \beta_{32}) = 1/2 - \gamma \end{array} \right. \\
3^{rd} \text{ order} & \left\{ \begin{array}{l} b_2 \alpha_2^2 + b_3 \alpha_3^2 = 1/3 \\ \alpha_2 \beta_{32} b_3 = 1/6 - \gamma/2 \\ \beta_{32} \beta_{21} b_3 = 1/6 - \gamma + \gamma^2 \end{array} \right.
\end{array}
\tag{3.25}
$$

For a method with lower number of stages one can use the same equation (3.25) neglecting the coefficients with higher subscripts.

The conditions to obtain third order accuracy for a three stages Rosenbrock method applied to an autonomous system can be computed neglecting the contribution of the derivative $\frac{\partial f}{\partial t}$ and read:

$$
\begin{array}{ll}
1^{st} \text{ order} & \{ b_1 + b_2 + b_3 = 1 \\
2^{nd} \text{ order} & \{ b_2 \beta_{21} + b_3 (\beta_{31} + \beta_{32}) = 1/2 - \gamma \\
3^{rd} \text{ order} & \left\{ \begin{array}{l} b_2 \alpha_2^2 + b_3 \alpha_3^2 = 1/3 \\ b_3 \beta_{32} \beta_{21} = 1/6 - \gamma + \gamma^2 \end{array} \right.
\end{array}
\tag{3.26}
$$

The first equation in (3.26) justifies the first coefficient in (3.19); the other order conditions have been used to set the coefficients of the Rosenbrock methods with two and three stages introduced in Subsection 5.1.

Due to relevance of the error (3.20) in the performance of an integration algorithm, several papers deal with error estimators. For brevity, we mention among others, Wiberg and Li (1993) and Lacoma and Romero (2007). For instance, Lacoma and Romero (2007) developed an algorithm for estimating the error when the *HHT*-α method is used both in the linear and non-linear regime. The correct formulation of the estimator $\boldsymbol{\theta}_{i+1}$, is based on the acceleration vector

$$\mathbf{a}_{i+1}^{new} = \frac{1}{\alpha}[\mathbf{a}_{i+1} - (1-\alpha)\mathbf{a}_i^{new}] \tag{3.27}$$

which is a third-order approximation to the exact acceleration $\ddot{\mathbf{u}}(t_{i+1})$ of the local problem, since the one estimated by the *HHT*-α method is not sufficiently accurate as shown in Subsection 1.2. The pseudo-code to evaluate the new error estimator for the *HHT*-α method is summarized here.

Given the solution at t_i, that is, the vectors $\mathbf{d}_i, \mathbf{v}_i, \mathbf{a}_i$:

1. Compute the solution at $t_{i+1}, \mathbf{d}_{i+1}, \mathbf{v}_{i+1}, \mathbf{a}_{i+1}$ using the *HHT*-α method.
2. Compute $\mathbf{a}_{i+1}^{new}$ at t_{i+1} by (3.27).
3. Compute the displacement, velocity and acceleration at instant $t_{i+1/2}$:
   a) for displacements and velocities, use the following expressions:

$$\mathbf{d}_{i+1/2}^{*} = \tilde{\mathbf{d}}_{i+1/2}^{*} = \frac{3}{4}\mathbf{d}_i + \frac{1}{4}\mathbf{d}_{i+1} + \frac{\Delta t_i}{4}\mathbf{v}_i \quad \text{or} \tag{3.28}$$

$$\mathbf{d}_{i+1/2}^{*} = \tilde{\mathbf{d}}_{i+1/2}^{*} = \frac{1}{2}(\mathbf{d}_i + \mathbf{d}_{i+1}) \tag{3.29}$$

$$+\frac{\Delta t_n}{4}(\mathbf{v}_i - \mathbf{v}_{i-1}) + \frac{\Delta t_n^2}{16}(\mathbf{a}_i^{new} + \mathbf{a}_{i+1}^{new})$$

$$\mathbf{v}_{i+1/2}^{*} = \mathbf{v}_i + \frac{3\Delta t_i}{8}\mathbf{a}_i^{new} + \frac{\Delta t_i}{8}\mathbf{a}_{i+1}^{new}]; \tag{3.30}$$

   b) for the acceleration, first compute the vector of internal forces at $t_{n+1/2}$ and, subsequently, solve the following linear system of equations:

$$\mathbf{M}\mathbf{a}_{i+1/2}^{*} = \mathbf{f}(t_{i+1/2}) + \mathbf{r}(\mathbf{d}_{i+1/2}^{*} + \mathbf{v}_{i+1/2}^{*}) \tag{3.31}$$

4. Compute the quadrature rule

$$\mathbf{Q}_n^{n+1} = \frac{\Delta t_n}{6}\begin{bmatrix} \mathbf{v}_i \\ \mathbf{a}_i^{new} \end{bmatrix} + \frac{4\Delta t_i}{6}\begin{bmatrix} \mathbf{v}_{i+1/2}^{*} \\ \mathbf{a}_{i+1/2}^{*} \end{bmatrix} + \frac{\Delta t_i}{6}\begin{bmatrix} \mathbf{v}_{i+1} \\ \mathbf{a}_{i+1}^{new} \end{bmatrix} \tag{3.32}$$

5. Compute the error estimate and its energy norm

$$\boldsymbol{\theta}_{i+1} = \tilde{z}_{i+1} - \mathbf{Q}_i^{i+1} - \tilde{z}_i \Longrightarrow \| \boldsymbol{\theta}_{i+1} \|_E \tag{3.33}$$

6. Set $i \leftarrow i + 1$.

Absolute stability and the phenomenon of stiffness Two general types of stability can be considered for time integration algorithms: i) the so-called *external stability*, conditioned by the external excitation; ii) the so-called *equilibrium stability* related to the free evolution of the discrete dynamical system starting from some initial conditions as with the Lyapunov stability (Nawrotzki and Eller, 2000). With reference to the *external stability*, different studies on the algorithmic behaviour with a forcing term were performed by (Preumont, 1982). He derived the transfer functions of the discretized SDoF equation and the comparison between the numerical and the exact transfer function provided a more complete picture of the behaviour of the Newmark method detecting spurious resonance conditions. Similar analyses were carried out on different schemes by others (Cannillo and Mancuso, 2000; Pegon, 2001; Mugan and Hulbert, 2001; Baldo et al., 2006).

Herein we develop *equilibrium stability*, which assures that initial distortions and round-off errors do not increase artificially in the computation process, *i.e.* the Lyapunov stability, or eventually damp out, *viz.* the asymptotical and global stability. We start introducing the definition of A-stability as the properties of an integrator of preserving the Lyapunov stability of the numerical solution when applied to a stable system, *i.e.* a system with s-plane representation on the left half-plane, described by the Dahlquist differential equation problem (Dahlquist, 1963)

$$\begin{aligned} y'(t) &= \lambda y(t) \\ y(0) &= 1 \end{aligned} \tag{3.34}$$

This is a desirable property for a numerical method and the domain $\boldsymbol{A}$ that corresponds to an A-stable method for any value of $\lambda\Delta t < 0$, is

$$\boldsymbol{A} = \{\lambda\Delta t \in C; |\boldsymbol{R}(\lambda\Delta t)| \leqslant 1\} \tag{3.35}$$

where $\boldsymbol{R}$ is named the *stability function* defined as $y_{k+1} = \boldsymbol{R}(\lambda\Delta t)\, y_k$.

For a particular type of system, called a stiff system (Lambert, 1991), the A-stability property might not be sufficient and we invoke global stability. In detail, a system is said to be stiff in a given interval $[0, t_f]$, if the neighboring solution curves approach the solution curve at a rate which is very large in comparison with the rate at which the solution varies in that interval. Thus, when we have a large value of $|\lambda\Delta t|$ with $|R|$ really close to 1, the high-frequency (stiff) components of the response are damped out very slowly. As a consequence, the L-stability property was introduced as the condition to be satisfied, in addition to the A-stability property, for stiff problems. It can be summarized as

$$\lim_{\lambda\Delta t\rightarrow-\infty} R(\lambda\Delta t) = 0 \tag{3.36}$$

and entails that an L-stable integration method nearly annihilates the response of a high-frequency mode in a single time step.

For instance and with reference to the *LSRT1* method introduced in Subsection 3.1, we investigate the stability property when λ is large. The solution with the Rosenbrock method provides:

$$\begin{aligned} y_{k+1} &= y_k + k_1 b_1, \quad k_1 = \frac{\lambda\Delta t}{(1-\lambda\Delta t\gamma)} y_k \\ y_{k+1} &= y_k + \frac{\lambda\Delta t}{(1-\lambda\Delta t\gamma)} y_k \\ R(\lambda\Delta t) &= \frac{y_{k+1}}{y_k} = \frac{1-\lambda\Delta t\gamma+\lambda\Delta t}{1-\lambda\Delta t\gamma} \end{aligned} \tag{3.37}$$

To analyse the A-stability property, we introduce the variable $z = -\lambda\Delta t$, imposing that $|R(z)| \leqslant 1$ with $z \geq 0$

$$\begin{aligned} R(z) &= \frac{1+z\gamma-z}{1+z\gamma} \\ |R(z)| &= \left|\frac{1+z(\gamma-1)}{1+z\gamma}\right| < 1 \\ \gamma &> 1/2 \end{aligned} \tag{3.38}$$

To induce the L-stability property, we impose that $\lim_{z\rightarrow+\infty} R = 0$, *i.e.*

$$\begin{aligned} R(z) &= \frac{1+z(\gamma-1)}{1+z\gamma} \\ \lim_{z\rightarrow+\infty} R &= \frac{z(\gamma-1)}{z\gamma} = \frac{\gamma-1}{\gamma} = 0 \\ \gamma &= \gamma_{11} = 1 \end{aligned} \tag{3.39}$$

This value agrees with (3.19). The plot of the relevant stability function $R(z)$ for $\gamma_{11} = 1$ is shown in Figure 10(a).

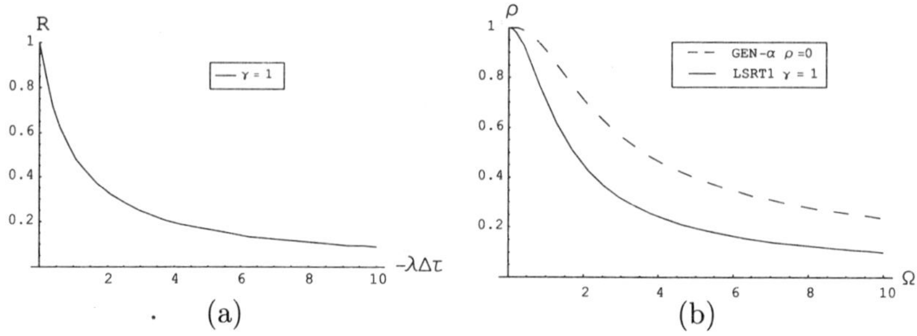

Figure 10. (a) Stability function plot for a first-order system with $\gamma = \mathbf{1}$ and with a single-stage Rosenbrock method assuming $\boldsymbol{b_1} = \mathbf{1}$;(b) stability function plot for a second-order system with $\gamma = \mathbf{1}$ and with a single-stage Rosenbrock method assuming $\boldsymbol{b_1} = \mathbf{1}$.

To analyse the properties of the *LSRT1* method for a mechanical second-order system, we consider the model problem

$$\begin{aligned} \boldsymbol{y}' &= \boldsymbol{p} \\ \boldsymbol{p}' &= -\boldsymbol{\omega}^2 \boldsymbol{y} \end{aligned} \tag{3.40}$$

As a result, $\boldsymbol{\lambda}^2 = -\boldsymbol{\omega}^2$ and the spectral radius $\boldsymbol{\rho} = |\boldsymbol{R}(\boldsymbol{z})|$ of the *LSRT1* method is simply obtained by substituting $\boldsymbol{\lambda}_{1,2} = \pm \boldsymbol{i\omega}$ into the stability function $\boldsymbol{R}(\boldsymbol{z})$. The corresponding evolution of $\boldsymbol{\rho}$ is plotted in Figure 10b. For brevity, it is compared only with the L-stable *CH-$\boldsymbol{\alpha}$* method characterized by $\boldsymbol{\rho_\infty} = \mathbf{0}$ (Chung and Hulbert, 1993); one can observe the property of the *LSRT1* method.

Both the property of accuracy dealt with in Subsection 3.2.1 and stability considered in this subsection entail convergence of a time integration algorithm applied to a linear system (Lambert, 1991). Nonetheless, the integration of non-linear dynamic systems may impair numerical stability and convergence of computed responses for several reasons (Soroushian et al., 2005). Some of them are listed here: i) computed responses affected by both truncation errors, see (3.22) and residuals of non-linearity solutions, *e.g.* (1.22); ii) approximations in the formulation of time integration methods, like the *CH-$\boldsymbol{\alpha}$* method that replaces the equations of motion with balance equations, see (1.12), where the acceleration-like variables are approximate as detected by (1.28).

Dissipation and dispersion Other important measures which are useful in the choice of time integration algorithms are the algorithmic damping ratio $\bar{\xi}$ and the frequency error $(\Omega - \bar{\Omega})/\Omega$, where $\bar{\Omega} = \bar{\omega}\Delta t$ (Hughes, 1987). For instance, with reference to the *LSRT1* method applied to **(3.40)**, the response can be expressed as

$$\begin{aligned} y_{k+1} &= \sum_{s=1}^{2} \bar{c}_s \lambda_s^{k+1} \\ &= e^{\left(-\bar{\xi}\bar{\Omega}(k+1)\right)} \left[d_1 \cos\left(\bar{\Omega}\,(k+1)\right) + d_2 \sin\left(\bar{\Omega}\,(k+1)\right)\right] \end{aligned} \tag{3.41}$$

being $\lambda_{1,2}$

$$\lambda_{1,2} = C \pm iD \tag{3.42}$$

the eigenvalues of the amplification matrix of the method and d_i constants determined from initial conditions. As a result, $\bar{\xi}$ and $\bar{\omega} = \bar{\Omega}/\Delta t$ read

$$\bar{\xi} = -\frac{\ln\left(C^2 + D^2\right)}{2\bar{\Omega}} \tag{3.43}$$

$$\bar{\omega}\Delta t = \bar{\Omega} = \arctan\left(\frac{D}{C}\right) \tag{3.44}$$

Both $\bar{\xi}$ and the frequency error $(\Omega - \bar{\Omega})/\Omega$ are depicted in Figure 11, respectively, and compared to the corresponding values of the *CH-α* method with $\rho_\infty = 0$. The favourable properties of the *LSRT1* method in terms of high algorithm damping and limited frequency error are evident.

These measures can be expressed also in an alternative form using a digital filter theory as presented in Subsection 4.3.

Error propagation in PsD testing In Subsection 3.2 we have reported an algorithm proposed by Lacoma and Romero (2007) able to perform error estimate when the *HHT-α* method is used to solve structural dynamic problems. The efficiency of a time-stepping scheme applied, for instance, to a *PsD* testing is connected to its reduced sensitivity to errors of different nature. In a *PsD* environment several errors arise (Combescure and Pegon, 1997) and the integrator should mitigate their effects. In fact, the time integration algorithm computes the displacement to be imposed on the structure and the digital board of the actuators generates an approach function to the desired target. As the approach process has to be stable and gradual, it is inevitable a certain difference between the effective displacement applied to the tested structure (superscript c) and the target displacement computed from the *PsD* algorithm, which is called stepwise control error

$$\mathbf{e}_{i+1}^{\mathbf{d,c}} = \mathbf{d}_{i+1}^{\mathbf{c}} - \mathbf{d}_{i+1}. \tag{3.45}$$

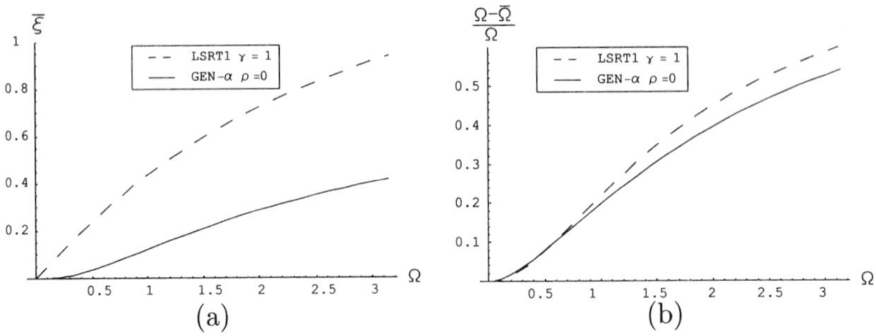

Figure 11. (a) Equivalent damping for a second-order system with a single-stage Rosenbrock method assuming $\boldsymbol{\gamma = 1}$ and $\boldsymbol{b_1 = 1}$; (b) frequency error for a second-order system with a single-stage Rosenbrock method assuming $\boldsymbol{\gamma = 1}$ and $\boldsymbol{b_1 = 1}$.

Accurate displacement control has been found to be of prime importance to ensure reliable response. In fact owing to undershoot, the controller generates error forces which are in phase with the structure under testing; if undershooting errors depend on the sign of the displacement increment, they add energy to the test piece (Combescure and Pegon, 1997). An additional displacement error is the stepwise measurement error (superscript m) caused by the precision of the displacement transducers which reads

$$\mathbf{e}_{i+1}^{\mathbf{d},m} = \mathbf{d}_{i+1}^{m} - \mathbf{d}_{i+1}^{\mathbf{c}}. \tag{3.46}$$

Another source of errors is the stepwise measurement error associated with the reaction force

$$\mathbf{e}_{i+1}^{\mathbf{r},m} = \mathbf{r}_{i+1}^{m} - \mathbf{r}_{i+1}^{\mathbf{c}}. \tag{3.47}$$

By considering a first order approximation, the total error on the reaction force can be recast in the following form

$$\mathbf{e}_{i+1}^{\mathbf{r}} = \mathbf{r}_{i+1}^{m} - \mathbf{r}_{i+1} = \mathbf{r}_{i+1}^{m} - \mathbf{r}_{i+1}^{\mathbf{c}} + \mathbf{r}_{i+1}^{\mathbf{c}} - \mathbf{r}_{i+1} \simeq \mathbf{e}_{i+1}^{\mathbf{r},m} + \mathbf{K}_{a,i+1}\mathbf{e}_{i+1}^{\mathbf{d},\mathbf{c}} \tag{3.48}$$

where $\mathbf{K}_{a,i+1}$ represents the actual stiffness of a structure. The off line analysis of the error effects on the structural response can easily be performed considering a linear elastic structure with a single DoF (Bursi and Shing, 1996; Combescure and Pegon, 1997; Bonelli and Bursi, 2004; Chang, 2005). More specifically, a single error in displacement or in force is equivalent to an external force applied to the structure leading to different effects depending on the scheme employed. The analysis described herein regards the effect of uncorrelated errors. Nonetheless, some errors like displacement control errors are correlated, and therefore analyses based on the discrete-time frequency response function of the scheme have been performed too (Combescure and Pegon, 1997; Bonelli and Bursi, 2004). Also on line analyses can be performed based on error monitors, which can provide on line early warnings of unacceptable levels of experimental errors (Mosqueda et al., 2007a, and Mosqueda et al., 2007b). These error monitors based on estimates of energy added to the simulation allow to pause the test, and correct the problem preferably before the specimen is irreparably damaged.

For brevity, we summarize herein the off line analysis applied to the $\boldsymbol{IPC - \rho_\infty}$ method (Bonelli and Bursi, 2004). Let's analyse the effect of uncorrelated errors by means of an error excitation $\boldsymbol{f_j^e}$ of impulsive type, such that

$$\boldsymbol{f_j^e} = \begin{cases} \boldsymbol{f^e} & \text{if } \boldsymbol{j = i - 1} \\ \mathbf{0} & \boldsymbol{j \neq i - 1} \end{cases} \tag{3.49}$$

Under this assumption, the response of a SDoF system evaluated with the $\boldsymbol{IPC - \rho_\infty}$ algorithm amounts to

$$\mathbf{Y}_i^e = \left(\left(1 - \alpha_f\right) \mathbf{A} + \alpha_f \mathbf{I}\right) \mathbf{B} \begin{bmatrix} \boldsymbol{f_{i-1}^e} \\ \mathbf{0} \end{bmatrix} \tag{3.50}$$

and hence

$$\mathbf{Y}_{i+r}^e = \mathbf{A}^r \mathbf{Y}_i^e = \mathbf{A}^r \left(\left(1 - \alpha_f\right) \mathbf{A} + \alpha_f \mathbf{I}\right) \mathbf{B} \begin{bmatrix} \boldsymbol{f_{i-1}^e} \\ \mathbf{0} \end{bmatrix} \tag{3.51}$$

where $\mathbf{A}$ is the amplification matrix of the $\boldsymbol{IPC - \rho_\infty}$, $\mathbf{Y} = \{\boldsymbol{d_i, \Delta t v_i, \Delta t^2 a_i, \Delta t^2 \frac{r_i^r}{m}}\}^T$ denotes its state vector and $\mathbf{B}$ is related to the forcing terms. Equation (3.51) can be represented in terms of eigenvalues $\boldsymbol{\lambda_k}$ of $\mathbf{A}$ and more specifically, the displacement assumes the following form

$$\boldsymbol{d_j} = \sum_{k=1}^{n} c_k \boldsymbol{\lambda_k^j}. \tag{3.52}$$

$\mathbf{A}$ has four eigenvalues and, retaining only the complex conjugate ones, (3.52) can be recast as

$$\boldsymbol{d_j} = e^{\left(-\xi \bar{\omega} i \Delta t\right)} \left(c_1 \cos\left(\bar{\omega} i \Delta t\right) + c_1^* \cos\left(\bar{\omega} i \Delta t\right)\right). \tag{3.53}$$

The error amplification factor $\boldsymbol{C_1}$ reads

$$\boldsymbol{C_1} = c_1 + c_1^*, \tag{3.54}$$

where $\boldsymbol{c}_1$ and $\boldsymbol{c}_1^*$ are complex conjugate. $\boldsymbol{C}_1$ and $\boldsymbol{e}^{(-\bar{\xi}\bar{\omega}i\Delta t)}$, allow the error propagation characteristics of a PsD algorithm to be evaluated. As lengthy expressions are associated with $\boldsymbol{c}_1$ and $\boldsymbol{c}_1^*$, only a graphical representation of $\boldsymbol{C}_1$ is provided.

Consider a control error $\boldsymbol{e}_{i+1}^{d,c}$ in the actuator movement with no measurement error. By substituting $\mathbf{Y}_i = \left\{ \boldsymbol{d}_i^r,\ \Delta t \boldsymbol{v}_i^r,\ \Delta t^2 \boldsymbol{a}_i^r,\ \Delta t^2 \frac{\boldsymbol{r}_i^r}{\boldsymbol{m}} \right\}^T$, which denotes the exact numerical solution evaluated from experimental feedback quantities, in (3.50) and using (3.48), one obtains the following error force

$$\boldsymbol{f}_i^e = \frac{\overline{\boldsymbol{K}} + \boldsymbol{K}}{\overline{\boldsymbol{K}} + 2\boldsymbol{K} - \boldsymbol{K}_a} (\boldsymbol{K}_a - \boldsymbol{K})\, \boldsymbol{e}_{i+1}^{d,c} \tag{3.55}$$

for the $\boldsymbol{IPC} - \boldsymbol{\rho}_\infty$ scheme where $\overline{\boldsymbol{K}}$ is found in (1.45). Conversely, the error force related to the measurement error $\boldsymbol{e}_{i+1}^{d,m}$ without control errors reads

$$\boldsymbol{f}_i^e = \boldsymbol{K}_a \boldsymbol{e}_{i+1}^{d,m}. \tag{3.56}$$

The performance of the $\boldsymbol{OSM} - \boldsymbol{\rho}_\infty$ scheme and of the $\boldsymbol{C} - \boldsymbol{\rho}_\infty$ method is presented, where these acronyms stand for the $\boldsymbol{OSM}$ and the $\boldsymbol{C} - \boldsymbol{\alpha}$ scheme implemented in the $\boldsymbol{CH} - \boldsymbol{\alpha}$ method, for a fair comparison (Bonelli and Bursi, 2004). In these cases, the error force caused by a stepwise control error $\boldsymbol{e}_{i+1}^{d,c}$ reads

$$\boldsymbol{f}_i^e = (\boldsymbol{K} - \boldsymbol{K}_a)\, \boldsymbol{e}_{i+1}^{d,c}. \tag{3.57}$$

The corresponding error amplification factors are depicted in Figure 12. The favourable behaviour of the $\boldsymbol{IPC} - \boldsymbol{\rho}_\infty$ method at high frequencies is evident. Moreover, the $\boldsymbol{IPC} - \boldsymbol{\rho}_\infty$ and the $\boldsymbol{C} - \boldsymbol{\rho}_\infty$ method do not show any unfavourable effect when $\boldsymbol{R} > \mathbf{1}$, where $\boldsymbol{R} = \frac{\boldsymbol{K}_a}{\boldsymbol{K}}$ is the stiffness ratio. Conversely, the $\boldsymbol{OSM} - \boldsymbol{\rho}_\infty$ scheme exhibits bifurcation, i.e. the complex conjugate roots $\boldsymbol{\lambda}_{1,2}$ become real and distinct, a clear warning of instability.

A complementary way to characterize the time-stepping scheme in the presence of errors consists in the analysis of its discrete-time frequency response function, assuming that the error force $\boldsymbol{f}_i^e$ acts at each time step (Combescure and Pegon, 1997; Bonelli and Bursi, 2004). It is clear that errors derived from experimental data can be random but also systematic owing to significant bias or calibration errors. Therefore, probabilistic techniques based on vectorized Monte Carlo simulations or the like should be used to find cumulative probability distributions for output variables of analyses and to investigate the uncertainty propagation in numerical/physical models. Analyses of this type were performed by Vasquez and Whiting (2006).

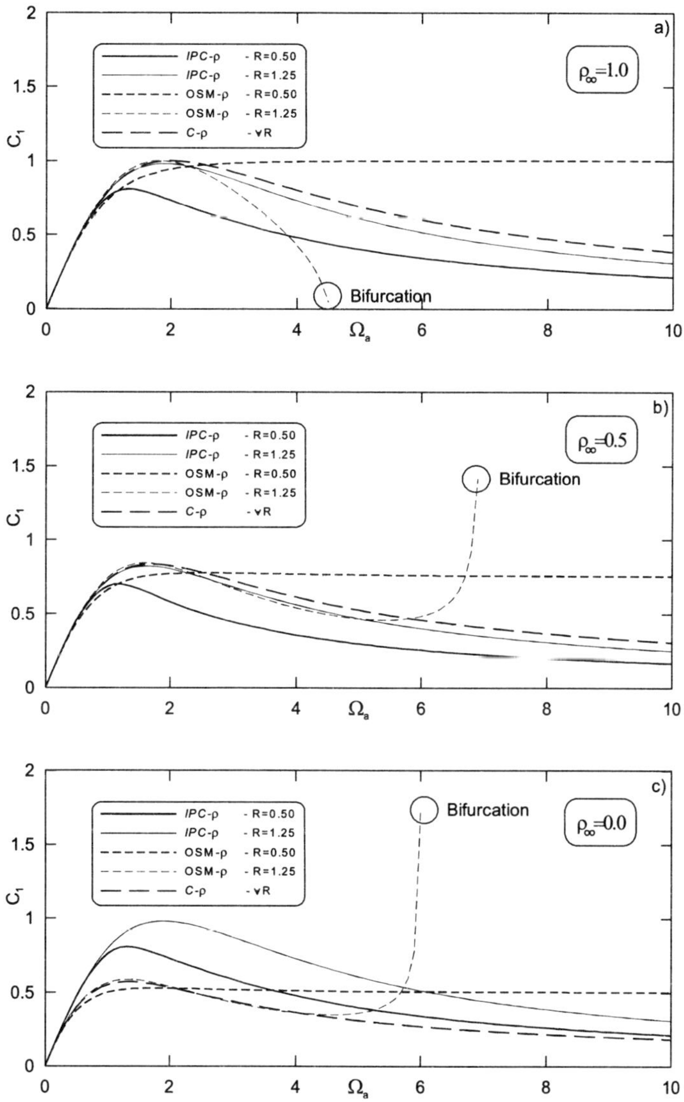

Figure 12. Control error amplification factor for various algorithms implemented in the CH-α method for $\boldsymbol{\xi} = \mathbf{0}$

4 Coupled integration and control engineering issues

In this section we introduce problems and analysis techniques that arise when coupling of numerical and physical substructures occurs. Real time situations are challenging in coupling and relevant requirements are dealt with in Subsection 4.1. As *SHSDS* require the use of actuators and controllers to control them, in Subsections 4.2 we deal with numerical aspects relevant to the discretization of an adaptive controller and its linearization. Subsection 4.3 concludes this Section applying concepts from digital filtering to time integrators.

4.1 Coupling algorithms

In this subsection, we study the simulation of coupled dynamic systems putting emphasis on partition and interconnection. The subject emerged over 30 years ago (Felippa and Park, 1980), but impulse was given to formulations and implementations rather recently (Park and Felippa, 2000 and Felippa et al., 2001) through the availability of inexpensive computer algebra systems. Borrowing the terminology from the area of *FSI*, see Zhang and Hisada (2004) among others, we distinguish between strong coupling or direct coupling when the variable vector of the discretized coupled system is solved and corrected simultaneously; conversely, in the weak coupling method an iterative method is employed to solve the same discretized equations. Moreover, we distinguish between monolithic approaches where solution methods are developed *ad hoc* to solve coupled applications; and partitioned methods where separate solvers are used for the fluid and the structure, respectively. Moreover, large systems of non-linear equations originated from subsystems are treated with iterative solvers, *i.e.* Jacobi or Gauss–Seidel solvers.

Typically, the coupled system can be expressed in the following form:

$$\dot{\mathbf{y}}_1 = \boldsymbol{f}_1(\mathbf{y}_1, \mathbf{y}_2, \mathbf{u}_1) \tag{4.1}$$

$$\mathbf{0} = \boldsymbol{g}_1(\mathbf{y}_1, \mathbf{y}_2, \mathbf{u}_1), \tag{4.2}$$

and

$$\dot{\mathbf{y}}_2 = \boldsymbol{f}_2(\mathbf{y}_1, \mathbf{y}_2, \mathbf{u}_2) \tag{4.3}$$

$$\mathbf{0} = \boldsymbol{g}_2(\mathbf{y}_1, \mathbf{y}_2, \mathbf{u}_2), \tag{4.4}$$

with the global coupling condition

$$\mathbf{0} = \boldsymbol{h}(\mathbf{y}_1, \mathbf{y}_2, \mathbf{u}_1, \mathbf{u}_2), \tag{4.5}$$

In these conditions, an algebraic loop might exist among some of the variables of the models. We assume that we have integrators for each subsystem separately. Hence, for Subsystem 1, we may say that given the values $(\mathbf{y}_1^{(i)}, \mathbf{u}_1^{(i)})$ and the functions $(\mathbf{y}_2, \mathbf{u}_2)$ over the interval $[t_i, t_{i+1}]$, with an implicit method we obtain

$$(\mathbf{y}_1^{(i+1)}, \mathbf{u}_1^{(i+1)}) = \mathbf{\Phi}_1(\mathbf{y}_1^{(i+1)}, \mathbf{u}_1^{(i+1)}, \mathbf{y}_1^{(i)}, \mathbf{u}_1^{(i)}, \mathbf{y}_2, \mathbf{u}_2) \tag{4.6}$$

Similarly, for Subsystem 2 we have a similar function,

$$(\mathbf{y}_2^{(i+1)}, \mathbf{u}_2^{(i+1)}) = \mathbf{\Phi}_2(\mathbf{y}_2^{(i+1)}, \mathbf{u}_2^{(i+1)}, \mathbf{y}_2^{(i)}, \mathbf{u}_2^{(i)}, \mathbf{y}_1, \mathbf{u}_1) \tag{4.7}$$

From these single system integrators, different options may be used to integrate the global system. The simplest strategy is a staggered scheme, also referred to as weak coupling. The scheme then is for Subsystem 1 to assume the time evolution of the state variables $(\mathbf{y}_2, \boldsymbol{z}_2)$ to be constant over the interval $[t_i, t_{i+1}]$, i.e. identically equal to $(\mathbf{y}_2^{(i)}, \mathbf{u}_2^{(i)})$, and then solve (4.6). Similarly for Subsystem 2, we assume $(\mathbf{y}_1, \boldsymbol{z}_1)$ over the time increment to be constant and identically equal to the initial values $(\mathbf{y}_1^{(i)}, \boldsymbol{z}_1^{(i)})$, and with this we can solve (4.7). To sum up, information is only exchanged between the subsystems after the increment has been performed; inside the increment, each subsystem is solved in parallel independently of each other. This scheme will only give first order accuracy in time, no matter how accurate the single system integrators are. This could be improved by a higher order extrapolation, but this might have a negative impact on its stability characteristics (Felippa et al., 2001). Although the integrators for the individual subsystems may be implicit, the staggered strategy is explicit and therefore conditionally stable. Moreover, global constraints are not satisfied exactly and energy conservation is not assured.

Implicit methods which use partitioned software but satisfy an implicit condition globally are desired and are referred to as strong coupling. Nonetheless, implicit methods usually need iterations and starting values to extrapolate the solution into the new time step. Thus, both Subsystems (4.6) and (4.7) can be independently solved and if wanted in parallel until convergence.

It is evident from Subsection 1.1 that integration algorithms employed in *SHSDS* must take into account the strategy to couple *NS* and *PS*. Firstly, the real-time requirement for numerical integrators must be fulfilled during the simulation, and therefore iterations during coupling must be avoided: this leads to a weak coupling. As a result, we need to resort to prediction-substitution procedures which can be based either on static or kinematic control at the interface. The relevant coupling possibilities are shown in

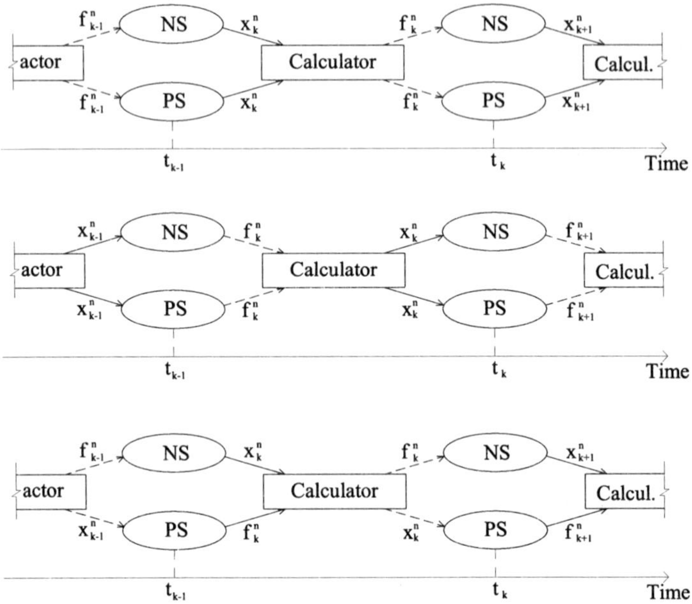

Figure 13. Coupling strategies with **f** force quantity vector and **X** kinematic quantity vector

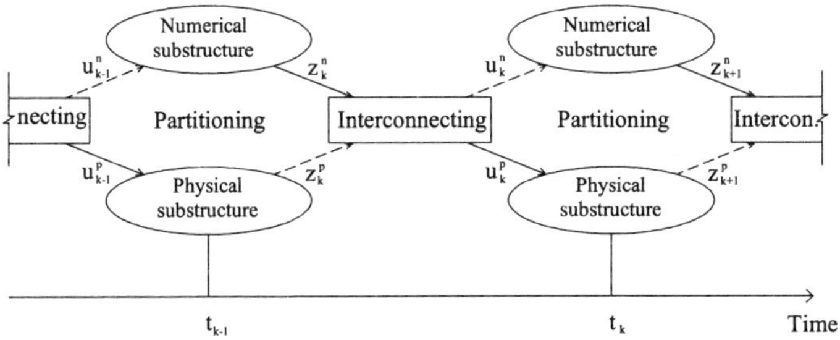

Figure 14. Coupled integration in displacement control

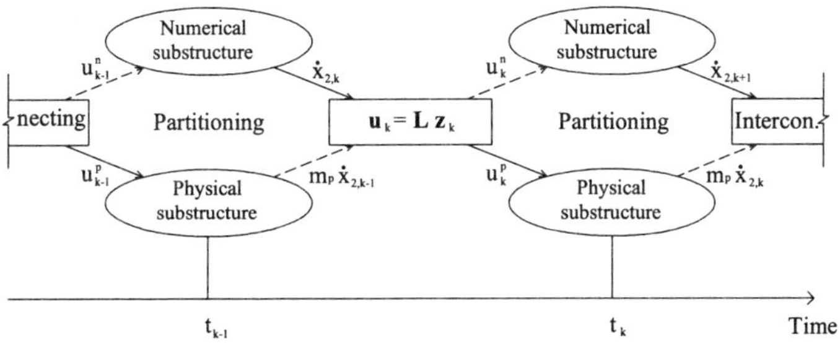

Figure 15. Coupled integration in displacement control with the algebraic coupling equation

Figure 13 where $\mathbf{f}$ represents the vector of force quantities and $\mathbf{X}$ the vector of kinematic quantities, respectively.

SHSDS does not require computations on the *PS* which behaves as a black box and the Calculator has to enforce an interconnecting condition between the *NS* and *PS*; thus, the mixed $\boldsymbol{f}-\boldsymbol{X}$ coupling strategy illustrated in Figure 13c is preferred. In detail, the interface kinematic vector $\mathbf{X}_{\boldsymbol{k}}^{\boldsymbol{n}}$ of the *NS* and the force quantity vector $\mathbf{f}_{\boldsymbol{k}}^{\boldsymbol{p}}$ of the *PS* are used as input to the Calculator, while the force quantity vector $\mathbf{f}_{\boldsymbol{k}}^{\boldsymbol{n}}$ of the *NS* and the interface kinematic vector $\mathbf{X}_{\boldsymbol{k}}^{\boldsymbol{p}}$ of the *PS* become output from the Calculator. Clearly, the mixed strategy can be set as a $\boldsymbol{X}-\boldsymbol{f}$ coupling strategy, but the disadvantage in operating the transfer system in force control favours the $\boldsymbol{f}-\boldsymbol{X}$ coupling strategy. This strategy is schematically depicted in Figure 14 were the interconnecting function of the Calculator is emphasized.

With reference to a splitted-mass SDoF system illustrated in Figure 1(c) with $\boldsymbol{m}_{\mathbf{2}}^{\mathrm{p}} = \boldsymbol{k}_{\mathbf{2}}^{\mathrm{p}} = \boldsymbol{c}_{\mathbf{2}}^{\mathrm{p}} = \mathbf{0}$, an algebraic coupling is shown in Figure 15 where we assumed that the application of kinematic quantities and the relevant measurements entail a delay of Δt (Bursi et al., 2008). All the inputs $\mathbf{u}_k^{\mathrm{n}}$ and $\mathbf{u}_k^{\mathrm{p}}$, corresponding to the *NS* and the *PS* inputs are shown and collected in a vector $\mathbf{u}_k$. The same is done with the outputs, using the vector $\mathbf{z}_k$. The interconnection enforces the algebraic condition

$$\mathbf{u}_k = \mathbf{L}\mathbf{z}_k, \tag{4.8}$$

where $\mathbf{L}$ is a time invariant Boolean localization matrix representing the input/output interconnection of the block diagram and containing only zeros and ones. As a result, one gets

$$\mathbf{L} = \begin{bmatrix} \mathbf{0} & \boldsymbol{L}^{\mathrm{p}} \\ \boldsymbol{L}^{\mathrm{n}} & \mathbf{0} \end{bmatrix}, \quad \mathbf{z}_k = \begin{Bmatrix} \dot{\boldsymbol{x}}_{2,k} \\ m^{\mathrm{p}}\dot{\boldsymbol{x}}_{2,k-1} \end{Bmatrix}, \quad \mathbf{u}_k = \begin{Bmatrix} \mathbf{u}_k^{\mathrm{n}} \\ \mathbf{u}_k^{\mathrm{p}} \end{Bmatrix}. \tag{4.9}$$

When using this coupling strategy, the set of differential equations (1.2) corresponding to the numerical subsystem are integrated to advance from t_{k-1} to t_k as shown in Figure 15. The proposed procedure is rather different from a truly parallel scheme (Felippa et al., 2001), which would entail the coupling force as predictor at t_k. In fact, in the actual prediction-substitution procedure the coupling force is taken at t_{k-1} owing to the nature of the *PS*. The prediction-substitution scheme used to enforce the interconnection is summarized in algorithmic form as follows:

- **Initialization procedure**:
  (a) $t = 0$; $k = 0$
  (b) Assuming $\mathbf{M}^n$, $\mathbf{C}^n$, $\mathbf{K}^n$ and the initial conditions $\mathbf{x}_0$, $\dot{\mathbf{x}}_0$ known, estimate $\ddot{\mathbf{x}}_0 = (\mathbf{M}^n)^{-1}(-\mathbf{C}^n\dot{\mathbf{x}}_0 - \mathbf{K}^n\mathbf{x}_0 + \mathbf{f}_{\mathrm{e},0})$
  (c) initial coupling forces: $\mathbf{f}_{\mathrm{s},0} = \mathbf{M}^{\mathrm{n}}\ddot{\mathbf{x}}_0 + \mathbf{C}^{\mathrm{n}}\dot{\mathbf{x}}_0 + \mathbf{K}^{\mathrm{n}}\mathbf{x}_0 - \mathbf{f}_{\mathrm{e},0}$
  (d) initial input vector: $\mathbf{f}_{\mathrm{e},0} \quad \mathbf{f}_{\mathrm{s},0}$
- **For each time step**:
  1. perform the numerical integration: $\mathbf{x}_k^n$;
  2. update the *PS* input vector: $\mathbf{x}_k^{\mathrm{p}} = \mathbf{x}_k^{\mathrm{n}}$;
  3. apply $\mathbf{x}_k^{\mathrm{p}}$ onto the *PS* while measuring $\mathbf{f}_{\mathrm{s,k}}$;
  4. update the *NS* input vectors: $\mathbf{f}_{\mathrm{e,k}} \quad \mathbf{f}_{\mathrm{s,k}}$;
  5. $k = k + 1$; $t_k = t_k + k\Delta t$.

Note that to perform the numerical integration of Step 1, it is necessary to employ explicit methods like the ones introduced in Subsection 1.1 and 3.1; or linearly implicit methods like the *LSRT1* method introduced in Subsection 3.1; or companion methods of the same class which will be introduced in Subsection 5.1. Moreover along the lines of Burgermeister et al.

(2006), the injection of $\mathbf{M}^p$, $\mathbf{C}^p$ and $\mathbf{K}^p$ of the *PS* in the Jacobian matrix $\mathbf{J}$ could improve the stability properties of linearly implicit methods. As shown in Subsection 5.2, these methods might loose their stability property when applied to coupled problems in *HiL*.

4.2 Delay and Compensation Techniques

Being the *SHSDS* based on the *PS* feedback, slightest inaccuracies translate into a feedback error which, in turn propagate into an actuator command signal error on the following time interval. To characterise this complex dynamics in terms that can be analysed and acted upon during a *SHSDS*, a pragmatic approach is to define a time delay and an amplitude error, assumed to vary continuously. The time delay approach of actuator dynamics has been used by numerous researchers in the field of *SHSDS* (Horiuchi et al., 1999 and Darby et al., 2002). As the delay can be of the order of 5-15ms, it is not negligible with respect to the $\boldsymbol{\Delta t}$ considered. In general, the total time delay in a control system can be partitioned into two contributions. The first part is referred to as fixed time delay owing to on-line data acquisition, filtering, manipulation of digital data inside the digital control processor, calculation of the required control force and signal transmission from the computer to the actuators. The second part usually depends on the particular dynamics of the actuators interacting with the controlled plant, which is referred to as floating delay time; it can be adjusted by explicitly combining the actuator dynamics with that of the plant (Chu et al., 2002). For instance in real-time *PsD* tests a fixed time delay of the order of $3\boldsymbol{\Delta t}$ was measured by Jung et al. (2007). In detail, the time delay between the computation of displacements in the target PC and the initiation of the computation of control signals in the digital controller was of the order of $2\boldsymbol{\Delta t}$. An additional $\boldsymbol{\Delta t}$ was required to execute the computation of control signals and sending signals to the valve drives. Bonnet (2006) classified the compensation methods devoted to the delay into two main categories: i) the schemes that, from known, assumed or calculated delay and amplitude error estimates, aim at directly correcting the command signal accordingly: they were labelled as *direct compensation schemes*; ii) the algorithms that correct the command signal from the knowledge of the real-time error to the desired output using more standard control engineering techniques, referred to as *outer-loop controllers*. Within the former category fall the forward prediction schemes proposed by Wallace et al. (2005) and Horiuchi et al. (1999), which assume that the delay and amplitude factor estimates are known and accurate; and the performance estimation algorithms suggested by Bonnet et al. (2007), which compute the delay and amplitude factor more precisely

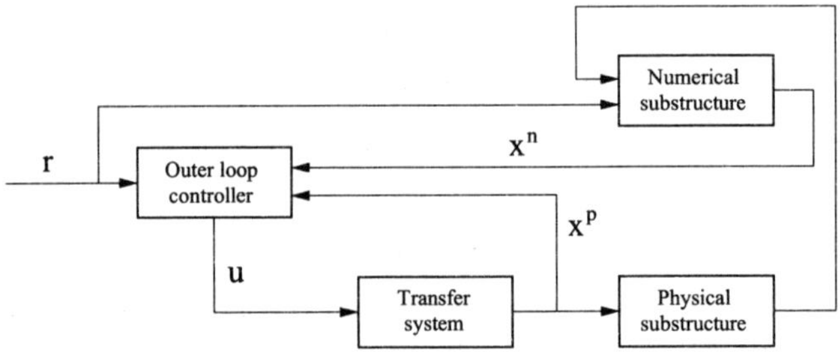

Figure 16. Block diagram representation of the outer loop controller concept

along a test starting from rough estimates. To the second category belong the minimal control synthesis family of outer-loop control algorithms (Stoten and Neild, 2003); instead of trying to compensate explicitly for the actuator delay, this approach considers the synchronisation error as a whole and minimises it. This is obtained by including a model of the transfer system as illustrated in Figure 16. With this approach all information goes through the controller, which computes the transfer system command signal accordingly. Eliminating the direct influence from the *NS* to the *PS* is the aim of the *outer loop controller.* This concept was implemented for simple *SHSDS* at the Bristol University using the shaking table as transfer system for the *PS* (Neild et al., 2005).

Different researchers also proposed to adopt adaptive control algorithms. In detail, adaptive control techniques provide a mechanism to continuously update the controller parameters with respect to the variation of the dynamics. Therefore, a change in a substructure would not necessarily imply the retuning of the control system. Researchers from Bristol University worked on the minimal control synthesis *(MCS)* family of adaptive controllers (Stoten and Neild, 2003) towards their use in *SHSDS* testing (Neild et al., 2005). Variants of this controller were proposed: the *MCSmd algorithm,* by introducing some previous knowledge of the transfer system to initialise the gains to typical values and an inverse of the reference model applied to the external demand before the application of the controller gains (Lim et al., 2005). The *MCSmd* outer loop controller was discussed and tested by Bonnet et al. (2007) through a series of *SHSDS* including nonlinearities in the *PS* and multi-axis tests. The stability and accuracy properties of the *MCS* algorithm by applying techniques from numerical analyses and time-delay effects usually present in the inner dynamics are considered hereafter.

4.3 The ZOH and the MCS controller

The *MCS* algorithm mentioned in the previous subsection belongs to dissipative controllers, as its properties are based on Popov's theory of hyperstability (Popov, 1973), which implies the positive real property of control systems and the dissipative (passive) property of systems. If one considers a square stable plant $(\mathbf{A}, \mathbf{B}, \mathbf{C})$, i.e. a linear system with the number of inputs equal to the number of outputs, an open-loop square system with simple poles is dissipative, if there exists a symmetric positive definite matrix $\mathbf{P}$ and a matrix $\mathbf{Q}$ that satisfy the following equations

$$\mathbf{A}^T\mathbf{P} + \mathbf{P}\mathbf{A} = -\mathbf{Q}^T\mathbf{Q} \tag{4.10}$$
$$\mathbf{B}^T\mathbf{P} = \mathbf{C} \tag{4.11}$$

This system is strictly dissipative if $\mathbf{Q}^T\mathbf{Q}$ is positive definite. The equations above allow for the simple determination of a dissipative system. Given $\mathbf{A}$ and $\mathbf{B}$ we select the matrix $\mathbf{Q}$. Next we solve (4.10) for $\mathbf{P}$ and we find the output matrix $\mathbf{C}$ from (4.11). Some matrices introduced above are indicated in Figure 17.

Stoten and Benchoubane (1990) proposed the *MCS* algorithm which derives from the *MRAS* formulation: it assumes null constant gains $\mathbf{K}$ and $\mathbf{K}_r$ and unknown plant parameters. Quantities to be known are the reference model parameters $\mathbf{A}_m$ and $\mathbf{B}_m$, the structure of the plant with the degree of freedom and order and the sign of the coefficients of $\mathbf{B}$, that usually is assumed to be positive. Therefore the control law becomes

$$\mathbf{u}(t) = \delta\mathbf{K}(t)\,\mathbf{x}(t) + \delta\mathbf{K}_r(t)\,\mathbf{r}(t) \tag{4.12}$$

and the output reads

$$\mathbf{y}_e(t) = \mathbf{C}_e\mathbf{x}_e(t) \tag{4.13}$$

where

$$\mathbf{C}_e = \mathbf{P} \tag{4.14}$$

and $\mathbf{P}$ is the solution of the Lyapunov equation (4.10). Stoten and Neild (2003) proposed a pragmatic solution for first order, *e.g.* $\boldsymbol{C_e} = [\boldsymbol{4/t_s}]$ and second order $\mathbf{C_e} = [\boldsymbol{4/t_s}\ \mathbf{1}]$ *SDoF* systems that induces an exact pole-zero cancellation. The block diagram of a standard MCS controller is shown in Figure 17.

The *MRAS* formulation contains a linear controller whose coefficients are determined from the plant model. In order to be implemented, it needs the development of a linearized plant model, a reference model, a linear model reference control and a hyperstable block. The *MCS* controller does not need all the above elements and provides excellent responses which match

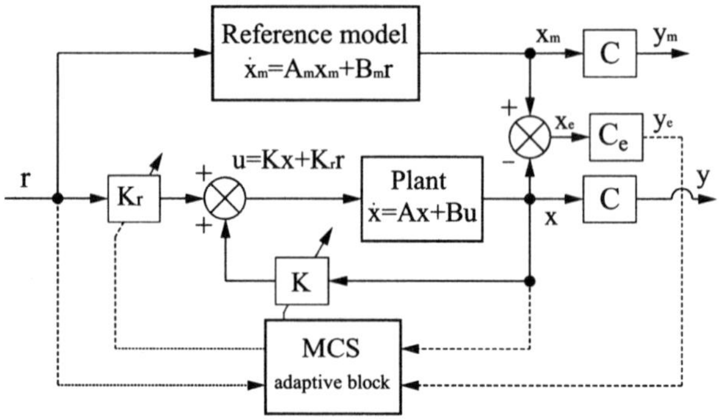

Figure 17. The standard Minimal Control Synthesis algorithm

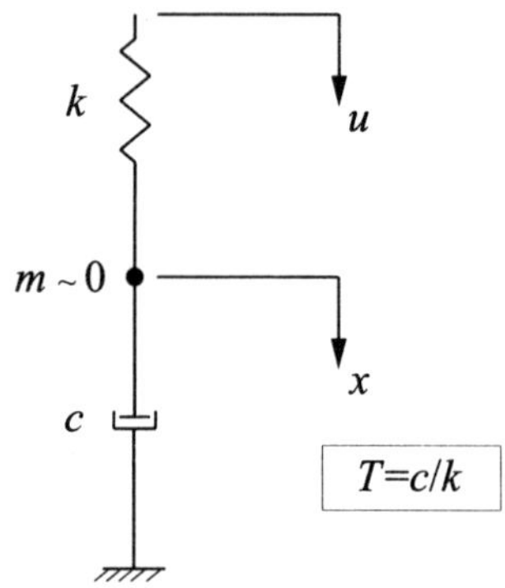

Figure 18. Plant model

those of the *MRAS* controller. Therefore, the comparable performance of the *MCS* algorithm with the *MRAS* algorithm permits to understand the advantages of the first one, since allows accurate system identification to be avoided.

The analysis of the *MCS* controller allows us to introduce: (i) the discrete control law of the *MCS* sampled by the Zero Order Hold (*ZOH*) method (Vaccaro, 1995); (ii) two linearization techniques based on *physical insight* and Taylor series expansion; (iii) a stability analysis. For brevity, both accuracy and frequency domain analyses are not dealt with here; they can be found in Bursi et al. (2007). Moreover, first-order linear time-invariant systems both for the plant and the reference model are considered.

The plant is a first-order system which can be represented by means of

Figure 18. Its behaviour is described by the continuous-time equation

$$\begin{aligned} (u(t) - x(t))\,k - c\dot{x}(t) &= 0 \\ \dot{x}(t) + x(t)/T &= u(t)/T \end{aligned} \tag{4.15}$$

where $u(t)$ is the input, *i.e.* the control variable, $x(t)$ is the output, *i.e.* the displacement of the mass m, and $T = c/k$ is the time constant of the system. Considering a state-space approach, (4.15) can be rewritten as

$$\dot{x}(t) = Ax(t) + Bu(t) \tag{4.16}$$

where $A = -1/T$ and $B = 1/T$. One can note that every term is scalar. By applying a *ZOH* sampling, (4.16) becomes

$$x[k+1] = A'x[k] + B'u[k] \tag{4.17}$$

where

$$A' = e^{A\Delta t} = e^{-\Delta t/T} \tag{4.18}$$

$$B' = \int_0^{\Delta t} e^{A\tau} d\tau B = \frac{1}{A}\left(e^{A\Delta t} - 1\right) B = 1 - e^{-\Delta t/T} \tag{4.19}$$

In a similar fashion, the discrete equivalent of the reference model becomes

$$x_m[k+1] = A'_m x_m[k] + B'_m r[k] \tag{4.20}$$

where

$$A'_m = e^{-\Delta t/T_m} = e^{-4\Delta t/t_s} \tag{4.21}$$

$$B'_m = 1 - e^{-\Delta t/T_m} = 1 - e^{-4\Delta t/t_s} \tag{4.22}$$

with T_m time constant of the reference model. It is assumed that $T_m = t_s/4$, where t_s is the settling time, used as a measure of the time taken for the oscillations to die away (Vaccaro, 1995).

Figure 17 shows the block diagram of the *MCS* controller whose control law is

$$u(t) = K(t)\,x(t) + K_r(t)\,r(t) \tag{4.23}$$

where $K(t)$ and $K_r(t)$ read

$$K(t) = \int_0^t \alpha y_e(\tau)\,x(\tau)\,d\tau + \beta y_e(t)\,x(t) \tag{4.24}$$

$$K_r(t) = \int_0^t \alpha y_e(\tau)\,r(\tau)\,d\tau + \beta y_e(t)\,r(t) \tag{4.25}$$

The error output is given by

$$y_e(t) = C_e x_e(t) \tag{4.26}$$

where $C_e = P$. In detail, P is the solution of the Lyapunov's equation and $x_e(t) = x_m(t) - x(t)$ is the error between the reference model and plant states.

To generalize the numerical treatment of the controller, we introduce an *ad hoc* delay in the control law. Thus, (4.23) becomes

$$u(t) = K(t - \gamma\delta t)\, x(t - \gamma\delta t) + K_r(t - \gamma\delta t)\, r(t - \gamma\delta t) \tag{4.27}$$

where $\gamma\delta t$ can assume any value from zero to infinity to represent typical sources of delay in a controlled system (Chu et al., 2002). The general formulation of the discrete control law equation provided by the *ZOH* for the *MCS* controller considering a fixed delay time $\gamma\delta t = \gamma\Delta t$ in (4.27) reads:

$$u[k] = K[k-\gamma]\, x[k-\gamma] + K_r[k-\gamma]\, r[k-\gamma] \tag{4.28}$$

where the adaptive gains read

$$\begin{aligned} K[k-\gamma] =& K[k-1-\gamma] + \beta y_e[k-\gamma] x[k-\gamma] \\ & - \sigma y_e[k-1-\gamma] x[k-1-\gamma] \end{aligned} \tag{4.29}$$

$$\begin{aligned} K_r[k-\gamma] =& K_r[k-1-\gamma] + \beta y_e[k-\gamma] r[k-\gamma] \\ & - \sigma y_e[k-1-\gamma] r[k-1-\gamma] \end{aligned} \tag{4.30}$$

with $\sigma = \beta - \alpha\Delta t$ (Vulcan, 2006).

By assuming no delay, *i.e.* $\gamma = 0$, the initial conditions of the *MCS* controller read $K[-1] = 0$ and $K_R[-1] = 0$, respectively. To complete the characterization of the controller, the eigenvalues of the reference model, α and β from (4.24) and (4.25) and C_e must be selected. In addition, the reference model must be stable. The values of α and β are arbitrary positive numbers and are selected by trial and error. However, an increase of β entails a reduction of the settling time t_s of the adaptation, while a reduction of the β/α ratio improves damping (Bursi et al., 2007). The ratio β/α can be set to 0.1 as a compromise between the speed of the adaptation mechanism and the stability limit (Vulcan, 2006). The relatively arbitrary selection of the error vector weighting matrix C_e is like the one proposed for continuous time systems, *i.e.* $C_e = 4/t_s$. Hereafter, the analysis is carried out by considering two limit cases: i) $\gamma = 0$; we implicitly assume that the processing time δ is negligible with respect to the time step Δt; and therefore, the control signal $\boldsymbol{u}$ generated by the *MCS* algorithm is obtained instantaneously; ii) the fixed delay time is equal to $\boldsymbol{\Delta t}$. The second case could be an arguably assumption, since in reality the delay may be a fraction of $\boldsymbol{\Delta t}$, *i.e.* $\mathbf{0} \leq \gamma \leq \mathbf{1}$; however, it is always possible to manage the delay to be equal to $\boldsymbol{\Delta t}$, such that this assumption is correct. Therefore, the analyses presented hereafter allow the effect of one $\boldsymbol{\Delta t}$ in the control law computation on the system to be understood. The reference input signal used for the stability analysis reads

$$r\left[k\right] = \left\{ \begin{array}{ll} \mathbf{0}, & k < \mathbf{0} \\ \mathbf{1}, & k \geq \mathbf{0} \end{array} \right. \tag{4.31}$$

As the overall system is non-linear owing to the adaptation mechanism, the stability analysis is carried out by linearizing the system at steady state in a neighbourhood of an operating point (Vulcan, 2006). Thus, the comparison between the linearization by *physical insight* and by the more rigorous Taylor series expansion is presented herein; this comparison shows the advantages of the first approach both in terms of computational burden and of an *a priori* selection of the control parameters $\boldsymbol{\alpha}$ and $\boldsymbol{\beta}$. In this first case, the analysis of the system is developed without delay and therefore $\gamma = \mathbf{0}$ in (4.28).

The linearization by *physical insight* is summarized herein. The discrete-time equations that rule the system can be arranged as

$$\left\{ \begin{array}{c} x_e\left[k+1\right] \\ x_e\left[k\right] \\ x\left[k+1\right] \\ K\left[k\right] \\ K_r\left[k\right] \end{array} \right\} = \mathbf{C}_{IT} \left\{ \begin{array}{c} x_e\left[k\right] \\ x_e\left[k-1\right] \\ x\left[k\right] \\ K\left[k-1\right] \\ K_r\left[k-1\right] \end{array} \right\} + \left\{ \begin{array}{c} B'_m r\left[k\right] \\ 0 \\ 0 \\ 0 \\ 0 \end{array} \right\}. \tag{4.32}$$

where $\mathbf{C}_{IT}$ reads,

$$\mathbf{C}_{IT} = \left[\begin{array}{ccccc} \mathbf{C}_{IT}^{11} & \mathbf{C}_{IT}^{12} & A'_m - A' & -B'x[k] & -B'r[k] \\ 1 & 0 & 0 & 0 & 0 \\ \mathbf{C}_{IT}^{31} & \mathbf{C}_{IT}^{32} & A' & B'x[k] & B'r[k] \\ \beta x[k]C_e & -\sigma x[k-1]C_e & 0 & 1 & 0 \\ \beta r[k]C_e & -\sigma r[k-1]C_e & 0 & 0 & 1 \end{array} \right] \tag{4.33}$$

with

$$\mathbf{C}_{IT}^{11} = A'_m - B'\beta C_e x^2[k] - B'\beta C_e r^2[k] \tag{4.34}$$
$$\mathbf{C}_{IT}^{12} = B'\sigma C_e r[k-1]r[k] + B'\sigma C_e x[k-1]x[k] \tag{4.35}$$
$$\mathbf{C}_{IT}^{31} = B'\beta C_e x^2[k] + +B'\beta C_e r^2[k] \tag{4.36}$$
$$\mathbf{C}_{IT}^{32} = -B'\sigma C_e r[k-1]r[k] - B'\sigma C_e x[k-1]x[k] \tag{4.37}$$

$\mathbf{C}_{IT}$ is not constant owing to the adaptation mechanism; thus, to reduce the complexity of the analysis, $\mathbf{C}_{IT}$ is locally linearized in a neighbourhood of an operating point, located at the steady-state condition, such that $\lim_{k\to+\infty} x_e[k] = 0$. The linearization by *physical insight* consists of substituting the values corresponding to the operating point, *i.e.*

$$x[k-1] = x[k] = r[k-1] = r[k] = 1 \text{ and } x_e[k-1] = x_e[k] = 0 \tag{4.38}$$

in the amplification matrix $\mathbf{C}_{IT}$, which in turn, is arranged to reduce the complexity of the analysis. As a result, the amplification matrix at steady state reads

$$\mathbf{C}_{IT,ss} = \begin{bmatrix} A'_m - 2B'\beta C_e & 2B'\sigma C_e & A'_m - A' & -B' & -B' \\ 1 & 0 & 0 & 0 & 0 \\ 2B'\beta C_e & -2B'\sigma C_e & A' & B' & B' \\ \beta C_e & -\sigma C_e & 0 & 1 & 0 \\ \beta C_e & -\sigma C_e & 0 & 0 & 1 \end{bmatrix} \tag{4.39}$$

where the eigenvalues of $\mathbf{C}_{IT,ss}$ read

$$\lambda_1 = 1,\ \lambda_2 = 0,\ \lambda_3 = A'_m \tag{4.40}$$
$$\lambda_4 = \tfrac{1}{2}(1 + A' - 2B'\beta C_e - \sqrt{(1 + A' - 2B'\beta C_e)^2 + 8B'\sigma C_e - 4A'})$$
$$\lambda_5 = \tfrac{1}{2}(1 + A' - 2B'\beta C_e + \sqrt{(1 + A' - 2B'\beta C_e)^2 + 8B'\sigma C_e - 4A'})$$

In the linearization by Taylor series expansion, the amplification matrix $\mathbf{C}_{IT,T,ss}$ is obtained by linearizing the system through Taylor series near the operating point at steady state conditions. Usually, the discrete linearization techniques follow those of continuous-time non-linear systems (Vaccaro, 1995). In detail, the system of state space equations of a non-linear discrete-time plant reads

$$\mathrm{x}[k+1] = \mathrm{f}(\mathrm{x}[k], u[k]) \tag{4.41}$$

As for continuous-time systems, it is advantageous to define variables centred about an operating point $(\mathbf{x}_0, \mathbf{u}_0)$

$$\begin{aligned} \mathrm{x}[k] &= \mathrm{x}_0 + \delta\mathrm{x}[k] \\ \mathrm{u}[k] &= \mathrm{u}_0 + \delta\mathrm{u}[k] \end{aligned} \tag{4.42}$$

By substituting (4.42) in (4.41), one gets

$$\delta \mathbf{x}\,[k+1] = \mathbf{f}\left(\delta \mathbf{x}\,[k] + \mathbf{x}_0, \delta \mathbf{u}\,[k] + \mathbf{u}_0\right) - \mathbf{x}_0 \tag{4.43}$$

Thus, expanding the rhs of (4.43) in Taylor series about an operating point and keeping only linear terms, one obtains

$$\delta \mathbf{x}\,[k+1] = \frac{\partial \mathbf{f}}{\partial \mathbf{x}\,[k]}\,|_{(x_0,\mathbf{u}_0)}\ \delta \mathbf{x}\,[k] + \frac{\partial \mathbf{f}}{\partial u\,[k]}\,|_{(x_0,\mathbf{u}_0)}\ \delta \mathbf{u}\,[k] \tag{4.44}$$

Therefore, the Jacobian matrix with respect to every state and error variable can be computed. When we apply this linearization to the system described by (4.32), the computation of the Jacobian matrix is needed as well as the evaluation of the resulting matrix at the operating point. Moreover, the vector related to the input is obtained by differentiating the equations with respect to $\boldsymbol{r}\,[\boldsymbol{k}]$ and then by evaluating every term at the operating point. The operating point is defined by

$$\begin{gathered} \boldsymbol{x}_{e0}\,[\boldsymbol{k}] = \boldsymbol{x}_{e0}\,[\boldsymbol{k}-1] = \mathbf{0},\ \ \boldsymbol{x}_0\,[\boldsymbol{k}] = \boldsymbol{x}_0\,[\boldsymbol{k}-1] = 1 \\ \boldsymbol{K}_0\,[\boldsymbol{k}-1] = \boldsymbol{K}_0\,[\boldsymbol{k}-2] = \boldsymbol{K}^*, \\ \boldsymbol{K}_{r0}\,[\boldsymbol{k}-1] = \boldsymbol{K}_{r0}\,[\boldsymbol{k}-2] = \boldsymbol{K}_r^*,\ \ \boldsymbol{r}_0\,[\boldsymbol{k}] = \boldsymbol{r}_0\,[\boldsymbol{k}-1] = 1 \end{gathered} \tag{4.45}$$

As the equations that rule the system do not depend on the variable $\boldsymbol{x}[\boldsymbol{k}-1], \boldsymbol{K}[\boldsymbol{k}-2]$ and $\boldsymbol{K}_r[\boldsymbol{k}-2]$, one can simply reduce the problem to a 5x5 system. As a result, one gets the following relationship

$$\left\{ \begin{array}{c} \boldsymbol{x}_e\,[\boldsymbol{k}+1] \\ \boldsymbol{x}_e\,[\boldsymbol{k}] \\ \boldsymbol{x}\,[\boldsymbol{k}+1] \\ \boldsymbol{K}\,[\boldsymbol{k}] \\ \boldsymbol{K}_r\,[\boldsymbol{k}] \end{array} \right\} = \mathbf{C}_{IT,T,ss} \left\{ \begin{array}{c} \boldsymbol{x}_e\,[\boldsymbol{k}] \\ \boldsymbol{x}_e\,[\boldsymbol{k}-1] \\ \boldsymbol{x}\,[\boldsymbol{k}] \\ \boldsymbol{K}\,[\boldsymbol{k}-1] \\ \boldsymbol{K}_r\,[\boldsymbol{k}-1] \end{array} \right\} + \mathbf{D}_{IT,T,ss}\,\{\boldsymbol{r}\,[\boldsymbol{k}]\} \tag{4.46}$$

where $\mathbf{C}_{IT,T,ss}$ and $\mathbf{D}_{IT,T,ss}$ at steady state read

$$\mathbf{C}_{IT,T,ss} =$$
$$\begin{bmatrix} A'_m - 2B'\beta C_e & 2B'\sigma C_e & A'_m - A' - B'K^* & -B' & -B' \\ 1 & 0 & 0 & 0 & 0 \\ 2B'\beta C_e & -2B'\sigma C_e & A' + B'K^* & B' & B' \\ \beta C_e & -\sigma C_e & 0 & 1 & 0 \\ \beta C_e & -\sigma C_e & 0 & 0 & 1 \end{bmatrix} \quad (4.47)$$

$$\mathbf{D}_{IT,T,ss} = \begin{bmatrix} -B'K^*_r + B'_m & 0 \\ 0 & 0 \\ B'K^*_r & 0 \\ 0 & 0 \\ 0 & 0 \end{bmatrix} \quad (4.48)$$

The eigenvalues of $\mathbf{C}_{IT,T,ss}$ are listed here

$$\lambda_1 = 1,\ \lambda_2 = 0,\ \lambda_3 = A'_m \quad (4.49)$$
$$\lambda_4 = \frac{1}{2} + \frac{1}{2}A' - B'\beta C_e + \frac{1}{2}B'K^* + \\ -\frac{1}{2}\sqrt{(1 + A' - 2B'\beta C_e + B'K^*)^2 - 4\left(A' + B'K^* - 2B'\sigma C_e\right)}$$
$$\lambda_5 = \frac{1}{2} + \frac{1}{2}A' - B'\beta C_e + \frac{1}{2}B'K^* + \\ +\frac{1}{2}\sqrt{(1 + A' - 2B'\beta C_e + B'K^*)^2 - 4\left(A' + B'K^* - 2B'\sigma C_e\right)}$$

One can observe that $\mathbf{C}_{IT,T,ss}$ is equal to $\mathbf{C}_{IT,ss}$ of (4.39), apart from the term $B'K^*$. Therefore, the results provided by *physical insight* corresponds to the case $B'K^*$ equal to zero. This approximation entails that either K^* or B' must be very small. As $B' = 1 - e^{-\Delta t/T}$, the two matrices meet if $\Delta t \rightarrow 0$, or T approaches large values. The term K^* appearing in (4.48) is not known a priori; therefore, the analysis can be performed only with the support of preliminary numerical simulations (Vulcan, 2006). Hereafter, the amplification matrices obtained at steady-state are analysed by introducing the following non-dimensional variables: $\frac{\Delta t}{T_m}, \frac{\beta}{4T_m}, \frac{T}{T_m}, \frac{\beta}{\alpha}\frac{1}{4T_m}$.

The absolute stability of the discrete system is assured when the modulus of the eigenvalues of the amplification matrix at the steady-state regime is less than one. For the linearization by Taylor series expansion only 4 and 5 of (4.49) are relevant for stability analysis; so these eigenvalue moduli and the eigenvalue moduli 4 and 5 of (4.40) obtained by means of the linearization by physical insight can be compared in Figure 19. In the figure, the

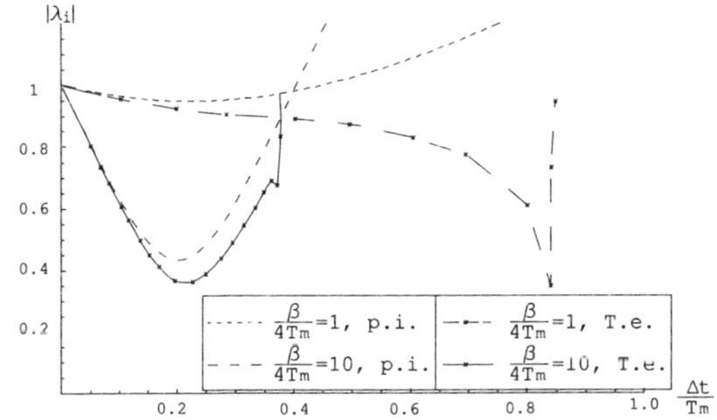

Figure 19. Eigenvalue moduli for a unit step input, $\beta/(\alpha 4TM) = \mathbf{0.1}$ and $T/T_M = \mathbf{10}$: comparison between the linearizations by *physical insight* and by Taylor series expansion

physical insight linearization is characterized by p.i., whereas the Taylor series expansion by T.e.; further, it is assumed , $t_s = 1$s, $T_m = \mathbf{0.25}$s, $\beta/\alpha = \mathbf{0.1}$, and consequently, $\frac{\beta}{\alpha}\frac{1}{4T_m} = \mathbf{0.1}$. Looking at the eigenvalues, they can be complex conjugates or real: from the plots one can underline that as long as the eigenvalues are complex conjugates, the system remains stable; but when they become real, the system gets quickly unstable as usual in discrete-dynamic analysis. Moreover, it is possible to determine the Sampling interval - Gain (SG) space, i.e. the stability region, as a function of $\frac{\Delta t}{T_m}$ and $\frac{\beta}{4T_m}$, for $\frac{\beta}{\alpha}\frac{1}{4T_m} = \mathbf{0.1}$ and $r = \mathbf{1}$; this domain is depicted in Figure 20a. This space may be interpreted as operational tool that allows the optimal combination of $\Delta t - \beta/\alpha$ values to be selected. In fact being the T/T_m ratio unknown, it is usually set by trial and errors. The advantage of this space is that it permits operators to determine *a priori* information on the behaviour of the sampled-data control system.

From the previous analysis we have demonstrated the effectiveness of the linearization by *physical insight* for most of the practical cases, i.e. for very short Δt or T_m approaches large values. Herein we employ this linearization introducing a fixed time step delay, i.e. $\gamma = \mathbf{1}$ in (4.28); thus, the control law reads

$$u[k] = K[k-1]\,x[k-1] + K_r[k-1]\,r[k-1] \tag{4.50}$$

By repeating the same type of analysis performed for the case without delay, we can plot the SG space for $\frac{\beta}{\alpha}\frac{1}{4T_m} = \mathbf{0.1}$ in Figure 20(b). From the figure it is evident that, if the control law u is computed with one time step delay, the stability limit reduces. This finding underlines the need of reducing the discrete time of the control law or to assume small $\frac{\Delta t}{T_m}$ ratios,

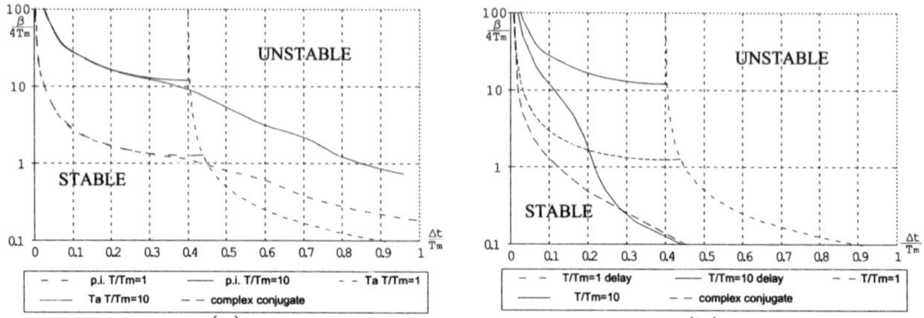

(a) (b)

Figure 20. (a) Absolute stability regions in the SG space for a unit step input and $\boldsymbol{\beta/(\alpha 4TM) = 0.1}$: comparison between the two linearizations;(b) Comparison of the SG space for the MCS algorithm with and without a fixed one time step delay and $\boldsymbol{\beta/(\alpha 4TM) = 0.1}$.

to increase the stability of the time-delayed system. Moreover, it highlights a greater reduction of the stability regions when the ratio $\boldsymbol{T/T_m}$ increases, i.e. when the plant is slow. Simulations and tests which confirm these and other analytical findings can be found in Vulcan (2006).

4.4 Bilinear transformation applied to time integration methods

From a signal processing point of view, each time integration formula introduced in Sections 2 and 3. can be viewed as a *process of digitally filtering* the *input* to produce the *output* at a discrete time $\boldsymbol{k\Delta t}$, the mechanical system acting as an *analog filter* (Preumont, 1982). Therefore, it is worthwhile to analyse how the poles of a time integrator, like the novel *LSRT1* method appears in the discrete counterpart of the $\boldsymbol{s}$-plane; and to compare these representations with the relevant properties of the *CH-*$\boldsymbol{\alpha}$ method introduced in Section 2.1. The location of the poles in the $\boldsymbol{s}$-plane influences and determines the absolute and the relative stability of a continuous time-invariant closed loop control system. In fact, complex closed-loop poles, close to the imaginary axis $\boldsymbol{j\omega}$, determine an oscillatory behaviour of the system; conversely, real closed-loop poles that lie on the left-hand side of the $\boldsymbol{s}$-plane cause an exponential decay behaviour of the system. It is noted that a closed-loop pole in the right-hand side of the $\boldsymbol{s}$-plane entails an unstable behaviour of the system (Sherrick, 2004).

Since the complex variables $\boldsymbol{s}$ and $\boldsymbol{z}$ are related to each other by the following relation

$$z = e^{\Delta t s} \tag{4.51}$$

the location of poles and zeros in the z-plane will depend on the pole and zero locations in the s-plane. This means that the stability of a linear discrete-time time-invariant closed-loop control system can be determined in terms of poles and zeros of the closed-loop pulse transfer function. Moreover, the dynamic behaviour of a discrete-time system depends on the sampling period Δt as well. As a result, a change of Δt causes a different location of the poles and zeros and, consequently, it also influences the stability of the system.

If we express the complex s variable in the following form

$$s = \sigma + j\omega \tag{4.52}$$

then z reads

$$z = e^{\Delta t(\sigma + j\omega)} = e^{\sigma \Delta t} e^{j\omega \Delta t} = e^{\sigma \Delta t} \left(\cos \omega \Delta t + j \sin \omega \Delta t\right) \tag{4.53}$$

i.e. the combination of an exponential term and a periodic term. Due to the presence of the periodic term, poles and zeros that differ in frequency of the sampling frequency $2\pi/\Delta t$ in the s-plane, are mapped into the same locations in the z-plane, see Figure 21, a clear manifestation of *aliasing*. Moreover, the imaginary axis $j\omega$ in the s-plane corresponds to the unit circle in the z-plane, because $\sigma = 0$ corresponds to $|z| = 1$. Since $\sigma < 0$ in the left half of the s-plane, the latter is mapped inside the unit circle in the z-plane, which therefore represents the stability area depicted in Figure 21 (Sherrick, 2004).

From a signal processing point of view, one popular way of creating digital filters from analog filters is to use a bilinear transform design (Sherrick, 2004). In detail, it can be used to obtain a one-to-one mapping from the z-plane into the w-plane with the relation:

$$w = \frac{1}{\Delta t} \frac{z - 1}{z + 1} \tag{4.54}$$

obtained taking the first term of the expansion of the natural logarithm from (4.51) into continuous fractions. The factor 2 at the numerator of (4.54) has been omitted to avoid any assumption. That factor is generally used to properly obtain the poles by using the trapezium rule (Tustin's method, in Franklin and Powell (1980)). The w-plane with the relevant stable region is sketched in Figure 21; and in this plane, the roots of the digital system have a similar representation as the roots of the continuous system in the

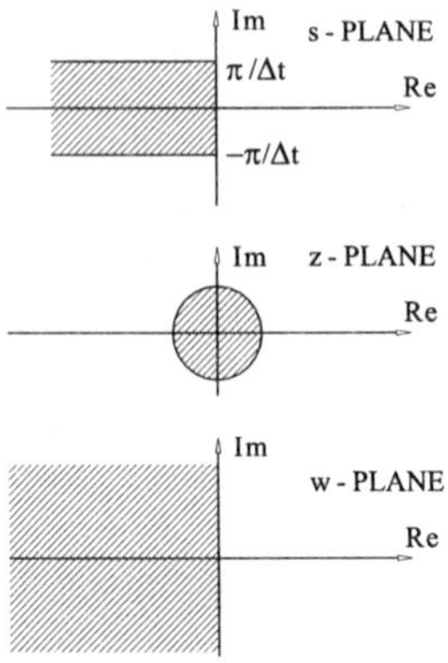

Figure 21. The s-, z- and w-plane and the relevant stability regions

prototype analog s-plane. Anew, from Figure 21 one can observe that every point on the w-plane is the image of an infinite number of points on the s-plane, with frequencies differing by integer multiples of ω_r=$2\pi/\Delta t$. By plugging (4.53) in (4.54), and defining

$$\Omega_{w,d} = \omega_{w,d}\Delta t, \quad \Omega_{s,d} = \omega_{s,d}\Delta t \tag{4.55}$$

one obtains

$$\Omega_{w,d} = \tan\frac{\Omega_{s,d}}{2} \tag{4.56}$$

In detail, there is a tendency of the digitization process to shorten the structure eigenvalues, generating a period elongation. Moreover, the very low structure periods are transformed by the integration process to periods $2\Delta t$: this phenomenon is called spurious high frequency oscillations.

If we consider a SDoF system, for the *LSRT1* method with one stage, we get,

$$\left\{ \begin{array}{c} k_{01} \\ k_{11} \end{array} \right\} = \left[\left[\begin{array}{cc} 1 & 0 \\ 0 & 1 \end{array} \right] - \gamma\Delta t \left[\begin{array}{cc} 0 & 1 \\ -m^{-1}k & -m^{-1}c \end{array} \right] \right]^{-1} \cdot \left[\begin{array}{c} p(t) \\ -m^{-1}kx(t) - m^{-1}cp(t) + m^{-1}f_e(t) \end{array} \right] \Delta t \tag{4.57}$$

$$\mathbf{y}_{k+1} = \left\{ \begin{array}{c} x[k+1] \\ p[k+1] \end{array} \right\} = \mathbf{y}_k + b_1 \left\{ \begin{array}{c} k_{01} \\ k_{11} \end{array} \right\} \tag{4.58}$$

$$x[k+1] = q_{00}x[k] + q_{01}p[k] + q_{02}f_e[k] \tag{4.59}$$

$$p[k+1] = q_{10}x[k] + q_{11}p[k] + q_{12}f_e[k] \tag{4.60}$$

where the coefficients of (4.59) and (4.60) can be easily derived. Then, we apply the z-transform to (4.59) and (4.60),

$$\begin{aligned} z\boldsymbol{X}[z] - z\boldsymbol{X}[0] &= \boldsymbol{q}_{00}\boldsymbol{X}[z] + \boldsymbol{q}_{01}\boldsymbol{P}[z] + \boldsymbol{q}_{02}\boldsymbol{F}_e[z] \quad (4.61) \\ z\boldsymbol{P}[z] - z\boldsymbol{P}[0] &= \boldsymbol{q}_{10}\boldsymbol{X}[z] + \boldsymbol{q}_{11}\boldsymbol{P}[z] + \boldsymbol{q}_{12}\boldsymbol{F}_e[z] \quad (4.62) \end{aligned}$$

and we can write

$$\boldsymbol{X}[z] = \boldsymbol{n}_0\boldsymbol{F}_e[k] + \boldsymbol{r}_d\boldsymbol{X}[0] + \boldsymbol{r}_v\boldsymbol{P}[0] \quad (4.63)$$

Then, we apply the bilinear transformation

$$z = \frac{-1 - w\Delta t}{-1 + w\Delta t} \quad (4.64)$$

and we obtain

$$\boldsymbol{X}[w] = \boldsymbol{n}_{0w}\boldsymbol{F}_e[w] + \boldsymbol{r}_{dw}\boldsymbol{X}[0] + \boldsymbol{r}_{vw}\boldsymbol{P}[0] \quad (4.65)$$

Each of the aforementioned coefficients, for instance $\boldsymbol{r}_{dw}$, can be expressed as

$$r_{dw} = \frac{N_{rdw}}{D_{rdw}} \quad (4.66)$$

where in this case, $\boldsymbol{D}_{rdw}$ represents the denominator of the transfer function between $\boldsymbol{X}[w]$ and $\boldsymbol{F}_e[w]$. By using the relations

$$k = m\omega^2,\ c = 2\xi m\omega,\ \Omega = \omega\Delta t \text{ and } \xi = 0 \quad (4.67)$$

$\boldsymbol{D}_{rdw}$ reads

$$\boldsymbol{D}_{rdw} = b_1\Omega_w^2 - 2b_1\Omega_w^2(b_1 - 2\gamma)w + (4 + b_1^2\Omega_w^2 - 4b_1\Omega_w^2\gamma + 4\Omega_w^2\gamma^2)w^2 \quad (4.68)$$

and, by assuming $b_1 = 1$ $\gamma = 1$ for the *LSRT1* method, one gets the poles

$$w_{1,2} = \frac{-2\Omega_w^2 \pm \sqrt{4\Omega_w^2 - 4\Omega_w^2\left(4 + \Omega_w^2\right)}}{2\left(4 + \Omega_w^2\right)} = \frac{\Omega_w}{\mp 2i - \Omega_w}. \quad (4.69)$$

w_2 can be represented in the w-plane as depicted in Figure 22(a).

Note that $\lim_{\Omega_w \to 0} Re[w_2] = 0$, $\lim_{\Omega_w \to 0} Im[w_2] = 0$ identify the continuous condition; while $\lim_{\Omega_w \to +\infty} Re[w_2] = -1$, $\lim_{\Omega_w \to +\infty} Im[w_2] = 0$ indicate that the method entail no oscillations *i.e.* the L-stability condition. This result can be compared in Figure 22(b) with the representation of the L-stable *CH-α* method with the assumption $\mathbf{f}_{e,k+1} =$

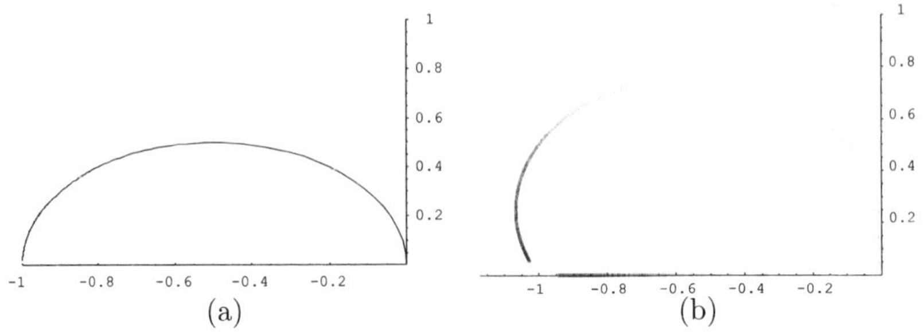

Figure 22. (a) Representation of one pole of the LSRT1 method for $\boldsymbol{\xi} = \mathbf{0.}$;(b) representation of one complex and one spurious pole of the CH-$\boldsymbol{\alpha}$ method for $\boldsymbol{\rho_\infty} = \mathbf{0.}$

$\mathbf{f}_{e,k}$. One can observe a similar trend for one of the complex conjugate poles as well as the presence of the so-called spurious root which does not entail oscillations. The favourable performance of the *LSRT1* method in terms of damping is evident from the comparison.

5 Novel schemes for simulations of monolithic and heterogeneous stiff systems in real time

For a specific number of stages introduced in Subsection 3.1, an important observation is that the Runge-Kutta formulae is not unique. This means that exist free parameters which can be chosen for specific needs. These novel L-stable real-time *(LSRT)* compatible algorithms derived from the Rosenbrock methods (Rosenbrock, 1963) can be introduced for *SHSDS*. In detail, two and three-stage methods, which are second- and third-order accurate, respectively, for uncoupled and linear problems, will be developed in a greater detail in Subsection 5.1 and applied to linear coupled problems in Subsection 5.2.

5.1 L-stable methods for monolithic systems

The two-stage L-stable real-time (LSRT2) method when applied to (2.3), exploits the estimates:

$$\mathbf{k}_1 = [\mathbf{I} - \gamma \Delta t \mathbf{J}]^{-1} \mathbf{f}(t_k, \mathbf{y}_k) \Delta t \qquad (5.1)$$

$$\mathbf{k}_2 = [\mathbf{I} - \gamma \Delta t \mathbf{J}]^{-1} (\mathbf{f}(t_{k+\alpha_2}, \mathbf{y}_{k+\alpha_2}) + \mathbf{J}\gamma_{21}\mathbf{k}_1) \Delta t \qquad (5.2)$$

where $\mathbf{y}_{k+\alpha_2}$ represents the value of the state vector at the α_2 fraction of Δt; the external force $\mathbf{f}_{e,k+\alpha_2}$ is defined at time $t_{k+\alpha_2} = t_k + \alpha_2 \Delta t$ and the measured substructuring force $\mathbf{f}_{s,k+\alpha_2}$ is obtained when applying $\mathbf{y}_{k+\alpha_2}$ at the α_2 fraction of Δt. Thus,

$$\mathbf{y}_{k+\alpha_2} = \mathbf{y}_k + \alpha_{21}\mathbf{k}_1 \tag{5.3}$$

and the state vector reads

$$\mathbf{y}_{k+1} = \mathbf{y}_k + b_1\mathbf{k}_1 + b_2\mathbf{k}_2 \tag{5.4}$$

The accuracy order of the method can be obtained by comparing the derivative terms of the Taylor series expansion of the exact solution with those of the numerical solution according with the developments in Subsection 3.2. For a two-stage method, the conditions needed to achieve second-order accuracy are

$$b_1 + b_2 = 1, \quad \alpha_2 b_2 = \tfrac{1}{2}, \quad b_2(\alpha_{21} + \gamma_{21}) = \tfrac{1}{2} - \gamma \tag{5.5}$$

Then, we impose the condition to assure L-stability, *viz.* $lim_{\lambda \to -\infty} |R(\lambda \Delta t)| = 0$, and the relevant γ values read

$$\gamma = 1 \pm \frac{\sqrt{2}}{2} \tag{5.6}$$

To enforce the compatibility of the algorithm with the real-time dSPACE software (dSP, 2001) the intermediate substeps must have a sample rate which is integer multiple of the base sample rate. As a result, α_2 and α_{21} must be rational numbers and, for instance,

$$\gamma = 1 - \frac{\sqrt{2}}{2}, \ \alpha_2 = \alpha_{21} = \frac{1}{2}, \ \gamma_{21} = -\gamma, \ b_1 = 0 \text{ and } b_2 = 1 \tag{5.7}$$

Figure 23(a) shows the spectral radius of the *LSRT2* method for a second-order system $\ddot{y} = -\omega^2 y$; the spectral radius of the *CH-α* with $\rho_\infty = 0$ is included for comparison. One can observe how a user can achieve a favourable behaviour of the algorithm in the low-frequency range, i.e. for $\gamma = 1 - \frac{\sqrt{2}}{2}$; or a favourable dissipation in the high-frequency range for $\gamma = 1 + \frac{\sqrt{2}}{2}$.

Formulae and parameter values relevant to the three-stage L-stable real-time *LSRT3* method read

$$\mathbf{k}_1 = [\mathbf{I} - \gamma \Delta t \mathbf{J}]^{-1} \mathbf{f}(t_k, \mathbf{y}_k) \Delta t \tag{5.8}$$

$$\mathbf{k}_2 = [\mathbf{I} - \gamma \Delta t \mathbf{J}]^{-1} \left(\mathbf{f}(t_{k+\alpha_2}, \mathbf{y}_{k+\alpha_2}) + \mathbf{J}\gamma_{21}\mathbf{k}_1\right) \Delta t \tag{5.9}$$

$$\mathbf{k}_3 = [\mathbf{I} - \gamma \Delta t \mathbf{J}]^{-1} \left(\mathbf{f}(t_{k+\alpha_3}, \mathbf{y}_{k+\alpha_3}) + \mathbf{J}\gamma_{31}\mathbf{k}_1 + \mathbf{J}\gamma_{32}\mathbf{k}_2\right) \Delta t \tag{5.10}$$

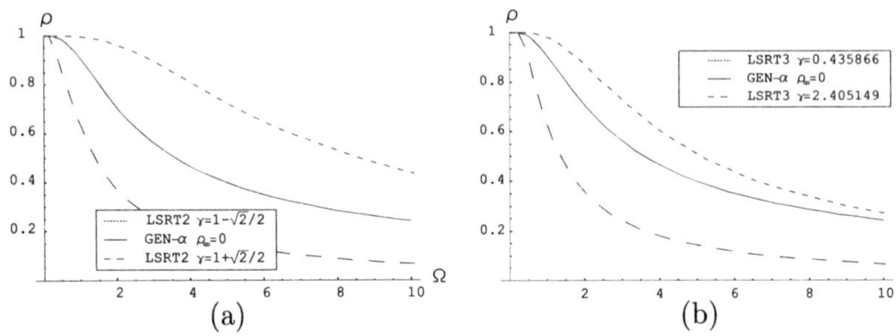

Figure 23. Spectral radius of the *LSRT* methods for a second-order system: (a) LSRT2 integrator; (b) LSRT3 integrator

where: $\mathbf{y}_{k+\alpha_2}$ represents the estimate of the state vector at the $\boldsymbol{\alpha}_2$ fraction of $\boldsymbol{\Delta t}$:

$$\mathbf{y}_{k+\alpha_2} = \mathbf{y}_k + \alpha_{21}\mathbf{k}_1; \tag{5.11}$$

$\mathbf{y}_{k+\alpha_3}$ represents the estimate of the state vector at the $\boldsymbol{\alpha}_3$ fraction of $\boldsymbol{\Delta t}$,

$$\mathbf{y}_{k+\alpha_3} = \mathbf{y}_k + \alpha_{31}\mathbf{k}_1 + \alpha_{32}\mathbf{k}_2 \tag{5.12}$$

Both the external forces $\mathbf{f}_{e,k+\alpha_2}$ and $\mathbf{f}_{e,k+\alpha_3}$ are defined at time $t_{k+\alpha_2} = t_k + \alpha_2\Delta t$ and $t_{k+\alpha_3} = t_k + \alpha_3\Delta t$, while the measured substructuring forces $\mathbf{f}_{s,k+\alpha_2}$ and $\mathbf{f}_{s,k+\alpha_3}$ are obtained when applying $\mathbf{y}_{k+\alpha_2}$ and $\mathbf{y}_{k+\alpha_3}$, respectively, to the PS. The state vector value reads

$$\mathbf{y}_{k+1} = \mathbf{y}_k + b_1\mathbf{k}_1 + b_2\mathbf{k}_2 + b_3\mathbf{k}_3 \tag{5.13}$$

The accuracy order of the method can be obtained comparing the derivative terms of the Taylor series expansion of the exact solution with those of the numerical solution (Bursi et al., 2008). For a three-stage method the conditions to achieve third-order accuracy read

$$\begin{gathered} b_1 + b_2 + b_3 = 1 \\ b_2(\alpha_{21} + \gamma_{21}) + b_3(\alpha_{31} + \gamma_{31} + \alpha_{32} + \gamma_{32}) = \frac{1}{2} - \gamma \\ \alpha_{21}b_2 + (\alpha_{31} + \alpha_{32})b_3 = \frac{1}{2},\ \alpha_{21}^2 b_2 + (\alpha_{31} + \alpha_{32})^2 b_3 = \frac{1}{3} \\ \alpha_{21}(\alpha_{32} + \gamma_{32})b_3 = \frac{1}{6} - \gamma/2 \\ (\alpha_{32} + \gamma_{32})(\alpha_{21} + \gamma_{21})b_3 = \frac{1}{6} - \gamma - \gamma^2 \end{gathered} \tag{5.14}$$

The γ values that assure L-stability can be derived from

$$\gamma^3 - 3\gamma^2 + \frac{3}{2}\gamma - \frac{1}{6} = 0 \tag{5.15}$$

To ensure full compatibility with dSPACE (dSP, 2001) α_{21} and $\alpha_{31} + \alpha_{32}$ must be rational numbers and $\alpha_{21} < \alpha_{31}+\alpha_{32}$. Therefore, we assume

$$\begin{gathered} \gamma = 0.4359 \text{ or } 2.4052, \ \gamma_{21} = 0.1818, \ \gamma_{31} = -0.0428 \\ \gamma_{32} = -0.5384, \ \alpha_2 = \frac{1}{3}, \ \alpha_3 = \frac{2}{3}, \ \alpha_{21} = \alpha_{31} = \alpha_{32} = \frac{1}{3} \\ b_1 = \frac{1}{4}, b_2 = 0 \text{ and } b_3 = \frac{3}{4} \end{gathered} \tag{5.16}$$

The comparison of the *LSRT3* method with the *CH*-α method with $\rho_\infty = 0$ in terms of spectral radius is illustrated in Figure 23(b). One can observe the capabilities offered by the proposed methods through the selection of γ.

5.2 L-stable methods for heterogeneous coupled systems

From Subsection 1.1 and the related ones a real-time simulation of a heterogeneous system can be performed with a linearly implicit algorithm without iteration and based on kinematic control. For the system of Figure 15 reduced for simplicity to a SDoF with $k_2^p = c_2^p = m_2^p = 0$, the coupling algorithm is schematically depicted in Figure 14 and further detailed in Figure 15. By assuming $\mathbf{y} = \{x\ \dot{x}\}^T = \{y_1\ y_2\}^T$, it entails: (i) the integration of the differential equation of the NS using the coupling force $m^p\dot{y}_{2,k-2} = u_{k-1}^n$ available at t_{k-1}; (ii) the motion of the actuator controlled to follow the predicted kinematic quantity $\dot{y}_{2,k} = u_k^p$ available at t_k; (iii) the measure of the coupling force $m^p\dot{y}_{2,k}$. In a greater detail, the *SHSDS* with the linearly implicit *LSRT2* method is summarized in algorithmic form as follows:

(a) Compute the Jacobian matrix $\mathbf{J}$
(b) Compute $\mathbf{k}_1$ from (5.1) and evaluate $\mathbf{y}_{k+\alpha_2}$ from (5.3)
(c) Impose $\mathbf{y}_{k+\alpha_2}$ to the PS, measure the coupling force $f_{s,k+\alpha_2}$, compute $\mathbf{k}_2$ from (5.2) and evaluate $\mathbf{y}_{k+1}$ from (5.4)
(d) Impose $\mathbf{y}_{k+1}$ to the PS and measure the coupling force $f_{s,k+1}$
(e) Set $k = k + 1$ and go to (**a**) for a non-linear system or to(**b**) for a linear system

This actual prediction-substitution procedure is rather different from a truly parallel scheme (Felippa et al., 2001), which at t_{k-1} would entail also a prediction of the coupling force $m^p\dot{y}_{2,k}$. This might be unpractical owing to the nature of the PS.

SHSDS tests can lose stability owing to: (i) control stability loss, see Horiuchi et al. (1999) among others; (ii) stability loss owing to the integration of heterogeneous systems. In this subsection we focus on point (ii), *i.e.* the stability of coupling integrators, taking for granted that the integrators considered herein and applied to monolithic linear elastic systems are L-stable. To avoid possible unwanted effects owing to the integration process, we present herein both a zero-stability analysis for the interconnection shown in Figure 14 and an absolute stability analysis of the discretized coupled system.

The zero-stability concept is nothing more than the minimal demand that the coupled difference system is well posed. In other words, the necessary and sufficient conditions for the proposed LSRT integrators to be convergent when an algebraic loop exists are that they are both consistent and zero-stable (see Lambert, 1991, p. 31). In this approach, the key to analyse the zero-stability of the coupling integrator is based on the fact that a one-step integrator uses a vanishing step size $\boldsymbol{\Delta t} \rightarrow \mathbf{0}$; this leads to a time invariant state vector $\mathbf{y}_{k+1} = \mathbf{y}_k = \mathbf{y}_0$ and, as a result, the stability condition of the coupled structure does not depend on the integration scheme. To perform this analysis, all the inputs $\mathbf{u}_k^n$ and $\mathbf{u}_k^p$, corresponding to the NS and the PS inputs shown in Figure 14, are collected in a vector $\mathbf{u}_k$. The same is done with the outputs, using the vector $\mathbf{z}_k$. The first step of this discrete analysis consists in relating inputs and outputs taken at the same instant at the interconnection, see Figure 15, through (4.8). For instance, for the structure of Figure 1 reduced for simplicity to a SDoF with $k_2^p = c_2^p = m_2^p = 0$, where $\mathbf{y}_k = \{x_k \quad \dot{x}_k\}^T = \{y_{1,k} \quad y_{2,k}\}^T$, one obtains,

$$\mathbf{L} = \begin{bmatrix} 0 & L^p \\ L^p & 0 \end{bmatrix}, \quad \mathbf{z}_k = \begin{Bmatrix} \dot{y}_{2,k} \\ m^p \dot{y}_{2,k-1} \end{Bmatrix}, \quad \mathbf{u}_k = \begin{Bmatrix} u_k^n \\ u_k^p \end{Bmatrix} \tag{5.17}$$

where $L^p = L^n = 1$. Anew for simplicity, we exploit the LSRT1 method and the SDoF structure neglecting the gravity force on the PS, thus we get

$$\mathbf{y}_{k+1} = \mathbf{A}_1 \mathbf{y}_k + \mathbf{B}_1 \mathbf{f}_k \tag{5.18}$$

where

$$\mathbf{A}_1 = \frac{1}{m^n + c^n \Delta t + k^n \Delta t^2} \begin{bmatrix} m^n + c^n \Delta t & m^n \Delta t \\ -k^n \Delta t & m^n \end{bmatrix},$$
$$\mathbf{B}_1 = \frac{\Delta t}{m^n + c^n \Delta t + k^n \Delta t^2} \begin{bmatrix} \Delta t & -\Delta t \\ 1 & -1 \end{bmatrix}, \quad \mathbf{f}_k = \begin{Bmatrix} f_{e,k} \\ u_k^n \end{Bmatrix} \tag{5.19}$$

Note that $\mathbf{A_1}$ is a 2×2 matrix; hence only complex eigenvalues characterize the behaviour of this class of algorithms. If the output equations are time invariant and linearly dependent on the inputs, see Figure 15, one can relate the input at time t_{k+1} to the input at time t_k through the following relation:

$$\mathbf{u}_{k+1} = \mathbf{C}_1 \mathbf{u}_k + \widetilde{\mathbf{f}} \tag{5.20}$$

where the matrix $\mathbf{C}_1$ and the vector $\widetilde{\mathbf{f}}$ contain constant terms as a result of the vanishing time step. In detail, for the SDoF structure one obtains

$$\mathbf{C}_1 = \frac{m^p}{m^n}\begin{bmatrix} 0 & m^n \\ 0 & -1 \end{bmatrix}, \quad \widetilde{\mathbf{f}} = \left\{ \begin{array}{c} 0 \\ \dfrac{1}{m^n}(f_{e,k+1} - k^n y_{1,k} - c^n y_{2,k}) \end{array} \right\} \tag{5.21}$$

In this condition, the coupled system is said to be zero stable if the spectral radius of the coupling matrix $\mathbf{C}_1$ lies within the unit circle in the Argand-Gauss plane. As a result, the way in which a complete system can be partitioned into an NS and a PS must follow certain rules. In practice, this stability condition applies to the splitting of any mass m_i present at the coupling interface. The condition results to be:

$$m_i^p \leq m_i^n \tag{5.22}$$

which means that an interface mass corresponding to the PS cannot be greater than its NS counterpart. This condition arises from the fact that dealing with feed-through systems, an algebraic loop arises between the acceleration $u_k^p = \dot{y}_{2,k}$ and the interface force $u_k^n = m^p \dot{y}^n_{2,k-1}$, see Figure 15, as their magnitudes are both dependent on each other; the algebraic loop can be eliminated together with condition (5.22) through an iterative process or the presence of filters (Kübler and Schiehlen, 2004). Nonetheless, iteration is not possible here owing to the real-time need for *SHSDS*; and the use of filters can modify the dynamics of the resulting emulated system; as a result, the algebraic loop is usually not eliminated owing to the presence of unintended delay (Kyrychko et al., 2006). In this proof, a delay of Δt between the acceleration at the interface u_k^p and the coupling force u_k^n has been considered:

$$\dot{y}_{2,k+1} = \frac{1}{m^n}(f_{e,k+1} - f_{s,k} - c^n y_{2,k+1} - k^n y_{1,k+1}) \tag{5.23}$$

This assumption and the stability condition (5.22) for vanishing Δt have been confirmed by numerical simulations and tests (Bursi et al., 2008).

Differently from the zero-stability analysis, herein we perform an absolute stability analysis relaxing the condition $\Delta t \to 0$. Being the system coupled and the *LSRT* integrators applied to the NS only, we need to introduce a state vector $\tilde{\mathbf{y}} = \left\{ \begin{array}{ccc} x & \dot{x} & f_s \end{array} \right\}^T = \left\{ \begin{array}{ccc} y_1 & y_2 & f_s \end{array} \right\}^T$ which comprises also the coupling force f_s acting on the mass m^n as illustrated in Figure 1(c).

For the *LSRT1* method and the SDoF system of Figure 1 with $m_2^p = k_2^p = c_2^p = 0$, one obtains

$$\tilde{\mathbf{y}}_{k+1} = \tilde{\mathbf{A}}_1 \tilde{\mathbf{y}}_k + \tilde{\mathbf{B}}_1 \tilde{\mathbf{f}}_{1,k} \tag{5.24}$$

where for stability, one is only interested to the amplification matrix $\tilde{\mathbf{A}}_1$; for $\gamma = 1$ it reads

$$\tilde{\mathbf{A}}_1 = \frac{1}{m^n + c^n \Delta t + k^n \Delta t^2} \begin{bmatrix} m^n + c^n \Delta t & m^n \Delta t & -\Delta t^2 \\ -k^n \Delta t & m^n & -\Delta t \\ -k^n m^p & -c^n m^p - \Delta t k^n m^p & -m^p \end{bmatrix} \tag{5.25}$$

Note that, differently from the previous zero-stability analysis, $\tilde{\mathbf{A}}_1$ is a 3×3 matrix owing to the presence of the coupling force $f_{s,k}$. The investigation of the spectral radius $\rho(\tilde{\mathbf{A}}_1)$ indicates that the *LSRT1* method with the zero-stability condition (5.22) satisfied is always L-stable for the coupled system under exam. Nonetheless the same analysis both with the *LSRT2* and the *LSRT3* method applied to the coupled system shows that the *LSRT2* is not A-stable for $\gamma = 1 - \sqrt{2}/2$. This result illustrated in Figure 24(a) was not expected and must warn the analyst dealing with coupled systems. Conversely, the *LSRT2* method recovers at least its A-stability for $\gamma = 1 + \sqrt{2}/2$ and any value of ξ, at a price of adding numerical damping in the low-frequency range components of the response. Again, the loss of stability of the *LSRT3* method for $\gamma = 0.43586$, $\xi = 0$ and small values of Ω is shown in Figure 24(b). Simulations and tests concerning both the stability and accuracy of *LSRT* methods can be found in Bursi et al. (2008).

6 Partitioned Methods and PM-alpha Methods

Partitioned methods were originally developed for the analysis of coupled problems, such as *FSI*. Based on Schur complements to split the coupled mechanical systems into subsystems, these partitioned analysis procedures create a coarse problem with a reduced number of unknowns by the elimination of internal subsystems unknowns. Only the original, *i.e.* primal unknowns -displacements, velocities or accelerations- are considered in the

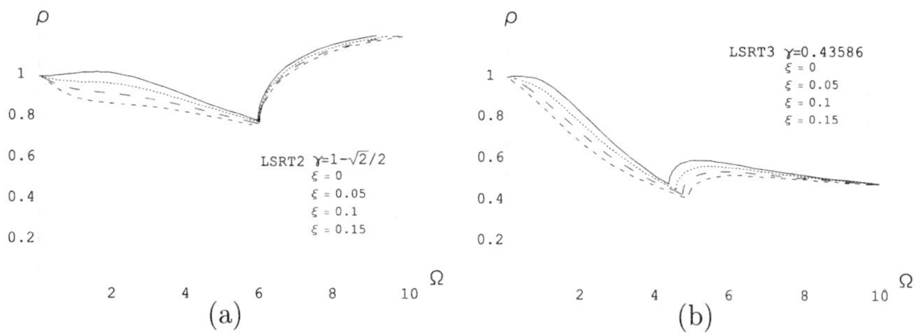

Figure 24. Spectral radius of the LSRT method applied to a SDOF system with $m^n/m^p = 3$ and different values of ξ: (a) *LSRT2* integrator; (b) *LSRT3* integrator

computation. Since the late 1980s, the development of parallel algorithms have been motivated, emphasizing the communication and synchronization between subsystems which is vital to the improvement of the performance of parallel computing. A distinct feature of these parallel algorithms is the use of dual unknowns, such as Lagrange multipliers, to enforce the continuity between subsystems.

In the last decade, the Finite Element Tearing and Interconnecting (*FETI*) method emerged as one of the most powerful parallel algorithms for quasi-static structural mechanics problems (Farhat and Roux, 1991); and it was later extended to dynamic problems (Farhat et al., 1995). In the *FETI* method, a given spatial domain is torn into non-overlapping substructures where an incomplete solution of the primal field is evaluated using a direct solver, and intersubstructure field continuity is enforced via Lagrange multipliers applied at substructure interfaces. The latter gluing phase generates a small size symmetric dual problem where the unknowns are Lagrange multipliers, and which is best solved with a preconditioned conjugate gradient algorithm. Gravouil and Combescure (2001) proved that imposing velocity continuity at the interface leads to a stable algorithm. In detail, they conceived a multi-time-step coupling method, labelled the *GC* method, able to couple arbitrary Newmark schemes with different time steps in different subdomains. In this context, they proved that the *GC* method is unconditionally stable as long as all individual subdomains satisfy their own stability requirements. Moreover, they showed that for multi-time-step cases the *GC* method entails energy dissipation at the interface, while for the case of a single time step in all subdomains the *GC* method is energy preserving. The *GC* method is very appealing for *HiL* and in particular for the continuous

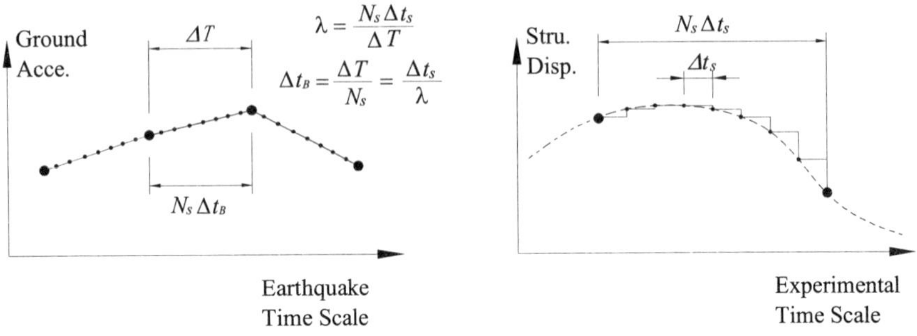

Figure 25. The continuous PsD testing.

PsD testing as heterogeneous numerical and physical substructures can be solved with different implicit/explicit Newmark schemes in different subdomains according to their complexity and characteristics. The possibility of performing a large amount of small time steps on a reduced number of DoFs at the laboratory, at about 1kHz frequency, while computing a large time step on a large number of DoFs on a remote computer, is mandatory for the proper implementation of the continuous *PsD* technique with *DS* (Pegon and Magonette, 2002, 2005). Figure 25 shows how $\boldsymbol{N_s}$ intermediate inputs are created by linear interpolation between two consecutive discrete ground acceleration records. The displacement command to the actuator is then generated in $\boldsymbol{\Delta t_s}$, *i.e.* the sampling time of controller. Thus, an original accelerogram span $\boldsymbol{\Delta T}$ is completed by $\boldsymbol{N_s}$ steps. The time expansion factor $\boldsymbol{\lambda}$ is defined as the ratio $\boldsymbol{\lambda} = \dfrac{\boldsymbol{N_s \Delta t_s}}{\boldsymbol{\Delta T}}$ between the experimental time and the earthquake time. For instance, $\boldsymbol{N_s} = \mathbf{500}$, $\boldsymbol{\Delta t_s} = \mathbf{2}$ms and $\boldsymbol{\Delta T} = \mathbf{5}$ms result in $\boldsymbol{\lambda} = \mathbf{200}$.

The *GC* method employed in the continuous *PsD* testing can maintain the smoothness of the displacement trajectory without using any extrapolation/interpolation assumption and can preserve the optimum signal/noise ratio of the continuous method. Unfortunately, the *GC* method as most of the available methods, is in essence a sequential staggered algorithm where the tasks in different subdomains are not concurrent. To solve the aforementioned problem, Pegon and Magonette (2002, 2005) developed and implemented an interfield parallel algorithm, the *PM* method, based on the *GC* algorithm, where the NS and the PS states advance simultaneously and possibly continuously. Though several studies point out the benefit of the parallelization of partitioned methods (Felippa et al., 2001) in view of real-time simulations of heterogeneous subsystems, there is still a paucity of publications devoted to this issue. Along these developments two novel methods are presented in Subsection 6.1 and 6.2, respectively: the *PM* and the *PM-α* method.

6.1 Analysis of the *PM* method

Formulation of the *PM* method The structure to be studied is divided into two subdomains: $\mathcal{A}$ and $\mathcal{B}$, respectively, to be discretized in time with time steps Δt_A and Δt_B assuming

$$\Delta t_A = ss\, \Delta t_B \tag{6.1}$$

where ss defines the number of substeps. Let $\mathbf{M}^A$, $\mathbf{M}^B$ denote the symmetric positive-definite mass matrices of the two subdomains, $\mathbf{R}^A$ and $\mathbf{R}^B$ the internal force vectors and $\mathbf{F}^A_{ext}$ and $\mathbf{F}^B_{ext}$ the vectors of applied forces, respectively. With this notation to hand, the equations of equilibrium on subdomain $\mathcal{A}$ at time t_{n+1} and subdomain $\mathcal{B}$ at time $t_{n+\frac{j}{ss}}$, $j = \{1, \ldots, ss\}$, together with the kinematic interface constraints between the subdomains can be written as

$$\mathbf{M}^A \ddot{\mathbf{u}}^A_{n+1} + \mathbf{R}^A\left(\mathbf{u}^A_{n+1}, \dot{\mathbf{u}}^A_{n+1}\right) = \mathbf{F}^A_{ext,n+1} + \mathbf{L}^{A^T} \Lambda_{n+1} \tag{6.2}$$

$$\mathbf{M}^B \ddot{\mathbf{u}}^B_{n+\frac{j}{ss}} + \mathbf{R}^B\left(\mathbf{u}^B_{n+\frac{j}{ss}}, \dot{\mathbf{u}}^B_{n+\frac{j}{ss}}\right) = \mathbf{F}^B_{ext,n+\frac{j}{ss}} + \mathbf{L}^{B^T} \Lambda_{n+\frac{j}{ss}} \tag{6.3}$$

$$\mathbf{L}^A \dot{\mathbf{u}}^A_{n+\frac{j}{ss}} + \mathbf{L}^B \dot{\mathbf{u}}^B_{n+\frac{j}{ss}} = \mathbf{0} \tag{6.4}$$

where the state variables $\mathbf{u}(t)$ are nodal quantities arising from a spatial discretization and their derivatives $\dot{\mathbf{u}}(t)$ and $\ddot{\mathbf{u}}(t)$ with respect to time t are indicated with superposed dots; Λ is the vector of Lagrange multipliers; $\mathbf{L}^A$ and $\mathbf{L}^B$ are the constraint matrices which express a linear relationship between the two connected boundaries (Combescure and Gravouil, 2002). In detail for a linear elastic system with classical damping, $\mathbf{R}^A\left(\mathbf{u}^A_{n+1}, \dot{\mathbf{u}}^A_{n+1}\right) = \mathbf{K}^A \mathbf{u}^A_{n+1} + \mathbf{C}^A \dot{\mathbf{u}}^A_{n+1}$ and $\mathbf{R}^B\left(\mathbf{u}^B_{n+\frac{j}{ss}}, \dot{\mathbf{u}}^B_{n+\frac{j}{ss}}\right) = \mathbf{K}^B \mathbf{u}^B_{n+\frac{j}{ss}} + \mathbf{C}^B \dot{\mathbf{u}}^B_{n+\frac{j}{ss}}$, where $\mathbf{K}^A$ and $\mathbf{K}^B$ denote the stiffness matrices of the two subdomains, respectively, and $\mathbf{C}^A$ and $\mathbf{C}^B$ are the damping matrices. In detail, the continuity relationship in (6.4) is enforced in terms of velocities.

The novel parallel procedure which was proposed by Pegon and Magonette (2002) is sketched in Figure 26. A time step equal to $2\Delta t_A$ is exploited in subdomain $\mathcal{A}$, to anticipate information on the subdomain $\mathcal{B}$ at the beginning of a new time step. Considering arbitrary numeric schemes of the

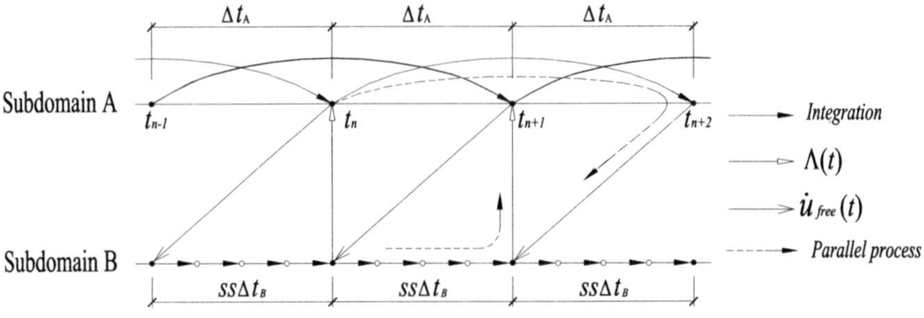

Figure 26. The *PM* method: a parallel procedure.

Newmark family with two characteristic parameters β_A, γ_A and β_B, γ_B for subdomain $\mathcal{A}$ and $\mathcal{B}$, respectively, the flow chart of the *PM* method reads:

1. solve the *free problem* in subdomain $\mathcal{A}$ using $2\Delta t_A$, thus advancing from t_{n-1} to t_{n+1}

$$\begin{aligned} \widetilde{\mathbf{M}}^A \ddot{\mathbf{u}}^A_{n+1_f} &= \mathbf{F}^A_{ext,n+1} - \mathbf{R}^A\left(\tilde{\mathbf{u}}^A_{n-1}, \tilde{\dot{\mathbf{u}}}^A_{n-1}\right) &(6.5)\\ \mathbf{u}^A_{n+1_f} &= \tilde{\mathbf{u}}^A_{n-1} + \alpha^A_1 \ddot{\mathbf{u}}^A_{n+1_f} &(6.6)\\ \dot{\mathbf{u}}^A_{n+1_f} &= \tilde{\dot{\mathbf{u}}}^A_{n-1} + \alpha^A_2 \ddot{\mathbf{u}}^A_{n+1_f} &(6.7) \end{aligned}$$

with

$$\widetilde{\mathbf{M}}^A = \mathbf{M}^A + \alpha^A_1 \mathbf{K}^A + \alpha^A_2 \mathbf{C}^A \quad (6.8)$$

$$\tilde{\mathbf{u}}^A_{n-1} = \mathbf{u}^A_{n-1} + 2\Delta t_A \dot{\mathbf{u}}^A_{n-1} + (2\Delta t_A)^2 \left(\frac{1}{2} - \beta_A\right)\ddot{\mathbf{u}}^A_{n-1} \quad (6.9)$$

$$\tilde{\dot{\mathbf{u}}}^A_{n-1} = \dot{\mathbf{u}}^A_{n-1} + 2\Delta t_A \left(1 - \gamma_A\right)\ddot{\mathbf{u}}^A_{n-1} \quad (6.10)$$

$$\alpha^A_1 = \beta_A \left(2\Delta t_A\right)^2 \quad \text{and} \quad \alpha^A_2 = \gamma_A \left(2\Delta t_A\right) \quad (6.11)$$

Here $\mathbf{R}^A\left(\tilde{\mathbf{u}}^A_{n-1}, \tilde{\dot{\mathbf{u}}}^A_{n-1}\right)$ is the internal force, that for a linear system reads $\mathbf{R}^A\left(\tilde{\mathbf{u}}^A_{n-1}, \tilde{\dot{\mathbf{u}}}^A_{n-1}\right) = \mathbf{K}^A \tilde{\mathbf{u}}^A_{n-1} + \mathbf{C}^A \tilde{\dot{\mathbf{u}}}^A_{n-1}$.

2. start the loop on ss substeps in subdomain $\mathcal{B}$

3. solve the *free problem* in subdomain $\boldsymbol{\mathcal{B}}$ at $t_{n+\frac{j}{ss}}$ with $j = \{1, \ldots, ss\}$

$$\widetilde{\mathbf{M}}^B \ddot{\mathbf{u}}^B_{n+\frac{j}{ss}_f} = \mathbf{F}^B_{ext,n+\frac{j}{ss}} - \mathbf{R}^B \left(\widetilde{\mathbf{u}}^B_{n+\frac{j-1}{ss}}, \widetilde{\dot{\mathbf{u}}}^B_{n+\frac{j-1}{ss}}\right) \tag{6.12}$$

$$\mathbf{u}^B_{n+\frac{j}{ss}_f} = \widetilde{\mathbf{u}}^B_{n+\frac{j-1}{ss}} + \alpha_1^B \ddot{\mathbf{u}}^B_{n+\frac{j}{ss}_f} \tag{6.13}$$

$$\dot{\mathbf{u}}^B_{n+\frac{j}{ss}_f} = \widetilde{\dot{\mathbf{u}}}^B_{n+\frac{j-1}{ss}} + \alpha_2^B \ddot{\mathbf{u}}^B_{n+\frac{j}{ss}_f} \tag{6.14}$$

with

$$\widetilde{\mathbf{M}}^B = \mathbf{M}^B + \alpha_1^B \mathbf{K}^B + \alpha_2^B \mathbf{C}^B \tag{6.15}$$

$$\widetilde{\mathbf{u}}^B_{n+\frac{j-1}{ss}} = \mathbf{u}^B_{n+\frac{j-1}{ss}} + \Delta t_B \dot{\mathbf{u}}^B_{n+\frac{j-1}{ss}} + \Delta t_B^2 (\frac{1}{2} - \beta_B) \ddot{\mathbf{u}}^B_{n+\frac{j-1}{ss}} \tag{6.16}$$

$$\widetilde{\dot{\mathbf{u}}}^B_{n+\frac{j-1}{ss}} = \dot{\mathbf{u}}^B_{n+\frac{j-1}{ss}} + \Delta t_B \left(1 - \gamma_B\right) \ddot{\mathbf{u}}^B_{n+\frac{j-1}{ss}} \tag{6.17}$$

$$\alpha_1^B = \beta_B \Delta t_B^2 \quad \text{and} \quad \alpha_2^B = \gamma_B \Delta t_B \tag{6.18}$$

In a *PsD* test, the internal resisting force $\mathbf{R}^B \left(\widetilde{\mathbf{u}}^B_{n+\frac{j-1}{ss}}\right)$ can be directly measured by imposing the displacement $\widetilde{\mathbf{u}}^B_{n+\frac{j-1}{ss}}$ on the substructure.

4. interpolate the *free velocity* in subdomain $\boldsymbol{\mathcal{A}}$

$$\dot{\mathbf{u}}^A_{n+\frac{j}{ss}_f} = \left(1 - \frac{j}{ss}\right) \dot{\mathbf{u}}^A_{n_f} + \left(\frac{j}{ss}\right) \dot{\mathbf{u}}^A_{n+1_f} \tag{6.19}$$

5. compute the Lagrange multipliers $\mathbf{\Lambda}_{n+\frac{j}{ss}}$ by solving the condensed interface problem

$$\mathbf{H}\mathbf{\Lambda}_{n+\frac{j}{ss}} = -(\mathbf{L}^A \dot{\mathbf{u}}^A_{n+\frac{j}{ss}_f} + \mathbf{L}^B \dot{\mathbf{u}}^B_{n+\frac{j}{ss}_f}) \tag{6.20}$$

where

$$\mathbf{H} = \alpha_2^A \mathbf{L}^A \widetilde{\mathbf{M}}^{A^{-1}} \mathbf{L}^{A^T} + \alpha_2^B \mathbf{L}^B \widetilde{\mathbf{M}}^{B^{-1}} \mathbf{L}^{B^T} \tag{6.21}$$

6. solve the *link problem* in subdomain $\boldsymbol{\mathcal{B}}$ at $t_{n+\frac{j}{ss}}$

$$\widetilde{\mathbf{M}}^B \ddot{\mathbf{u}}^B_{n+\frac{j}{ss}_l} = \mathbf{L}^{B^T} \mathbf{\Lambda}_{n+\frac{j}{ss}} \tag{6.22}$$

$$\mathbf{u}^B_{n+\frac{j}{ss}_l} = \alpha_1^B \ddot{\mathbf{u}}^B_{n+\frac{j}{ss}_l} \text{ and } \dot{\mathbf{u}}^B_{n+\frac{j}{ss}_l} = \alpha_2^B \ddot{\mathbf{u}}^B_{n+\frac{j}{ss}_l} \tag{6.23}$$

7. compute the kinematic quantities of subdomain $\boldsymbol{\mathcal{B}}$ at $t_{n+\frac{j}{ss}}$, by summing the *free quantities* (Point **3**) and *link quantities* (Point **6**)

$$(\cdot) = (\cdot)_f + (\cdot)_l \tag{6.24}$$

8. if $j = ss$, then end the loop in subdomain $\boldsymbol{\mathcal{B}}$
9. solve the *link problem* in subdomain $\boldsymbol{\mathcal{A}}$ using a time step $2\Delta t_A$ from t_{n-1} to t_{n+1}

$$\widetilde{\mathbf{M}}^A \ddot{\mathbf{u}}^A_{n+1_l} = \mathbf{L}^{A^T} \boldsymbol{\Lambda}_{n+1} \tag{6.25}$$

$$\mathbf{u}^A_{n+1_l} = \boldsymbol{\alpha}^A_1 \ddot{\mathbf{u}}^A_{n+1_l} \text{ and } \dot{\mathbf{u}}^A_{n+1_l} = \boldsymbol{\alpha}^A_2 \ddot{\mathbf{u}}^A_{n+1_l} \tag{6.26}$$

10. compute the kinematic quantities of subdomain $\boldsymbol{\mathcal{A}}$ at t_{n+1} by summing the *free problem* (Point **1**) and the *link problem* (Point **9**)

$$(\cdot) = (\cdot)_f + (\cdot)_l \tag{6.27}$$

The strategy described above drives the algorithm illustrated in Figure 26. The dashed line describes the ongoing parallel process in the two subdomains. Note that the process in subdomain $\boldsymbol{\mathcal{A}}$ is split into two independent parts, linked through the subdomain $\boldsymbol{\mathcal{B}}$, which enables the parallel computation and synchronized exchange of information. The *PM* method is referred to as a simplified interfield modification of the basic *GC* method. Extensions to the non-linear case can be found in Pegon and Magonette (2005); Pegon (2008).

Amplification matrix of the PM method The *PM* method uses a mesh partition with two different schemes on two or more subdomains. For this reason, the modal analysis approach to stability is inapplicable. As a result, the *GC* method as other methods of this class was analysed through the energy approach (see Hughes, 1987, p. 564). Although the *PM* method is spectrally stable as it will be shown herein, the well-known energy approach is not easily extendible to this algorithm, owing to the large number of system variables. For this reason, consistency and stability properties of the *PM* method are examined exploiting a standard approach for linear model problems (Bonelli et al., 2008).

In *PsD* applications, it is typically preferred to use an implicit scheme in subdomain $\boldsymbol{\mathcal{A}}$ for the NS and an explicit scheme in subdomain $\boldsymbol{\mathcal{B}}$ for the PS. Thus, it is avoided to have Newton iterations or correction formulations in the procedure devoted to the subdomain $\boldsymbol{\mathcal{B}}$, which require an estimate of the stiffness matrix of the PS. In what follows, we exploit the trapezoidal rule, i.e. $\beta = \frac{1}{4}, \gamma = \frac{1}{2}$ in subdomain $\boldsymbol{\mathcal{A}}$, and the *CD* method, i.e. $\beta =$

$\mathbf{0}, \gamma = \frac{1}{2}$ in subdomain $\mathcal{B}$. Nonetheless, as the original *GC* method, the *PM* scheme employed in this section is convergent with any Newmark scheme with $\gamma \neq \mathbf{1/2}$.

Time stepping schemes applied to linear problems can normally be recast in a recursive form as

$$\mathbf{X}_{n+1} = \mathbf{A}\ \mathbf{X}_n + \mathbf{L}_n \tag{6.28}$$

where $\mathbf{X}$ is an appropriate state vector depending on the formulation of the scheme, $\mathbf{A}$ is the amplification matrix and $\mathbf{L}$ is the load vector which depends on external forces. The main difficulty in the characterization of the *PM* method is the choice of sufficient state variables. To evaluate the state of the *PM* method at time t_{n+1}, not only the variables at time t_n but also some variables at t_{n-1} are needed. This feature transforms the scheme in a second-degree method, which can be analysed as proposed in Young, 1971, p. 486, to define the amplification matrix. To proceed to t_{n+1}, the knowledge of variables on subdomain $\mathcal{B}$ at time t_n is sufficient, whereas subdomain $\mathcal{A}$ needs also the knowledge of variables at time t_{n-1}. Hence, the following state vector is considered:

$$\mathbf{X}_n = \begin{pmatrix} \mathbf{X}^A_{n-1} & \mathbf{X}^A_n & \mathbf{X}^B_n \end{pmatrix}^T \tag{6.29}$$

where $\mathbf{X}^A_n$ collects the kinematic quantities of subdomain $\mathcal{A}$

$$\mathbf{X}^A_n = \begin{pmatrix} \mathbf{u}^A_n & \dot{\mathbf{u}}^A_n & \dot{\mathbf{u}}^A_{n_f} & \ddot{\mathbf{u}}^A_n \end{pmatrix}^T \tag{6.30}$$

and $\mathbf{X}^B_n$ those of subdomain $\mathcal{B}$

$$\mathbf{X}^B_n = \begin{pmatrix} \mathbf{u}^B_n & \dot{\mathbf{u}}^B_n & \ddot{\mathbf{u}}^B_n \end{pmatrix}^T \tag{6.31}$$

As a result, $\mathbf{X}_n$ has the dimension $\mathbf{8}\ n^A + \mathbf{3}\ n^B$ with n^A and n^B the DoFs in the two subdomains. For homogeneous linear model problems, considered without damping for the sake of simplicity, (6.28) turns out to be

$$\begin{pmatrix} \mathbf{X}^A_n \\ \mathbf{X}^A_{n+1} \\ \mathbf{X}^B_{n+1} \end{pmatrix} = \mathbf{A} \begin{pmatrix} \mathbf{X}^A_{n-1} \\ \mathbf{X}^A_n \\ \mathbf{X}^B_n \end{pmatrix} = \begin{bmatrix} \mathbf{0} & \mathbf{I} & \mathbf{0} \\ \bar{\mathbf{A}}^{AA} & \mathbf{A}^{AA} & \mathbf{A}^{AB} \\ \bar{\mathbf{A}}^{BA} & \mathbf{A}^{BA} & \mathbf{A}^{BB} \end{bmatrix} \begin{pmatrix} \mathbf{X}^A_{n-1} \\ \mathbf{X}^A_n \\ \mathbf{X}^B_n \end{pmatrix} \tag{6.32}$$

where $\mathbf{I}$ is the identity matrix, and $\mathbf{A}$ is expressed using the block matrix form determined through the symbolic analysis provided in Bonelli et al. (2008). For a general MDoF system, the state vector (6.29)-(6.31) can be reduced by removing $\dot{\mathbf{u}}^A_{n-1_f}$. Indeed (6.9)-(6.11) show that $\dot{\mathbf{u}}^A_{n-1_f}$ has no contribution from the scheme to pass from t_{n-1} to t_n. It also conforms to the fact that the corresponding row and column vectors are zero in $\mathbf{A}$ except the submatrix part in $\mathbf{I}$, that entails $\dot{\mathbf{u}}^A_{n_f} = \dot{\mathbf{u}}^A_{n_f}$. By summing up, the state vector $\mathbf{X}_n$ reads

$$\mathbf{X}_n = \begin{pmatrix} \mathbf{u}^A_{n-1} & \dot{\mathbf{u}}^A_{n-1} & \ddot{\mathbf{u}}^A_{n-1} & \mathbf{u}^A_n & \dot{\mathbf{u}}^A_n & \dot{\mathbf{u}}^A_{n_f} & \ddot{\mathbf{u}}^A_n & \mathbf{u}^B_n & \dot{\mathbf{u}}^B_n & \ddot{\mathbf{u}}^B_n \end{pmatrix}^T \tag{6.33}$$

with a dimension of $7n^A + 3n^B$.

Accuracy To begin the time integration process, the *PM* algorithm needs an estimate of free velocities and accelerations at an initial time as well as quantities' values of the previous or first step. As a result, an accuracy analysis needs to be addressed both for the initialization process and for the local truncation error as suggested in Hulbert and Hughes (1987). To obtain more seemingly convenient results along the lines of Hulbert and Hughes (1987), we introduce an alternative form of (6.28)

$$\bar{\mathbf{X}}_{n+1} = \bar{\mathbf{A}}\, \bar{\mathbf{X}}_n + \bar{\mathbf{L}}_n \tag{6.34}$$

where

$$\bar{\mathbf{X}}_n = \begin{pmatrix} \mathbf{u}^A_{n-1} & \dot{\mathbf{u}}^A_{n-1} & \ddot{\mathbf{u}}^A_{n-1}\Delta t_A & \mathbf{u}^A_n & \dot{\mathbf{u}}^A_n & \dot{\mathbf{u}}^A_{n_f} & \ddot{\mathbf{u}}^A_n \Delta t_A \\ \mathbf{u}^B_n & \dot{\mathbf{u}}^B_n & \ddot{\mathbf{u}}^B_n \Delta t_A \end{pmatrix}^T \tag{6.35}$$

$$\bar{\mathbf{A}} = \bar{\mathbf{M}}\mathbf{A}\bar{\mathbf{M}}^{-1} \quad , \quad \bar{\mathbf{L}}_n = \bar{\mathbf{M}}\mathbf{L}_n \tag{6.36}$$

$$\bar{\mathbf{M}} = \text{block diagonal}\left[\, \mathbf{I} \quad \mathbf{I} \quad \Delta t_A \mathbf{I} \quad \mathbf{I} \quad \mathbf{I} \quad \mathbf{I} \quad \Delta t_A \mathbf{I} \quad \mathbf{I} \quad \mathbf{I} \quad \Delta t_A \mathbf{I} \,\right] \tag{6.37}$$

with $\mathbf{I}$ being the identity matrix. In detail, Δt_A has been introduced in the acceleration terms of (6.35). The load vector $\bar{\mathbf{L}}_n$ in (6.34) is not included in the following analysis, assuming that the power of the loading error term of its approximation is greater than the order of accuracy of the method.

The local truncation error $\bar{\tau}_n$ is defined as

$$\bar{\tau}_n = \bar{\tau}\,(t_n) = \bar{\mathbf{A}}\bar{\mathbf{X}}\,(t_n) - \bar{\mathbf{X}}\,(t_{n+1}) \tag{6.38}$$

being $\bar{\mathbf{X}}\,(t_n)$ the corresponding exact solution of the state vector $\bar{\mathbf{X}}_n$ at t_n. The exact solution of the *free velocity*, $\dot{\mathbf{u}}^A_f(t_n)$ reads

$$\dot{\mathbf{u}}^A_f(t_n) = \dot{\mathbf{u}}^A(t_n) - \alpha^A_2 \widetilde{\mathbf{M}}^{A^{-1}} \left(\mathbf{M}^A \ddot{\mathbf{u}}^A(t_n) + \mathbf{K}^A \mathbf{u}^A(t_n)\right) \tag{6.39}$$

which is derived from (6.25) and (6.27).

With regard to the SDoF model problem described at length at the end of this subsection, from symbolic calculations one gets

$$\bar{\boldsymbol{\tau}}_n = \boldsymbol{O}\left(\Delta t_A^3\right) \quad \textbf{and} \quad \bar{\boldsymbol{\tau}}_n = \boldsymbol{O}\left(\Delta t_A^2\right) \tag{6.40}$$

for $\boldsymbol{ss} = \mathbf{1}$ and $\boldsymbol{ss} = \mathbf{2}$, respectively. Moreover, for the Two-DoF model problem one gets

$$\bar{\boldsymbol{\tau}}_n = \boldsymbol{O}\left(\Delta t_A^3\right) \quad \textbf{and} \quad \bar{\boldsymbol{\tau}}_n = \boldsymbol{O}\left(\Delta t_A^2\right) \tag{6.41}$$

for $\boldsymbol{ss} = \mathbf{1}$ and $\boldsymbol{ss} = \mathbf{2}$, respectively. In general, it has been obtained that the order of accuracy of $\bar{\boldsymbol{\tau}}_n$ for $\boldsymbol{ss} > \mathbf{2}$ is identical to the one for $\boldsymbol{ss} = \mathbf{2}$.

From (6.29), one observes that two starting points on subdomain $\boldsymbol{\mathcal{A}}$ are necessary to initialize the *PM* method. To obtain the additional point on subdomain $\boldsymbol{\mathcal{A}}$, one may use the *GC* method for the first step followed by the *PM* method in the remaining steps, *i.e.*

$$\mathbf{X}_0^A, \mathbf{X}_0^B \longrightarrow \mathbf{X}_1^A, \mathbf{X}_1^B, \text{ through the } GC \text{ method;} \tag{6.42}$$

which, entails the following initialization error

$$\bar{\mathbf{e}}_2 = \bar{\mathbf{X}}_2 - \bar{\mathbf{X}}\left(2\Delta t_A\right) = \bar{\mathbf{A}}\bar{\mathbf{X}}_1 - \bar{\mathbf{X}}\left(2\Delta t_A\right) \tag{6.43}$$

For the SDoF model problem, the initial condition of the exact solution entails $\boldsymbol{u}(\mathbf{0}) = \boldsymbol{d}_0$, $\dot{\boldsymbol{u}}(\mathbf{0}) = \boldsymbol{v}_0$. By taking

$$\mathbf{X}_0^A = \left\{d_0, v_0, v_{f,0}, a_0\right\}^T, \; \mathbf{X}_0^B = \left\{d_0, v_0, a_0\right\}^T \tag{6.44}$$

with $\boldsymbol{v}_{f,0} = \boldsymbol{v}_0$ and $\boldsymbol{a}_0$ determined by the initial equilibrium condition, one obtains, respectively, when $\boldsymbol{ss} = \mathbf{1}$ and $\boldsymbol{ss} = \mathbf{2}$

$$\bar{\mathbf{e}}_2 = \boldsymbol{O}\left(\Delta t_A^3\right) \quad \textbf{and} \quad \bar{\mathbf{e}}_2 = \boldsymbol{O}\left(\Delta t_A^1\right) \tag{6.45}$$

For the Two-DoF model problem, the results for $\bar{\mathbf{e}}_2$ are similar.

Stability Since the PM method couples two or more subdomains, a zero-stability analysis involving (6.20) has preceeded the absolute stability analysis (He, 2008) along the line of Subsection 5.2. This analysis does not entail any special system parameter condition.

With regard to a SDoF model problem, the absolute values of the eigenvalues of the amplification matrices are plotted in Figure 27 with respect to $\boldsymbol{\Omega}_B = \boldsymbol{\omega}_B \times \boldsymbol{\Delta t}_B$ employing few values of $\boldsymbol{ss}$. The number of nonzero eigenvalues is found to be $\mathbf{6}$. Among them only one pair of the complex conjugate eigenvalues are the principal eigenvalues while the remaining ones are

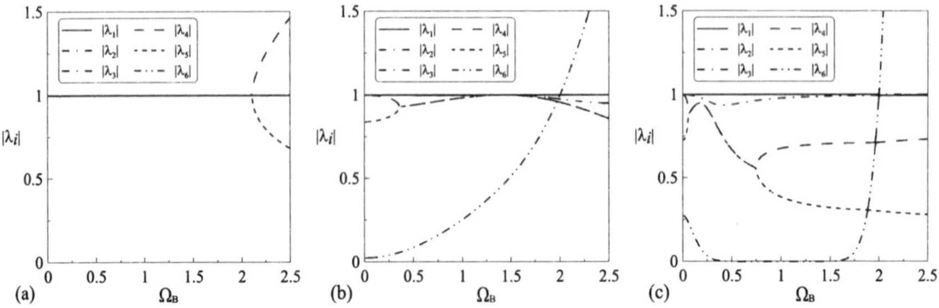

Figure 27. $|\lambda_i|$ of the SDoF problem for the *PM* method when $b_1 = 10$: (a) $ss = 1$; (b) $ss = 2$; and (c) $ss = 20$.

spurious. In detail, the non-zero eigenvalues λ_i $(i = 1, ..., 6)$ of the amplification $\bar{\mathbf{A}}$ in the limit $\Delta t_A = 0$ read

$$\begin{bmatrix} 1 & 1 & 1 & -1 & \frac{-b_1-2\sqrt{-1-b_1}}{2+b_1} & \frac{-b_1+2\sqrt{-1-b_1}}{2+b_1} \end{bmatrix} \tag{6.46}$$

if $ss = 1$ and

$$\begin{bmatrix} 1 & 1 & 1 & -1 & \frac{-8b_1-\sqrt{-256-256b_1+b_1^4}}{(4+b_1)^2} & \frac{-8b_1+\sqrt{-256-256b_1+b_1^4}}{(4+b_1)^2} \end{bmatrix} \tag{6.47}$$

if $ss = 2$. It can be shown that $|\lambda_5|$ and $|\lambda_6|$ are equal to unity in (6.46) and are strictly less than one unless $b_1 = \frac{m_A}{m_B} = \frac{K_B}{K_A}$ defined in (6.52) is either 0 or tends to $+\infty$ in (6.47). Linearly independent eigenvectors exist for the given nonzero eigenvalues in (6.46) and (6.47). In general, we verified that $\bar{\mathbf{A}}$ has linearly independent eigenvectors for all nonzero eigenvalues. As a result, the spectral radius of $\bar{\mathbf{A}}$, $\max_i |\lambda_i(\bar{\mathbf{A}})| \leq 1$ suffices the stability . The stability limit is $\Omega_B = 2$ when $ss > 1$. When $ss = 1$ the stability limit is higher, which is found to be related to the system parameters b_1. Nevertheless, the scheme is always stable when $\Omega_B \leq 2$. Similar results were found with respect to the Two-DoF model problem (Bonelli et al., 2008). In summary, the *PM* method preserves the same property as the *GC* method, *i.e.* it is stable if the stability requirement in each subdomain is satisfied.

Convergence With regard to the convergence of the *PM* method, the global error $\bar{\mathbf{e}}_n$ at t_n is affected by both the local truncation error and the

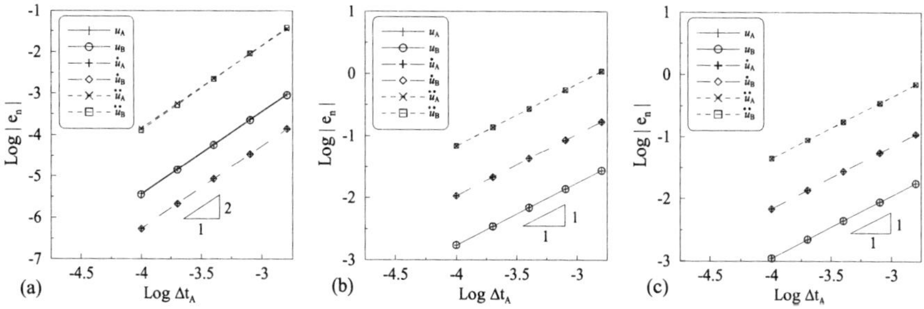

Figure 28. Global error of the SDoF model problem ($b_1 = 10$) for the *PM* method with: (a) $ss = 1$; (b) $ss = 2$; and (c) $ss = 20$.

initialization error,

$$\bar{\mathrm{e}}_n = \bar{\mathbf{A}}^{n-2}\bar{\mathrm{e}}_2 + \sum_{i=2}^{n-1} \bar{\mathbf{A}}^{n-(i+1)}\bar{\tau}_i \tag{6.48}$$

with

$$\bar{\mathrm{e}}_n = \bar{\mathrm{e}}(t_n) = \bar{\mathbf{X}}_n - \bar{\mathbf{X}}(t_n) \tag{6.49}$$

A method is convergent of the order k, if it is stable and consistent of the order k, i.e. $\bar{\tau}_i = O\left(\Delta t_A^{k+1}\right)$ and $\bar{\mathrm{e}}_2 = O\left(\Delta t_A^k\right)$. In detail, $k = 2$ when $ss = 1$ and $k = 1$ when $ss > 1$ for the *PM* method.

Figure 28 shows the global error $|\mathbf{e}_n| = |\mathbf{X}_n - \mathbf{X}(t_n)|$ versus Δt_A in a logarithmic scale for the SDoF model problem, being $\mathbf{X}(t_n)$ the corresponding exact solution of the state vector $\mathbf{X}_n$ at t_n. Simulations confirm analytical results: the *PM* scheme is second-order accurate for $ss = 1$ and first-order accurate for $ss > 1$.

To feature more the spectral properties of the *PM* method, we write the discrete solution as (Hulbert and Chung, 1994)

$$u_n^j = \sum_{i=1}^{m} c_i \lambda_i^n \, , \, \dot{u}_n^j = \sum_{i=1}^{m} \bar{c}_i \lambda_i^n \, , \, \ddot{u}_n^j = \sum_{i=1}^{m} \tilde{c}_i \lambda_i^n \tag{6.50}$$

where λ_i's are the nonzero eigenvalues of the amplification matrix, c_i, $\bar{c}_i$ and $\tilde{c}_i$ serve as generic constants and j represents the subdomain $\mathcal{A}$ or $\mathcal{B}$. Figure 29 shows both the $|\lambda_i|$'s and $|c_i|$'s of u_n^A, $\dot{u}_n^A$, $\ddot{u}_n^A$ of the SDoF model problem with $b_1 = 10, ss = 50$. The number of nonzero eigenvalues plotted in the figure is 6; among them, λ_2, λ_3 are the principal complex conjugate eigenvalues being the relevant constants different from zero at

$\Delta t_A \to \mathbf{0}$, while the others are spurious ones with the relevant constants being zero at $\Delta t_A \to \mathbf{0}$. In detail, a spurious real eigenvalue $|\lambda_1| = 1$ always exists. Figure 29 shows how $c_1 \neq \mathbf{0}$ and $\bar{c}_1 = \tilde{c}_1 = \mathbf{0}$ for $\Delta t_A \neq \mathbf{0}$. This spectral behaviour is responsible of a drift-off error of displacements for finite time steps, typically of the partitioned algorithms with constraints only on velocities. Conversely, the remaining spurious eigenvalues do not entail this behaviour being less than 1. Nonetheless, the drift-off errors tend to zero as $\Delta t_A \to \mathbf{0}$. For on-line simulations, the drift-off effects are found to be negligible by using a small time step which is typical in the application of continuous PsD tests with substructuring. Moreover, for $\ddot{u}_A$ the constants relevant to λ_5 and λ_6 do not vanish for $\Delta t_A \to \mathbf{0}$; and the same phenomenon is observed for $\ddot{u}_B$. As a result, if we analyse the local truncation error $\tau_n = \mathbf{AX}(t_n) - \mathbf{X}(t_{n+1})$, one order lower errors are found in the terms related to the acceleration compared to the ones related to displacement and velocity. We verified numerically that those are the effects of λ_5 and λ_6. As the moduli of λ_5 and λ_6 are less than the principal ones, these contributions will damped out and eventually will not effect the global convergence; on this basis, we introduced the alternative state vector $\bar{\mathbf{X}}_n$ and the amplification matrix $\bar{\mathbf{A}}$ in (6.34).

Representative Model Problems The single-DoF mass-spring emulated system considered herein is split into two parts as illustrated in Figure 30(a). The total mass m and the total stiffness k are the sum of the mass and the stiffness of each part, i.e.

$$m = m_A + m_B \quad , \quad k = k_A + k_B \tag{6.51}$$

The idea is to compare various cases keeping the natural frequency f of the emulated system unchanged. As a result, in the analyses and simulations it is assumed that

$$f = \frac{1}{2\pi}\sqrt{\frac{k}{m}} = 1Hz, \quad \frac{m_A}{m_B} = b_1 \quad \text{and} \quad \frac{k_A}{k_B} = \frac{1}{b_1} \tag{6.52}$$

A Two-DoF mass-spring emulated system can be split into two parts as depicted in Figure 30 (b). The natural frequencies f_1 and f_2 of the emulated system are left unchanged. As a result, for the different cases examined we have assumed

$$f_1 = \frac{1}{2\pi}\sqrt{\frac{k}{m}} = 1\text{Hz}, \;\; f_2 = \sqrt{3}f_1 \text{ and } \frac{m_A}{m_B} = r \text{ with } m_A + m_B = m \tag{6.53}$$

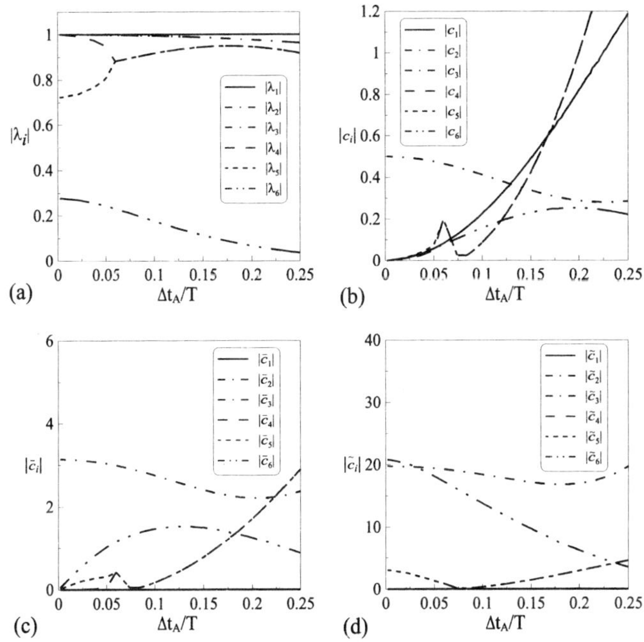

Figure 29. Discrete solution of the PM method: (a) $|\boldsymbol{\lambda}_i|$; (b) $|\boldsymbol{c}_i|$ of $\boldsymbol{u}_n^{\boldsymbol{A}}$; (c) $|\bar{\boldsymbol{c}}_i|$ of $\dot{\boldsymbol{u}}_n^{\boldsymbol{A}}$; and (d) $|\tilde{\boldsymbol{c}}_i|$ of $\ddot{\boldsymbol{u}}_n^{\boldsymbol{A}}$.

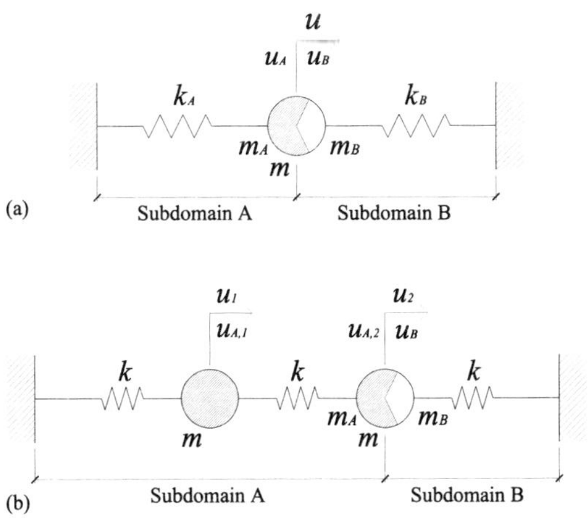

Figure 30. Model problems: (a) partitioned single-DoF system and (b) partitioned two-DoF system.

6.2 The PM-α method

Variants of the *CH*-$\boldsymbol{\alpha}$ Methods Arnold and Brüls (2007) proposed a new form of the *CH*-$\boldsymbol{\alpha}$ method introduced in Subsection 1.2 where the dynamic equilibrium for the time discrete response is enforced at the end of each time step, instead of using a weighted formulation of the equilibrium equation (Chung and Hulbert, 1993). Moreover, it was proved that second-order accuracy of the new algorithm holds also in the accelerations. By using a predictor-corrector form, the novel algorithm suggested by Arnold and Brüls for a linear problem reads
Predictors:

$$\widetilde{\mathbf{u}}_n = \mathbf{u}_n + \Delta t \dot{\mathbf{u}}_n + \Delta t^2 (1/2 - \frac{\beta}{1-\alpha_m})\mathbf{a}_n + \Delta t^2 \beta \frac{\alpha_f}{1-\alpha_m} \ddot{\mathbf{u}}_n \tag{6.54}$$

$$\widetilde{\dot{\mathbf{u}}}_n = \dot{\mathbf{u}}_n + \Delta t (1 - \frac{\gamma}{1-\alpha_m})\mathbf{a}_n + \Delta t \gamma \frac{\alpha_f}{1-\alpha_m} \ddot{\mathbf{u}}_n \tag{6.55}$$

Equilibrium equation:

$$\mathbf{M}\ddot{\mathbf{u}}_{n+1} + \mathbf{C}\dot{\mathbf{u}}_{n+1} + \mathbf{K}\mathbf{u}_{n+1} = \mathbf{F}_{n+1} \tag{6.56}$$

Correctors:

$$\mathbf{u}_{n+1} = \widetilde{\mathbf{u}}_n + \Delta t^2 \beta \frac{1-\alpha_f}{1-\alpha_m} \ddot{\mathbf{u}}_{n+1} \tag{6.57}$$

$$\dot{\mathbf{u}}_{n+1} = \widetilde{\dot{\mathbf{u}}}_n + \Delta t \gamma \frac{1-\alpha_f}{1-\alpha_m} \ddot{\mathbf{u}}_{n+1} \tag{6.58}$$

Recursive equation:

$$(1-\alpha_m)\mathbf{a}_{n+1} + \alpha_m \mathbf{a}_n = (1-\alpha_f)\ddot{\mathbf{u}}_{n+1} + \alpha_f \ddot{\mathbf{u}}_n \ , \ \mathbf{a}_0 = \ddot{\mathbf{u}}_0 \tag{6.59}$$

The state variables $\mathbf{u}$ are nodal quantities arising from a spatial discretization and their derivatives $\dot{\mathbf{u}}$ and $\ddot{\mathbf{u}}$ with respect to time t are indicated with superposed dots. $\mathbf{a}$ are the auxiliary acceleration-like variables, which are identical to the computational accelerations in Chung and Hulbert (1993). By contrast, $\ddot{\mathbf{u}}$, which are the true accelerations keep the property of second-order accuracy as stated in (1.28)(Erlicher et al., 2002).

For the solution procedure, we implemented the *CH*-$\boldsymbol{\alpha}$ method in a $\ddot{\mathbf{u}}$-form.

$$\widetilde{\mathbf{M}}^I \ddot{\mathbf{u}}_{n+1} = \mathbf{F}_{n+1} - \mathbf{C}\widetilde{\dot{\mathbf{u}}}_n - \mathbf{K}\widetilde{\mathbf{u}}_n \tag{6.60}$$

where

$$\widetilde{\mathbf{M}}^I = \mathbf{M} + \Delta t \gamma \frac{1-\alpha_f}{1-\alpha_m} \mathbf{C} + \Delta t^2 \beta \frac{1-\alpha_f}{1-\alpha_m} \mathbf{K} \tag{6.61}$$

After (6.60) is solved, (6.57) and (6.58) may then be used to compute $\mathbf{u}_{n+1}$ and $\dot{\mathbf{u}}_{n+1}$, respectively. In addition, the recursive equation (6.59) needs to be solved for $\mathbf{a}_{n+1}$. As a result four quantities, i.e. $\mathbf{u}$, $\dot{\mathbf{u}}$, $\ddot{\mathbf{u}}$ and $\mathbf{a}$ have to be solved for each time step. The new implementation of the *CH*-α method by Arnold and Brüls (2007) is no more a one-step three-stage method as indicated in (3.4) but a one-step four-stage method.

Based on the implicit form of the *CH*-α method by Arnold and Brüls (2007), we propose a new explicit form of the *CH*-α method enforcing the equilibrium at the end of each time step. Like its implicit progenitor, the explicit *CH*-α method preserves the properties of second-order accuracy in the accelerations, being a one-step four-stage method. Hence, the following equilibrium equation is proposed for the new explicit form of the *CH*-α method:

$$\mathbf{M}\ddot{\mathbf{u}}_{n+1} + \mathbf{C}\dot{\mathbf{u}}_{n+1} + \mathbf{K}\widetilde{\mathbf{u}}_n = \mathbf{F}_{n+1} \tag{6.62}$$

where $\widetilde{\mathbf{u}}_n$ is defined by (6.54). Equation (6.62) together with (6.54), (6.55) and (6.57) - (6.59) provide an explicit implementation of the *CH*-α method. For the solution procedure, one may use the following $\ddot{\mathbf{u}}$-form:

$$\widetilde{\mathbf{M}}^E\ddot{\mathbf{u}}_{n+1} = \mathbf{F}_{n+1} - \mathbf{C}\widetilde{\dot{\mathbf{u}}}_n - \mathbf{K}\widetilde{\mathbf{u}}_n \tag{6.63}$$

where

$$\widetilde{\mathbf{M}}^E = \mathbf{M} + \Delta t\gamma\frac{1-\alpha_f}{1-\alpha_m}\mathbf{C} \tag{6.64}$$

On each time step, after (6.63) is solved, $\mathbf{u}_{n+1}$, $\dot{\mathbf{u}}_{n+1}$ and $\mathbf{a}_{n+1}$ are calculated using (6.57), (6.58) and (6.59). Moreover, the contribution of $\mathbf{K}$ in (6.61) is neglected to keep an explicit form.

It has been proved by He (2008) that the new explicit form of the *CH*-α method is numerically equivalent to the explicit-α method proposed by Daniel (2003). Moreover, the integration parameters of the new algorithm can be obtained using those of the implicit *CH*-α method of Chung and Hulbert (1993), with the spectral radius ρ_∞ substituted by ρ_b, the spectral radius at the bifurcation point of the explicit algorithm. In terms of ρ, which is either ρ_∞ for the implicit algorithm or ρ_b for the explicit one, the algorithmic parameters of the new implicit and explicit *CH*-α method read

$$\alpha_m = \frac{2\rho-1}{\rho+1}\,,\ \alpha_f = \frac{\rho}{\rho+1} \tag{6.65}$$

$$\beta = \frac{1}{4}(\frac{1}{2}+\gamma)^2\,,\ \gamma = \frac{1}{2} - \alpha_m + \alpha_f \tag{6.66}$$

In detail, the bifurcation limit $\boldsymbol{\Omega}_b$ and the stability limit $\boldsymbol{\Omega}_s$ of the explicit algorithm, in the case of $\mathbf{C} = \mathbf{0}$ are defined as

$$\boldsymbol{\Omega}_b = (1+\rho_b)\sqrt{2-\rho_b} \tag{6.67}$$

$$\boldsymbol{\Omega}_s = \sqrt{\frac{12(1+\rho_b)^3(2-\rho_b)}{10+15\rho_b-\rho_b^2+\rho_b^3-\rho_b^4}} \tag{6.68}$$

where b and s subscript are used to denote bifurcation and stability.

Formulation of the algorithm By adopting the *CH-α* methods which enforce equilibrium at the end of each time step, a novel interfield parallel method is obtained, which is referred to as the *PM-α* method. We consider subdomain A to be integrated with a coarse time step Δt_A using the implicit *CH-α* method and subdomain B to be integrated with a small time step Δt_B using the explicit *CH-α* method. Hence, the flow-chart of the PM-α method reads:

1. solve the free problem in subdomain A using $2\Delta t_A$, thus advancing from t_{n-1} to t_{n+1}

$$\widetilde{\mathbf{M}}^A \ddot{\mathbf{u}}^A_{n+1_f} = \mathbf{F}^A_{n+1} - \mathbf{C}^A \tilde{\dot{\mathbf{u}}}^A_{n-1} - \mathbf{K}^A \tilde{\mathbf{u}}^A_{n-1} \tag{6.69}$$

$$\mathbf{u}^A_{n+1_f} = \tilde{\mathbf{u}}^A_{n-1} + \alpha^A_1 \ddot{\mathbf{u}}^A_{n+1_f} \tag{6.70}$$

$$\dot{\mathbf{u}}^A_{n+1_f} = \tilde{\dot{\mathbf{u}}}^A_{n-1} + \alpha^A_2 \ddot{\mathbf{u}}^A_{n+1_f} \tag{6.71}$$

where $\widetilde{\mathbf{M}}^A$, $\tilde{\mathbf{u}}^A_{n-1}$ and $\tilde{\dot{\mathbf{u}}}^A_{n-1}$ are given by (6.61), (6.54) and (6.55), respectively; and

$$\alpha^A_1 = 4\Delta t^2_A \beta_A \frac{1-\alpha^A_f}{1-\alpha^A_m} \text{ and } \alpha^A_2 = 2\Delta t_A \gamma_A \frac{1-\alpha^A_f}{1-\alpha^A_m} \tag{6.72}$$

2. start the loop on ss substeps in subdomain $\mathcal{B}$ at the j-th subsetp with $j = \{1, \ldots, ss\}$
3. solve the *free problem* in subdomain $\mathcal{B}$ at $t_{n+\frac{j}{ss}}$ by

$$\widetilde{\mathbf{M}}^B \ddot{\mathbf{u}}^B_{n+\frac{j}{ss}f} = \mathbf{F}^B_{n+\frac{j}{ss}} - \mathbf{C}^A \tilde{\dot{\mathbf{u}}}^B_{n+\frac{j-1}{ss}} - \mathbf{K}^B \tilde{\mathbf{u}}^B_{n+\frac{j-1}{ss}} \tag{6.73}$$

$$\mathbf{u}^B_{n+\frac{j}{ss}f} = \tilde{\mathbf{u}}^B_{n+\frac{j-1}{ss}} + \alpha^B_1 \ddot{\mathbf{u}}^B_{n+\frac{j}{ss}f} \tag{6.74}$$

$$\dot{\mathbf{u}}^B_{n+\frac{j}{ss}f} = \tilde{\dot{\mathbf{u}}}^B_{n+\frac{j-1}{ss}} + \alpha^B_2 \ddot{\mathbf{u}}^B_{n+\frac{j}{ss}f} \tag{6.75}$$

where $\widetilde{\mathbf{M}}^B$, $\widetilde{\mathbf{u}}^B_{n+\frac{j-1}{ss}}$ and $\widetilde{\dot{\mathbf{u}}}^B_{n+\frac{j-1}{ss}}$ are given by (6.64), (6.54) and (6.55), respectively; and

$$\alpha_1^B = \Delta t_B^2 \beta_B \frac{1-\alpha_f^B}{1-\alpha_m^B} \text{ and } \alpha_2^B = \Delta t_B \gamma_B \frac{1-\alpha_f^B}{1-\alpha_m^B} \tag{6.76}$$

In case of a PsD test, $\mathbf{R}^B\left(\widetilde{\mathbf{u}}^B_{n+\frac{j-1}{ss}}\right)$ can be directly measured by imposing the displacement $\widetilde{\mathbf{u}}^B_{n+\frac{j-1}{ss}}$ on the physical substructure B.

4. interpolate the free velocity in subdomain A by

$$\dot{\mathbf{u}}^A_{n+\frac{j}{ss}_f} = (1-\frac{j}{ss})\dot{\mathbf{u}}^A_{n_f} + \frac{j}{ss}\dot{\mathbf{u}}^A_{n+1_f} \tag{6.77}$$

5. compute the Lagrange multipliers $\mathbf{\Lambda}_{n+\frac{j}{ss}}$ by solving the condensed interface problem

$$\mathbf{H}\mathbf{\Lambda}_{n+\frac{j}{ss}} = -(\mathbf{L}^A\dot{\mathbf{u}}^A_{n+\frac{j}{ss}_f} + \mathbf{L}^B\dot{\mathbf{u}}^B_{n+\frac{j}{ss}_f}) \tag{6.78}$$

with

$$\mathbf{H} = \alpha_2^A \mathbf{L}^A \widetilde{\mathbf{M}}^{A^{-1}} \mathbf{L}^{A^T} + \alpha_2^B \mathbf{L}^B \widetilde{\mathbf{M}}^{B^{-1}} \mathbf{L}^{B^T} \tag{6.79}$$

6. solve the *link problem* in subdomain $\mathcal{B}$ at $t_{n+\frac{j}{ss}}$

$$\widetilde{\mathbf{M}}^B \ddot{\mathbf{u}}^B_{n+\frac{j}{ss}_l} = \mathbf{L}^{B^T}\mathbf{\Lambda}_{n+\frac{j}{ss}} \tag{6.80}$$

$$\mathbf{u}^B_{n+\frac{j}{ss}_l} = \alpha_1^B \ddot{\mathbf{u}}^B_{n+\frac{j}{ss}_l} \tag{6.81}$$

$$\dot{\mathbf{u}}^B_{n+\frac{j}{ss}_l} = \alpha_2^B \ddot{\mathbf{u}}^B_{n+\frac{j}{ss}_l} \tag{6.82}$$

7. Compute the kinematic quantities $\mathbf{u}$, $\dot{\mathbf{u}}$ and $\ddot{\mathbf{u}}$ of subdomain B at $t_{n+\frac{j}{ss}}$ by

$$(\cdot) = (\cdot)_{free} + (\cdot)_{link} \tag{6.83}$$

and then compute the auxiliary acceleration-like variables $\mathbf{a}^B_{n+\frac{j}{ss}}$ of subdomain B at $t_{n+\frac{j}{ss}}$ by

$$(1-\alpha_m^B)\mathbf{a}^B_{n+\frac{j}{ss}} + \alpha_m^B \mathbf{a}^B_{n+\frac{j-1}{ss}} = (1-\alpha_f^B)\ddot{\mathbf{u}}^B_{n+\frac{j}{ss}} + \alpha_f^B \ddot{\mathbf{u}}^B_{n+\frac{j-1}{ss}} \quad \text{with} \quad \mathbf{a}_0^B = \ddot{\mathbf{u}}_0^B \tag{6.84}$$

8. if $j = ss$, then end the loop in subdomain $\mathcal{B}$

9. solve the *link problem* in subdomain $\mathcal{A}$ using a time step $2\Delta t_A$ from t_{n-1} to t_{n+1}

$$\widetilde{\mathbf{M}}^A \ddot{\mathbf{u}}^A_{n+1_l} = \mathbf{L}^{A^T} \mathbf{\Lambda}_{n+1} \quad (6.85)$$

$$\mathbf{u}^A_{n+1_l} = \alpha^A_1 \ddot{\mathbf{u}}^A_{n+1_l} \quad (6.86)$$

$$\dot{\mathbf{u}}^A_{n+1_l} = \alpha^A_2 \ddot{\mathbf{u}}^A_{n+1_l} \quad (6.87)$$

10. Calculate the kinematic quantities $\mathbf{u}$, $\dot{\mathbf{u}}$ and $\ddot{\mathbf{u}}$ of subdomain A at t_{n+1} by

$$(\cdot) = (\cdot)_{free} + (\cdot)_{link} \quad (6.88)$$

and then compute the auxiliary acceleration-like variables $\mathbf{a}^A_{n+1}$ of subdomain A at t_{n+1} by

$$(1-\alpha^A_m)\mathbf{a}^A_{n+1} + \alpha^A_m \mathbf{a}^A_{n-1} = (1-\alpha^A_f)\ddot{\mathbf{u}}^A_{n+1} + \alpha^A_f \ddot{\mathbf{u}}^A_{n-1}\ , \ \mathbf{a}^A_0 = \ddot{\mathbf{u}}^A_0 \quad (6.89)$$

Regarding the initialization of the *PM*-α method, the problem is similar to that of the *PM* method as two starting points are required on subdomain A. We suggest to use a starting condition similar to that of the *PM* method, see (6.42)), which has been discussed before.

Amplification matrix of the PM-α method In the following analysis, the implicit *CH*-α method in subdomain A and the explicit *CH*-α method in subdomain B are characterized by ρ_∞ and ρ_b, respectively, according to (6.65) and (6.66).

When applied to linear problems, the *PM*-α method can be recast in a recursive form as

$$\mathbf{X}_{n+1} = \mathbf{A}\ \mathbf{X}_n + \mathbf{L}_n \quad (6.90)$$

with $\mathbf{X}$ an appropriate state vector depending on the formulation of the scheme, $\mathbf{A}$ the amplification matrix and $\mathbf{L}$ the load vector which depends on external forces. To proceed to the next time step the PM-α method requires not only the state variables at time t_n but also some ones at t_{n-1}. For this reason, we consider the following state vector:

$$\mathbf{X}_n = \begin{pmatrix} \mathbf{X}^A_{n-1} & \mathbf{X}^A_n & \mathbf{X}^B_n \end{pmatrix}^T \quad (6.91)$$

where $\mathbf{X}^A_n$ collects the kinematic quantities of subdomain $\mathcal{A}$

$$\mathbf{X}^A_n = \begin{pmatrix} \mathbf{u}^A_n & \dot{\mathbf{u}}^A_n & \dot{\mathbf{u}}^A_{n_{free}} & \ddot{\mathbf{u}}^A_n & \mathbf{a}^A_n \end{pmatrix}^T \quad (6.92)$$

.

and $\mathbf{X}_n^B$ those of subdomain $\mathcal{B}$

$$\mathbf{X}_n^B = \left(\mathbf{u}_n^B \quad \dot{\mathbf{u}}_n^B \quad \ddot{\mathbf{u}}_n^B \quad \mathbf{a}_n^B \right)^T \tag{6.93}$$

Consequently, $\mathbf{X}_n$ has the dimension $\mathbf{10}\ \boldsymbol{n^A} + \mathbf{4}\ \boldsymbol{n^B}$ with $\boldsymbol{n^A}$, $\boldsymbol{n^B}$ the DoFs in the two subdomains. For homogeneous linear model problems, considered without damping for the sake of brevity, (6.91) turns out to be

$$\begin{pmatrix} \mathbf{X}_n^A \\ \mathbf{X}_{n+1}^A \\ \mathbf{X}_{n+1}^B \end{pmatrix} = \mathbf{A} \begin{pmatrix} \mathbf{X}_{n-1}^A \\ \mathbf{X}_n^A \\ \mathbf{X}_n^B \end{pmatrix} = \begin{bmatrix} \mathbf{0} & \mathbf{I} & \mathbf{0} \\ \bar{\mathbf{A}}^{AA} & \mathbf{A}^{AA} & \mathbf{A}^{AB} \\ \bar{\mathbf{A}}^{BA} & \mathbf{A}^{BA} & \mathbf{A}^{BB} \end{bmatrix} \begin{pmatrix} \mathbf{X}_{n-1}^A \\ \mathbf{X}_n^A \\ \mathbf{X}_n^B \end{pmatrix} \tag{6.94}$$

where $\mathbf{I}$ is the identity matrix, and $\mathbf{A}$ is expressed using the block matrix.

Accuracy of the PM-α method In order to perform the accuracy analysis, we introduce an alternative form of (6.91)

$$\bar{\mathbf{X}}_{n+1} = \bar{\mathbf{A}}\ \bar{\mathbf{X}}_n + \bar{\mathbf{L}}_n \tag{6.95}$$

where

$$\bar{\mathbf{X}}_n = \left(\begin{matrix} \mathbf{u}_{n-1}^A & \dot{\mathbf{u}}_{n-1}^A & \dot{\mathbf{u}}_{n-1_f}^A & \ddot{\mathbf{u}}_{n-1}^A \Delta t_A & \mathbf{a}_{n-1}^A \Delta t_A & \mathbf{u}_n^A \\ & \dot{\mathbf{u}}_n^A & \dot{\mathbf{u}}_{n_f}^A & \ddot{\mathbf{u}}_n^A \Delta t_A & \mathbf{a}_n^A \Delta t_A & \\ & \mathbf{u}_n^B & \dot{\mathbf{u}}_n^B & \ddot{\mathbf{u}}_n^B \Delta t_A & \mathbf{a}_n^B \Delta t_A & \end{matrix} \right)^T \tag{6.96}$$

$$\bar{\mathbf{A}} = \bar{\mathbf{M}}\mathbf{A}\bar{\mathbf{M}}^{-1}, \ \bar{\mathbf{L}}_n = \bar{\mathbf{M}}\mathbf{L}_n \tag{6.97}$$

$$\bar{\mathbf{M}} = \text{block diagonal}\left[\begin{matrix} \mathbf{I} & \mathbf{I} & \mathbf{I} & \Delta t_A \mathbf{I} & \Delta t_A \mathbf{I} & \mathbf{I} & \mathbf{I} & \\ & \mathbf{I} & \Delta t_A \mathbf{I} & \Delta t_A \mathbf{I} & \mathbf{I} & \mathbf{I} & \Delta t_A \mathbf{I} & \Delta t_A \mathbf{I} \end{matrix} \right] \tag{6.98}$$

In particular, $\boldsymbol{\Delta t_A}$ has been introduced in the acceleration terms of (6.96).

Given the definition of local truncation error $\bar{\tau}_n$ in (6.38), the accuracy analysis results of the PM-$\boldsymbol{\alpha}$ method for the SDoF model problem introduced above turn out to be similar to those stated in (6.40) and (6.41). Moreover, given the definition of the initialization error $\bar{\mathbf{e}}_2$ in (6.43), the PM-$\boldsymbol{\alpha}$ method exhibits on the SDoF model problem similar results to those obtained in (6.45).

Stability The stability of the PM-$\boldsymbol{\alpha}$ method is investigated herein employing the spectral analysis approach on the SDoF model problem. It is obtained that $\bar{\mathbf{A}}$ defined in (6.90) is endowed with 9 non-zero eigenvalues.

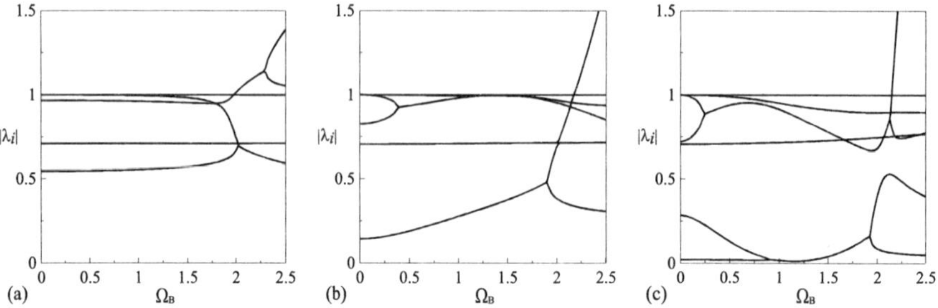

Figure 31. $|\lambda_i|$ of the SDoF problem for the *PM*-α method when $b_1 = 10$: (a) $ss = 1$; (b) $ss = 2$; and (c) $ss = 5$.

In detail, with $\rho_\infty = \rho_b = 1$ and $\Delta t_A = 0$ the non-zero eigenvalues read

$$\begin{bmatrix} 1 & 1 & 1 & -1 & -1 & i & -i & \frac{-b_1-2\sqrt{-1-b_1}}{2+b_1} & \frac{-b_1+2\sqrt{-1-b_1}}{2+b_1} \end{bmatrix} \tag{6.99}$$

if $ss = 1$, and

$$\left[1 \quad \frac{1}{\frac{-8b_1-\sqrt{-256-256b_1+b_1^4}}{(4+b_1)^2}} \quad \frac{1}{\frac{-8b_1+\sqrt{-256-256b_1+b_1^4}}{(4+b_1)^2}} \quad 1 \quad -1 \quad i \quad -i \right] \tag{6.100}$$

if $ss = 2$. The modulus of the eigenvalues are less than or equal to 1, provided any value of $b_1 = \frac{m_A}{m_B} = \frac{k_B}{k_A}$. Moreover, the repeated eigenvalues have linearly independent eigenvectors. We verified that the aforementioned findings are valid for any $\Delta t_A > 0$, $0 \le \rho_\infty \le 1$ and $0 \le \rho_b \le 1$. As a result, the method is stable if $\max_i |\lambda_i(\bar{\mathbf{A}})| \le 1$. The absolute values of the eigenvalues of $\bar{\mathbf{A}}$ are plotted in Figure 31 versus $\Omega_B = \omega_B \times \Delta t_B$ employing few values of ss. The global problem has the stability limit always equal to or higher than that of the explicit subdomain $\mathcal{B}$.

Convergence The global error $\bar{\mathbf{e}}_n$ at t_n of the PM-α method is also affected by both the local truncation error and the initialization error, as indicated by (6.48)). Analytical findings of He (2008) indicate that the PM-α method is second-order accurate when $ss = 1$ and first-order accurate when $ss > 1$. Figure 32 shows the global error $|\mathbf{e}_n| = |\mathbf{X}_n - \mathbf{X}(t_n)|$ versus Δt_A in a logarithmic scale for the SDoF model problem, which confirms the convergence analysis results.

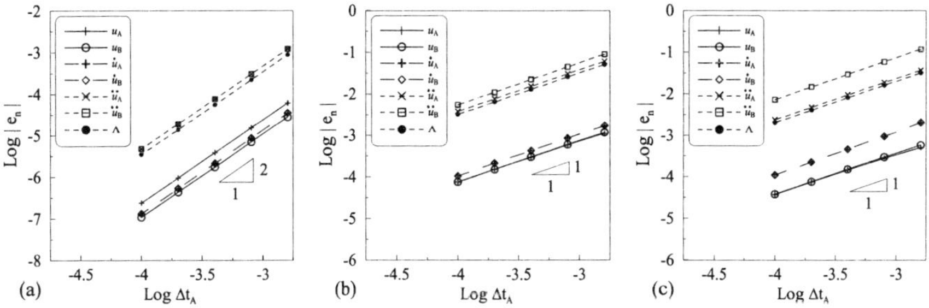

Figure 32. Global error of the SDoF problem ($b_1 = 10$) for the *PM-α* method with: (a) $ss = 1$; (b) $ss = 2$; and (c) $ss = 20$.

Numerical dissipation Figure 33 (a) shows the numerical damping ratio induced by the *PM-α* method on the SDoF model problem, where $\rho = \rho_\infty = \rho_b$. In detail, the algorithmic damping ratio $\bar{\xi}$ and the distorted angula frequency $\bar{\Omega}$ have been defined in (3.43) and (3.44), respectively. Figure 33(a) shows the high-frequency dissipation capabilities of the PM-α method which are controllable by the user. Conversely, 33(b) shows the spectral radius of the scheme, ignoring the real unity eigenvalue, which contributes to a constant drift-off displacement error as indicated in Figure 29) for the *PM* method. One can observe from Figure 33 that the equivalent damping generated at the shared node by the *PM* method is modest even for large time steps. In contrast with that, the *PM-α* method provides adequate numerical dissipation of the higher modes components of the response.

Some Numerical Simulations The Four-DoF emulated system considered herein can be split into two parts as shown in Figure 34(a). The displacement vector of subdomain $\mathcal{A}$ is chosen to be $\mathbf{U}_A = \{u_2, u_3, u_4\}^T$ while the displacement vector of subdomain $\mathcal{B}$ is chosen as $\mathbf{U}_B = \{u_1, u_2\}^T$. The mass matrices of each subdomain read

$$\mathbf{M}^A = m \begin{bmatrix} \frac{r}{1+r} & 0 & 0 \\ 0 & 1 & 0 \\ 0 & 0 & 1 \end{bmatrix} \text{ and } \mathbf{M}^B = m \begin{bmatrix} 1 & 0 \\ 0 & \frac{1}{1+r} \end{bmatrix} \tag{6.101}$$

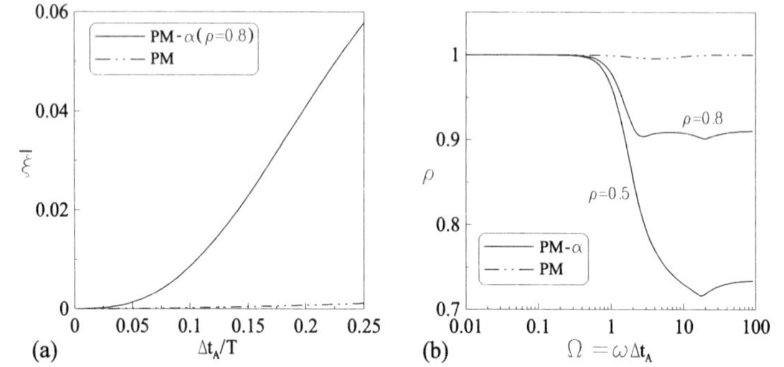

Figure 33. Numerical dissipation of the *PM-*α method for the SDoF problem when $b_1 = 0.01$ and $ss = 5$: (a) algorithmic damping ratio and (b) spectral radius.

while the stiffness matrices of each subdomain read

$$\mathbf{K}^A = 10^6 \begin{bmatrix} 35.5 & -32.6 & 9.7 \\ -32.6 & 32.3 & -5.6 \\ 9.7 & -5.6 & 7.4 \end{bmatrix} \text{ and } \mathbf{K}^B = 10^6 \begin{bmatrix} 12.4 & -6.6 \\ -6.6 & 3.9 \end{bmatrix} \tag{6.102}$$

where the condensed mass $m = 20,000kg$ for each DoF and $r = \frac{m_A}{m_B}$ being $m = m_A + m_B$. As a result, the natural frequencies for the emulated system read

$$f_1 = 0.47\ Hz,\ f_2 = 2.90\ Hz,\ f_3 = 4.18\ Hz \text{ and } f_4 = 9.47\ Hz \tag{6.103}$$

Figure 34(b) and (c) show the results provided by the *PM* and *PM-*α methods, respectively, using the same time step $\Delta t_A/T_1 = 1/40$ and $ss = 20$. It is evident that the unwanted high-frequency components of the response are traced by the *PM* method, while the *PM-*α method damps out them after a few time steps.

7 Conclusions and Future Perspectives

This section summarizes the main conclusions of the material presented in the chapter which is not intended to cover all of the numerical methods and techniques presented or all of the types of systems covered. Moreover, the section offers recommendations for future directions.

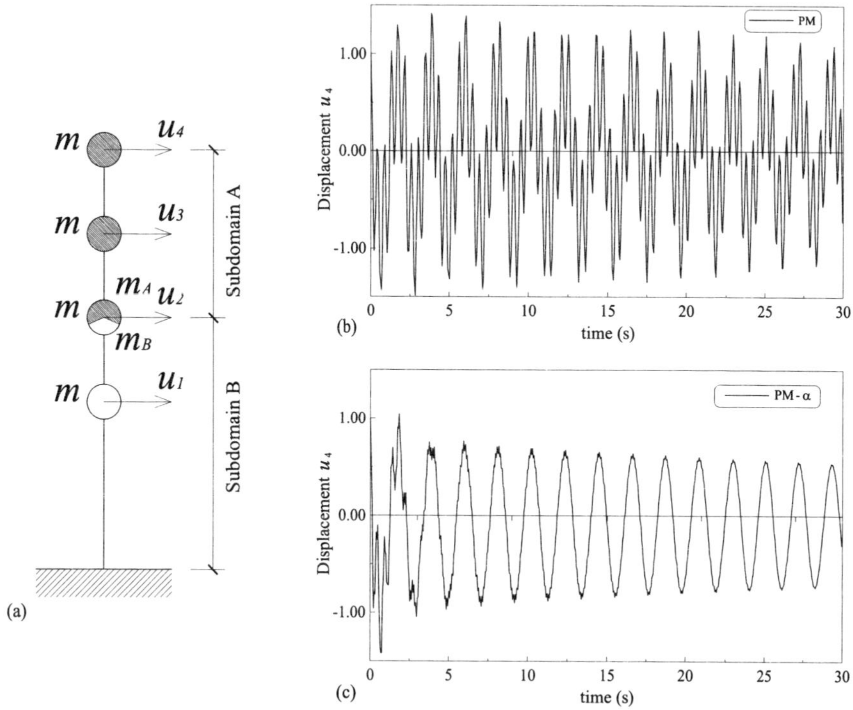

Figure 34. Partitioned four-DoF system: (a) the model problem; the displacement response in free vibration integrated by (b) the *PM* method and (c) the *PM-α* method.

7.1 Conclusions

Although analytical or symbolic solutions provide the greatest insight regarding the influence of various parameters in the transient dynamic response of monolithic structural systems as well as of heterogeneous (numerical/physical) subsystems, the researchers frequently resort to numerical studies or testing techniques or fusion of both in order to discern trends of their dynamic behaviour. Thus, long and expensive trials with simulations and experiments are often the only avenue left to understand the behaviour of complex non-linear mechanical and civil-engineering systems. Since this chapter has been the first, we have presented a brief state-of-the-art on real-time testing with *DS* in the context of *HiL* and *PsD* testing. We have introduced the *CH-α* methods whose special cases even today represent the most common structural integrators used in *FE* codes and testing. In detail, we have shown applications both in monolithic and innovative parallelized partitioned methods based on Lagrange multipliers able to guarantee fast pseudo-dynamic tests of complex substructures. We have also introduced few variants of the Rosenbrock methods conceived to be real-time compatible, linearly implicit and thus rather powerful for real-time *HiL* applications. These time integration methods have been used for the solution of coupled-field real-time problems, primarily for structure-structure interaction problems occurring in *HiL*. However, they can be easily extended to structure-control problems. Because *HiL* and *PsD* techniques require actuators and therefore controllers to control them, for this reason numerical analyses applied to integrate adaptive controllers and to linearize them in order to understand their properties have been illustrated.

7.2 Future perspectives

As fully illustrated in the previous sections, research on numerical integration is ongoing and continuing to develop and up to now it has studied in depth only a few aspects of real-time *HiL* and *PsD* with *DS*. Other aspects only mentioned in the chapter do deserve full scrutiny and in-depth research: we believe that this effort is worthy, being the possibilities offered by the simulation of heterogeneous systems with *DS* both unique and attractive. In fact, other *HiL* techniques such as shaking tables or *PsD* testing with *DS* are evolving towards it. Future developments of the *LSRT* integrators applied to the structure-structure interaction shall include the proper simulation of damping, especially for emulated structures with non-proportional damping. Moreover, the application of the Rosenbrock-based *LSRT* algorithms to structure-control interaction, including the exploitation of the velocity through a derivative-feedback control to assure a better control of the acceleration of the transfer system, deserves further studies. Basing on the current literature, it is highly probable that each future successful generalized control strategy is going to contain adaptive elements. Thus, the analysis described in this chapter that was performed on a first-order system controlled by a discrete *MCS* controller, could be further developed for second-order systems that model structural systems. Finally, since the effect of the delay reduces the stability limit and the effectiveness of the numerical damping generated by the controller or other numerical techniques, this important characteristic appears to be worth while in-depth study and research in the simulation of heterogeneous systems with *DS* even in view of the presence of a delay either multiple of the time step or variable.

Bibliography

M. Arnold and O. Brüls. Convergence of the generalized-α scheme for constrained mechanical systems. *Multibody Syst Dyn*, 18:185–202, 2007.

M. Arnold, B. Burgermeister, and A. Eichberger. Linearly implicit time integration methods in real-time applications: DAEs and stiff ODEs. *Multibody System Dynamics*, 17:99–117, 2007.

G. Baldo, A. Bonelli, O. Bursi, and S. Erlicher. The accuracy of the generalized method in the time integration of non linear single and two DoF forced systems. *Computational Mechanics*, 38:15–31, 2006.

V. Bayer, U. E. Dorka, U. Füllekrug, and J. Gschwilm. On real-time pseudo-dynamic sub-structure testing: algorithm, numerical and experimental results. *Aerospace Science and Technology*, 9:223–232, 2005.

B. Biondi and G. Muscolino. Component-mode synthesis method variants in the dynamics of coupled structures. *Meccanica*, 35:17–38, 2000.

A. Bonelli and O. S. Bursi. Generalized-α methods for seismic structural testing. *Earthquake Engineering and Structural Dynamics*, 33(10):1067–1102, 2004.

A. Bonelli, O. S. Bursi, and M. Mancuso. Explicit predictor-multicorrector time discontinuons Galerkin methods for linear dynamics. *Journal of Sound and Vibration*, 246(4):625–652, 2001.

A. Bonelli, O. S. Bursi, and M. Mancuso. Explicit predictor-multicorrector time discontinuons Galerkin methods for non-linear dynamics. *Journal of Sound and Vibration*, 256(4):659–724, 2002.

A. Bonelli, O. S. Bursi, L. He, P. Pegon, and G. Magonette. Convergence analysis of a parallel interfield method for heterogeneous simulations with dynamic substructuring. *International Journal for Numerical Methods in Engineering*, 2008. DOI: 10.1002/nme.2285.

P. A. Bonnet, C. N. Lim, M. S. Williams, A. Blakeborough, S. A. Neild, D. P. Stoten, and C. A. Taylor. Real-time hybrid experiments with Newmark integration, MCSmd outer-loop control and multi-tasking strategies. *Earthquake Engineering and Structural Dynamics*, 36(1):577–588, 2007.

P.A. Bonnet. *The development of multi-axis real-time substructure testing.* PhD thesis, University of Oxford, 2006.

O. Brüls and J.C. Golinval. On the numerical damping of time integrators for coupled mechatronic systems. *Computer Methods in Applied Mechanics and Engineering*, 197:577–588, 2008.

B. Burgermeister, M. Arnold, and B. Ester. DAE time integration for real-time applications in multi-body dynamics. *ZAMM Z. Angew. Math. Mech.*, 86(10):759–771, 2006.

O. S. Bursi and M. Mancuso. Analysis and performance of a predictor-multicorrector time discontinuous Galerkin method in non-linear elastodynamics. *Earthquake engineering and structural dynamics*, 31:1793–1814, 2002.

O. S. Bursi and P. B. Shing. Evaluation of some implicit time-stepping algorithms for pseudodynamic tests. *Earthquake engineering and structural dynamics*, 25:333–355, 1996.

O. S. Bursi, D. P. Stoten, and L. Vulcan. Convergence and frequency-domain analysis of a discrete first-order model reference adaptive controller. *Structural Control and Health Monitoring*, 14:777–807, 2007.

O. S. Bursi, B. A. Gonzalez, L. Vulcan, N. A. Neild, and D. J. Wagg. Novel coupling Rosenbrock-based algorithms for real-time dynamic substructure testing. *Earthquake Engineering and Structural Dynamics*, 37(3): 339–360, 2008.

J. C. Butcher. *Numerical Methods for Ordinary Differential Equations.* Wiley, 2003.

V. Cannillo and M. Mancuso. Spurious resonances in numerical time integration methods for linear dynamics. *Journal of Sound and Vibration*, 238(3):389–399, 2000.

S. Y. Chang. Error propagation in implicit pseudodynamic testing of nonlinear systems. *Journal of Engineering Mechanics*, 131(12):1257–1269, 2005.

S.-Y. Chang. An improved on-line dynamic testing method. *Engineering Structures*, 24:587–596, 2002.

S. Y. Chu, T. T. Soong, C. C. Lin, and Y. Z. Chen. Time-delay effect and compensation on direct output feedback controlled mass damper systems. *Earthquake Eng. and Struct. Dyn.*, 31(1):121–137, 2002.

S. Y. Chu, T. T. Soong, and A. M. Reinhorn. *Active, Hybrid, and Semi-active Structural Control: A Design and Implementation Handbook.* Wiley, Chichester, 2005.

J. Chung and G. M. Hulbert. A time integration algorithm for structural dynamics with improved numerical dissipation: the Generalized-alpha method. *Journal of Applied Mechanics*, 60:371–375, 1993.

A. Combescure and A. Gravouil. A numerical scheme to couple subdomains with different time-steps for predominantly linear transient analysis. *Computer Methods in Applied Mechanics and Engineering*, 191: 1129–1157, 2002.

D. Combescure and P. Pegon. α-Operator Splitting time integration technique for pseudodynamic testing error propagation analysis. *Soil Dynamics and Earthquake Engineering*, 16:427–443, 1997.

J. Cuadrado, J. Cardenal, and E. Bayo. Modeling and solution methods for efficient real-time simulation of multibody dynamics. *Multibody System Dynamics*, 1:1897–1913, 1997.

G. Dahlquist. A special stability problem for linear multistep methods. *BIT Numerical Mathematics*, 3(1):27–43, 1963.

W. J. T. Daniel. Explicit/implicit partitioning and a new explicit form of the generalized alpha method. *Communications in numerical methods in engineering*, 19:909–920, 2003.

A. P. Darby, M. S. Williams, and A. Blakcborough. Stability and delay compensation for real-time substructure testing. *Journal of Engineering Mechanics*, 128(12):1276–1284, 2002.

J. R. Dormand and P. J. Prince. A family of embedded Runge-Kutta formulae. *Journal of Computational and Applied Mathematics*, 6:19–26, 1980.

Real-time interface, Implementation guide, Release 3.2. dSPACE GmbH, Paderborn, Germany, 2001.

S. Erlicher, L. Bonaventura, and O. S. Bursi. The analysis of generalized-α methods for non-linear dynamic problems. *Computational Mechanics*, 28:83–104, 2002.

C. Farhat and F.-X. Roux. A method of finite element tearing and interconnecting and its parallel solution algorithm. *International Journal for Numerical Methods in Engineering*, 32:1205–1227, 1991.

C. Farhat, P.S. Chen, and J. Mandel. A scalable Lagrange multiplier based domain decomposition method for time dependent problems. *International Journal for Numerical Methods in Engineering*, 38:3831–3854, 1995.

C. Farhat, J. Cortial, C. Dastillung, and H. Bavestrello. Time-parallel implicit integrators for the near-real-time prediction of linear structural dynamic responses. *International Journal for Numerical Methods in Engineering*, 67:697–724, 2006.

C. A. Felippa and K. C. Park. Staggered transient analysis procedures for coupled dynamic systems: formulation. *Comp. Meths. Appl. Mech. Engrg.*, 24:61–112, 1980.

C. A. Felippa, K.C. Park, and C. Farhat. Partitioned analysis of coupled mechanical systems. *Computer Methods in Applied Mechanics and Engineering*, 190:3247–3270, 2001.

G. F. Franklin and J. D. Powell. *Digital Control of Dynamic Systems.* Addison-Wesley Publishing Company, New York, 1980.

A. Gravouil and A. Combescure. Multi-time-step explicit-implicit method for non-linear structural dynamics. *International Journal for Numerical Methods in Engineering*, 50:199–225, 2001.

E. Hairer and G. Wanner. *Solving Ordinary Differential Equations II. Stiff and Differential-Algebraic Problems.* Springer-Verlag, 1996.

E. Hairer, S. P. Norsett, and G. Wanner. *Solving Ordinary Differential Equations I. Nonstiff Problems.* Springer-Verlag, 1987.

L. He. *Dcvclopment of Partitioned Time Integration Schemes for Parallel Simulation of Heterogeneous Systems.* PhD thesis, University of Trento, 2008.

H. M. Hilber, T. J. R. Hughes, and R. L. Taylor. Improved numerical dissipation for time integration algorithms in structural dynamics. *Earthquake Engineering and Structural Dynamics*, 5:283–292, 1977.

T. Horiuchi, M. Inoue, T. Konno, and Namita Y. Real-time hybrid experimental system with actuator delay compensation and its application to a piping system with energy absorber. *Earthquake Eng. and Struct. Dyn.*, 28(10):1121–1141, 1999.

T. J. R. Hughes. *The Finite Element Method, Linear Static and Dynamic Finite Element Analysis.* Prentice-Hall, Englewood Cliffs, NJ, 1987.

T. J. R. Hughes and G. M. Hulbert. Space-time finite element methods for elastodynamics: formulations and error estimates. *Computer Methods in Applied Mechanics and Engineering*, 66:339–363, 1988.

T. J. R. Hughes, K. S. Pister, and R. L. Taylor. Implicit-explicit finite elements in non-linear transient analysis. *Computer Methods in Applied Mechanics and Engineering*, 17/18:159–182, 1979.

G. M. Hulbert. *Space-time finite element methods: 1970's and Beyond*, chapter An Overview of Variational Integrators, pages 116–123. CIMNE, Barcelona, Spain, 2004.

G. M. Hulbert and J. Chung. The unimportance of the spurious root of time integration algorithms for structural dynamics. *Communications in Numerical Methods in Engineering*, 10:591–597, 1994.

G. M. Hulbert and T. J. R. Hughes. An error analysis of truncated starting conditions in step-by-step time integration: Consequences for structural dynamics. *Earthquake Engineering and Structural Dynamics*, 15:901–910, 1987.

G.M. Hulbert and J. Chung. Explicit time integration algorithms for structural dynamics with optimal numerical dissipation. *Computer Methods in Applied Mechanics and Engineering*, 137:175–188, 1996.

R. Y. Jung, P. B. Shing, E. Stauffer, and B. Thoen. Performance of a real-time pseudodynamic test system considering nonlinear structural response. *Earthquake Engineering and Structural Dynamics*, 32:1785–1809, 2007.

R. Kübler and W. Schiehlen. Modular simulation in multibody system dynamics. *Multibody System Dynamics*, 4:107–127, 2004.

Y. N. Kyrychko, K. B. Blyuss, A. Gonzalez-Buelga, S. J. Hogan, and D. J. Wagg. Real-time dynamic substructuring in a coupled oscillator-pendulum system. *Proceedings of the Royal Society A*, 462:1271–1294, 2006.

L. M. Lacoma and I. Romero. Error estimation for hht method in non-linear solid dynamics. *Computers and structures*, 85:158–169, 2007.

J. D. Lambert. *Numerical Methods for Ordinary Differential Systems: The Initial Value Problem.* Wiley, 1991.

X. D. Li and N. E. Wiberg. Structural dynamic analysis by a time-discontinuous galerkin finite element method. *International Journal for Numerical Methods in Engineering*, 39:2131–2152, 1996.

C. N. Lim, S.A. Neild, D. P. Stoten, and D. Drury. Real-time dynamic substructure testing via an adaptive control strategy. *1st International Conference on Advances in Experimental Structural Engineering, Nagoya, Japan*, pages 393–400, 2005.

F. Lopez-Almansa, A. H. Barbat, and J. Rodellar. SSP algorithm for linear and non-linear dynamic response simulation. *International Journal for Numerical Methods in Engineering*, 26(12):2687 – 2706, 1998.

Simulink: simulation and model-based design. The MathWorks, Natick, MA, 2005.

W. E. Misselhorn, N. J. Theron, and P. S. Els. Investigation of hardware-in-the-loop for use in suspension development. *Vehicle System Dynamics*, 44:65–81, 2006.

C. Moler and C. Van Loan. Nineteen dubious ways to compute the exponential of a matrix. *SIAM Review*, 45:1–46, 1978.

G. Mosqueda, B. Stojadinovic, and S.A. Mahin. Real-time error monitoring for hybrid simulation. part I: Methodology and experimental verification. *Journal of Structural Engineering, ASCE*, 133(8):1100–1108, 2007a.

G. Mosqueda, B. Stojadinovic, and S.A. Mahin. Real-time error monitoring for hybrid simulation. part II: Structural response modification due to error. *Journal of Structural Engineering, ASCE*, 133(8):1109–1117, 2007b.

A. Mugan and G. M. Hulbert. Alternative characterization of time integration schemes. *International Journal for Numerical Methods in Engineering*, 51(3):351–376, 2001.

M. Nakashima, T. Kaminosomo, M. Ishida, and K. Ando. Integration techniques for substructuring pseudodynamic test. In *fourth U.S. National Conference on Earthquake Engineering*, volume 2, Palm Springs, California, 1990.

P. Nawrotzki and C. Eller. Numerical stability analysis in structural dynamics. *Comput. Methods Appl. Mech. Engrg.*, 189:915–929, 2000.

S. A. Neild, D. P. Stoten, D. Drury, and D. J. Wagg. Control issues relating to real-time substructuring experiments using a shaking table. *Earthquake Engineering and Structural Dynamics*, 34(9):1171 – 1192, 2005.

N. M. Newmark. A method of computation for structural dynamics. *Journal of the Engineering Mechanics Division ASCE*, 85:67–94, 1959.

K. C. Park and Carlos A. Felippa. A variational principle for the formulation of partitioned structural systems. *Int. J. Numer. Meth. Engng.*, 47:395–418, 2000.

P. Pegon. Alternative characterization of time integration schemes. *Computer Methods in Applied Mechanics and Engineering*, 190(20):2707–2727, 2001.

P. Pegon. Continuous psd testing with substructuring. In O. S. Bursi and D. J. Wagg, editors, *Modern Testing Techniques for Structural Systems, Dynamics and Control.* CISM, 2008.

P. Pegon and G. Magonette. Continuous PSD testing with non-linear substructuring: Presentation of a stable parallel inter-field procedure. Technical Report I.02.167, E.C., JRC, ELSA, Ispra, Italy, 2002.

P. Pegon and G. Magonette. Continuous PsD testing with non-linear substructuring: using the operator splitting technique to avoid iterative procedures. Technical Report SPI.05.30, E.C., JRC, ELSA, Ispra, Italy, 2005.

A. V. Pinto, P. Pegon, G. Magonette, and G. Tsionis. Pseudo-dynamic testing of bridges using non-linear substructuring. *Earthquake Engineering and Structural Dynamics*, 33:1125–1146, 2004.

V.M. Popov. *Hyperstability of Automatic control systems.* Springer-Verlag, New York, 1973.

A. Preumont. Frequency domain analysis of time integration operators. *Earthquake Engineering and Structural Dynamics*, 10(5):691–697, 1982.

AM. Reinhorn, MV. Sivaselvan, Z. Liang, and X. Shao. Real-time dynamic hybrid testing of structural systems. In *Proceeding of the 13th World Conference on Earthquake Engineering*, Vancouver, Canada, 2004.

H. H. Rosenbrock. Some general implicit processes for the numerical solution of differential equations. *Computer Journal*, 5:329–330, 1963.

J. C. Samin, O. Brüls, J. F. Collard, L. Sass, and P. Fisette. Multiphysics modeling and optimization of mechatronic multibody systems. *Multibody System Dynamics*, 18:345–373, 2007.

J. D. Sherrick. *Concepts In Systems and Signals.* Prentice Hall, 2004.

P. B. Shing. Real-time hybrid testing techniques. In O. S. Bursi and D. J. Wagg, editors, *In Modern Testing Techniques for Structural Systems, Dynamics and Control.* CISM, 2008.

P. B. Shing and M.T. Vannan. On the accuracy of an implicit time integration for pseudodynamic tests. *Earthquake Engineering and Structural Dynamics*, 19:631–651, 1990.

P. B. Shing, M.T. Vannan, and E. Cater. Implicit time integration for pseudodynamic tests. *Earthquake Engineering and Structural Dynamics*, 20:551–576, 1991.

P. B. Shing, O. S. Bursi, and M.T. Vannan. Pseudodynamic tests of a concentrically braced frame using substructuring techniques. *J. of Construct. Steel Research*, 29:121–148, 1994.

A. Soroushian, P. Wriggers, and J. Farjoodi. On practical integration of semi-discretized equations of motion part 1: reasons for probable instability and improper convergence. *Journal of Sound and Vibration*, 18: 705–731, 2005.

D. P. Stoten and H. Benchoubane. Empirical studies of an mrac algorithm with minimal controller synthesis. *International Journal of Control*, 51(4):823–849, 1990.

D. P. Stoten and S. A. Neild. The error-based minimal control synthesis algorithm with integral action. *Proc. of the Institetion of Mechanical Engineers, Part I: Journal of systems and Control Engineering*, 217: 187–201, 2003.

C. R. Thewalt and S. A. Mahin. Hybrid solution techniques for generalized pseudodynamic testing. Technical Report No. UCB/EERC-87/09, Earthquake Engineering Research Center, University of California Berkeley, CA, 1987.

R. J. Vaccaro. *Digital Control A State-Space Approach*. McGraw-Hill, New York, 1995.

V. R. Vasquez and W. B. Whiting. Accounting for both random errors and systematic errors in uncertainty propagation analysis of computer models involving experimental measurements with monte carlo methods. *Risk Analysis*, 25(6):1669–1681, 2006.

L. Vulcan. *Discrete-time analysis of integrator algorithms applied to S.I.S.O adaptive controllers with minimal control systhesis*. PhD thesis, University of Trento, Italy, 2006.

D. J. Wagg, S. A. Neild, and P. Gawthrop. Real-time testing with dynamic substructuring. In O. S. Bursi and D. J. Wagg, editors, *Modern Testing Techniques for Structural Systems, Dynamics and Control*. CISM, 2008.

M.I. Wallace, J. Sieber, S.A. Neild, D.J. Wagg, and B. Krauskopf. Stability analysis of real-time dynamic substructuring using delay differential equation models. *Earthquake Engineering and Structural Dynamics*, 34 (15):1817 – 1832, 2004.

M.I. Wallace, D.J. Wagg, and S.A. Neild. An adaptive polynomial based forward prediction algorithm for multi-actuator real-time dynamic substructuring. *Proceedings of the Royal Society A*, 2064(461):3807–3826, 2005.

N. E. Wiberg and X.D. Li. A post-processing technique and an a posteriori error estimate for the Newmark method in dynamic analysis. *Earthquake Engineering and Structural Dynamics*, 22(6):465–489, 1993.

M.S. Williams and A. Blakeborough. Laboratory testing of structures under dynamic loads: an introductory review. *Philosophical Transactions of the Royal Society of London, Series A*, 359(1786):1651–1670, 2001.

B. Wu, H. Bao, J. Ou, and S. Tian. Stability and accuracy analysis of the central difference method for real-time substructure testing. *Earthquake Engineering and Structural Dynamics*, 34(7):705–718, 2005.

B. Wu, G. Xu, Q. Wang, and M. S. Williams. Operator-splitting method for real-time substructure testing. *Earthquake Engineering and Structural Dynamics*, 35(3):293–314, 2006.

D. M. Young. *Iterative Solution of Large Linear System.* Academic Press, Orlando, 1971.

Q. Zhang and T. Hisada. Studies of the strong coupling and weak coupling methods in fsi analysis. *International Journal for Numerical Methods in Engineering*, 60:2013–2029, 2004.

Y. Zhang, R. Sause, J. M. Ricles, and C. J. Naito. Modified predictor-corrector numerical scheme for real-time pseudo dynamic tests using state-space formulations. *Earthquake Engineering and Structural Dynamics*, 34(3):271–288, 2005.

A State Space Approach to System Identification

Dionisio Bernal[†]

[†]Civil and Environmental Engineering, Center for Digital Signal Processing, Northeastern University, Boston MA 02115

Abstract System identification is the process of obtaining descriptions of dynamical systems based on the analysis of experimental observations. This chapter presents, in sufficient detail to allow a numerical implementation by the interested reader, an algorithm for identifying a state space description of a time invariant system that behaves linearly. The chapter begins by clarifying the connection between the discrete time models that are identified from sampled data and the underlying continuous time physical systems. The (formal) connection is shown to be strictly related to the assumed parameterization of the excitation in the inter-sample and it is noted that the commonly used zero order hold premise (constant load within the sampling time), ubiquitous in control applications, is usually suboptimal in off-line identification. Specifically, it is contended that accuracy in the transfer from discrete to continuous time can be promoted by operating on the premise that the input is a band limited function. The notions of controllability and observability, which relate the sensor deployment to the modes that can be identified from observations, are discussed following the discrete to continuous transfer section.

The central part of the chapter examines the extraction of pulse response functions for multiple input multiple output testing setups and the use of these functions in the formulation of a discrete time state space model by means of the Eigensystem Realization Algorithm. The chapter continues by illustrating the connection between the realization matrices and the modal properties of the system, i.e., modal frequencies, damping ratios and mode shapes. The need to separate system modes from modes that appear in the computations due to inevitable approximations and sensor noise is also discussed and some specific guidelines to discriminate between the two are given. The identification part of the chapter closes with an introduction to the identification of systems where the excitation signals are not deterministically known. An appendix presenting a

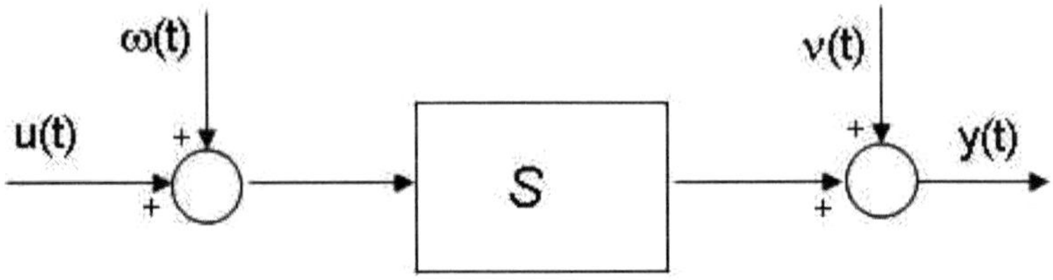

Figure 1. Schematic representation of a dynamical system.

technique to localize damage in structural or mechanical systems from changes in realization results concludes the chapter.

1 System Identification

System identification is the process of obtaining descriptions of dynamical systems based on the analysis of experimental observations. A diagrammatic representation of a dynamical system is illustrated in Figure 1. The system is driven by a set of known inputs $u(t)$ and (perhaps) a set of unmeasured disturbances $\omega(t)$ and there are some output measurements $y(t)$, typically equal to the true quantities plus some additive noise $\nu(t)$. The objective of identification is to arrive at some useful description of S by processing the measured inputs and outputs. The form chosen for the dynamical model is typically determined by the intended application. The model is said to be parametric when it belongs to a class whose members are differentiated by specific values of the parameter space. Non-parametric models are descriptions in terms of plots or functions, i.e., impulse response functions, frequency response functions, pulse response functions, spectral densities, etc.

An identification task involves: 1) collection of data, 2) selection of the type of model desired for S and 3) definition of a criterion to select the best model from the selected type. In the design of the experiment decisions are made regarding the number and location of inputs and output sensors, the type of input signals to be used (broad-band, narrow band, stationary, non-stationary) the sampling rate, etc., so the system characteristics are well reflected in the measurements. These decisions are made on the basis of knowledge about the system obtained either from physical inspection, experience, or from analysis. Models may be linear or non-linear, time invariant or allow for variation of the parameters in time, they can be deterministic or stochastic, or mixed, and may be formulated in frequency or in the time domain. This chapter does not provide a survey of identification techniques but presents, in sufficient detail to allow a numerical implementation by the interested reader, one particular algorithm for identification of a linear time invariant (LTI) state-space description of S. The reader can find comprehensive treatment of identification techniques in a number of references. On the state-space identification approach that is central to this chapter a basic reference is the book by Juang (2000). Subspace techniques are an alternative way to obtain state space descriptions that have received much attention in recent years. A first text on subspace techniques was written by Van Overschee and De Moor (1996) and a more recent excellent coverage appears in the book by Katayama (2005). Comprehensive discussion of classical identification techniques can be found in various sources, the books by Lung (1999), Ubenhauen and Rao (1987) and Soderstrom and Stoica (2001), are excellent references.

1.1 A Health Monitoring Application

System identification plays a central role in the design of algorithms used to identify and characterize undesirable changes (faults or damages) that may appear during the service life of a structure or mechanism. An approach that uses changes in an identified state-space description to locate the position of stiffness related damage is presented, for illustrative purposes, as an appendix to this chapter.

1.2 Dimensionality and State Space Description

Real systems are invariably infinitely dimensional. Models of real systems can be infinitely or finite dimensional. In finite dimensional models the state is a collection of variables that if known at a given time can be projected into the future given the model and the excitation. The mathematical structure that describes the dynamics in terms of the evolution of the state is known as a state-space representation. The state space representation provides a map from input to state and from state to output and, because of the flexibility afforded by the choice of state variables, infinite representations, each known as a realization, exist for a given input-output map. The extraction of a realization from measured input-output data is the central objective of the material in the subsequent sections. Once a realization is obtained the modal properties of the system, i.e., frequencies, damping ratios and eigenvectors (at the sensor locations) can be easily computed.

2 State Space Representations in Discrete and Continuous Time

In practice signals are digitized and processed as sequences and the resulting mathematical structures connecting them are said to be discrete time models. The terminology is not entirely accurate because discrete time models refer (strictly) to cases where the dependent quantities are discrete and this is not the case in structural identification. Notwithstanding that sampled-time is a more precise designation, we've opted here in favor of familiarity and will use the discrete time terminology throughout the chapter. When the objective of performing identification is to obtain an input-output map there is usually no need to convert the discrete time model to continuous time. If, however, the relation between identification results and the properties of the physical system are of interest the connection with the underlying continuous time model has to be established. This section examines this connection.

We begin by considering the state space description of a linear time invariant (LTI) finite dimensional system in continuous time and a measurement equation, namely

$$\dot{x} = A_c x + B_c u \tag{2.1}$$

$$y = C_c x + D_c u \tag{2.2}$$

where the output, y, input, u, and state, x, are vectors of dimensions m, r and n, and $\{A_c,\ B_c,\ C_c,\ D_c\}$ are matrices of appropriate dimensions. The solution of equation 2.1 is well known (Kailath 1990), namely

$$x_{(t)} = \mathrm{e}^{A_c t} x_0 + \int_0^t \mathrm{e}^{A_c(t-\tau)} B_c u(t) \mathrm{d}\tau \tag{2.3}$$

which, taking $t = k\Delta t$ can be expressed as

$$x_{k+1} = A_d x_k + A_d \int_0^{\Delta t} \mathrm{e}^{-A_c \tau} B_c u(t) \mathrm{d}\tau \tag{2.4}$$

where τ has the origin at the start of each time step and

$$A_d = \mathrm{e}^{A_c \Delta t} \tag{2.5}$$

To convert equation 2.4 into an algebraic form the integral on the right hand side has to be resolved. Consider the parameterization of the analog input given by

$$u(\tau) = \sum_{v} f_j(\tau) u_{k+j} \tag{2.6}$$

where $f_j(\tau)$ are basis functions defined over the time step and the summation is over a set of select integers j, that are collected in the vector v. The reason for choosing v instead of the more standard notation in the summation is because the integers that appear in this vector need not be sequential , for example, $v = \{0, 1, 4\}$ is a valid vector and it implies that j takes the values in the bracket. Substituting equation 2.6 into equation 2.4 one gets

$$x_{k+1} = A_d x_k + \sum_v B_j u_{k+j} \tag{2.7}$$

with

$$B_j = A_d \int_0^{\Delta t} \mathrm{e}^{-A_c \tau} B_c f_j(\tau) \mathrm{d}\tau \tag{2.8}$$

or

$$B_j = T_j B_c \tag{2.9}$$

where the definition of T_j is evident. The standard discrete time state space model that is fitted to observations is

$$z_{k+1} = A_d z_k + B_d u_k \tag{2.10}$$

$$y_k = C_d z_k + D_d u_k \tag{2.11}$$

Examination shows that equation 2.7 can be transformed to the form in equation 2.10 by taking

$$x_k = z_k + \sum_v \sum_{i=1}^{j} A_d^{j-i} B_j u_{k+i-1} \tag{2.12}$$

Indeed, substituting equation 2.12 into equation 2.7 gives

$$z_{k+1} = A_d z_k + \left(\sum_v A_d^j B_j \right) u_k \tag{2.13}$$

which has the desired form. The output equation obtained by substituting equation 2.12 into equation 2.2 is

$$y_k = C_d z_k + C_c \sum_v \sum_{i=1}^{j} A_d^{j-i} B_j u_{k+i-1} + D_d u_k \tag{2.14}$$

which, in contrast with the state recurrence, fits the standard form (equation 2.11) only when the integers j are in the set $[0, 1]$. One concludes, therefore, that the most general parameterization of the class in equation 2.6 that fits the standard state space model is

$$u(\tau) = f_0(\tau)u_k + f_1(\tau)u_{k+1} \tag{2.15}$$

The inputs that fit equation 2.15 are a subset of the finite dimensional set of equation 2.6 and we note that for these one has

$$B_d = B_0 + A_d B_1 \tag{2.16}$$

and

$$D_d = C_c B_1 + D_c \tag{2.17}$$

We close by noting that the key step of the previous analysis, i.e., the transformation in equation 2.12, is a generalization of a form introduced by Hanselmann (1987).

Summary

The expressions mapping the continuous and discrete time state space representations for any member of the parameterization in equation 2.15 are:

$$A_c = \frac{\log(A_d)}{\Delta t} \tag{2.18}$$

$$B_c = (T_0 + AT_1)^{-1} B_d \tag{2.19}$$

$$C_c = C_d \tag{2.20}$$

$$D_c = D_d - C_c T_1 B_c \tag{2.21}$$

where

$$T_j = A_d \int_0^{\Delta t} \mathrm{e}^{-A_c \tau} f_j(\tau) \mathrm{d}\tau \quad (j = 0 \text{ and } 1) \tag{2.22}$$

Not previously noted is the fact that the initial conditions in the physical system and those that hold in the discrete time model are not necessarily the same. Indeed, particularizing equation 2.12 to the parameterization of equation 2.15 one concludes that the initial conditions in continuous and discrete time are related by

$$(x_0)_{\text{disctime}} = (x_0)_{\text{conttime}} - B_1 u_0 \tag{2.23}$$

The result in equation 2.23 is relevant in simulation studies where one often starts a system from rest yet, if both B_1 and u_0 are non-zero, the identification algorithm should be instructed that initial conditions are not zero.

The Band Limited Approximation

A recent study on the discrete to continuous (D2C) transfer operation (Bernal 2007a) contends that when the input is not finite dimensional, accuracy in the continuous time model can be promoted by operating on the premise that the input is a band limited function. In this case (which is the common one in identification) the specific D2C formulas are: equation 2.18 and equation 2.20 for the system and the state to output matrices, and, for the input to state matrix

$$B_c = \frac{A_d^{-1}}{\Delta t} B_d \tag{2.24}$$

Furthermore, it is shown in the same reference that the direct transmission is most accurate if computed from constraints that relate it to the triplet $\{A_c, B_c, C_c\}$, specifically

$$D_c = C_c A_c^{1-b} B_c \tag{2.25}$$

where $b = \{0, 1, 2\}$ for displacement velocity or acceleration measurements respectively. Before moving to the identification problem it is necessary to examine how the location and the number of input and output sensors affects what part of the system "can be seen" when one looks at the data. To do this we examine briefly the concepts of controllability and observability and in doing so we choose the discrete time model for mathematical expediency.

2.1 Controllability

A state x_p is said to be controllable if there is an input that can transfer the system from this state to the origin in finite time; when the process is from the origin to x_p the state is said to be reachable. In practice controllability is commonly used to imply reachability because the concepts are equivalent when the system matrix A_d is not singular (Antsaklis and Michel 2006). In structural systems A_d is not singular so the distinction between controllability and reachability is unnecessary here. If all possible states are controllable the system is said to be *completely controllable* or simply, *controllable*. From the discrete time state space model one can see (using x as the state vector designation) that;

$$\begin{aligned} x_1 &= A_d x_0 + B_d u_0 \\ x_2 &= A_d^2 x_0 + A_d B_d u_0 + B_d u_1 \\ x_3 &= A_d^3 x_0 + A_d^4 B_d u_0 + A_d B_d u_1 + B_d u_2 \end{aligned} \tag{2.26}$$

or, in general, at time step p,

$$x_p = A_d^p x_0 + \sum_{i=0}^{p-1} A_d^i B_d u_{p-1-i} \tag{2.27}$$

or

$$x_p = A_d^p x_0 + [B_d \quad A_d B_d \quad A_d^2 B_d \quad \ldots \quad A_d^{p-1} B_d] \begin{bmatrix} u_{p-1} \\ u_{p-2} \\ \vdots \\ u_0 \end{bmatrix} \tag{2.28}$$

From equation 2.28, it follows that an arbitrary x_p can be reached only if $(x_p - A_d^p x_0)$ lies in the column space of $[B_d \quad A_d B_d \quad A_d^2 B_d \quad \ldots \quad A_d^{p-1} B_d]$, and, since $(x_p - A^p x_0)$ is arbitrary one concludes that controllability at step p is equivalent to requiring that this matrix be full column rank. In summary, a linear finite dimensional discrete-time system of order n is controllable at step p if the $[n \times pr]$ block matrix Q_p

$$Q_p = [B_d \quad A_d B_d \quad A_d^2 B_d \quad \ldots \quad A_d^{p-1} B_d] \tag{2.29}$$

known as the controllability matrix at step p, is rank n. Since the number of columns in the controllability matrix is $p \times r$ it is evident that controllability requires that $p \geq n/r$. A question that comes to mind is whether or not there is a limit for p after which there is no point checking if the rank is going to increase further. The answer is positive and is a consequence of the Cayley-Hamilton theorem (Golub and Van Loan 1989). In particular, examination shows that

$$\text{rank}[B_d \quad A_d B_d \quad \ldots \quad A_d^{n-1} B_d] = n \tag{2.30}$$

is a necessary and sufficient condition for controllability. A simply supported undamped beam where the dynamic excitation is a concentrated load in the center provides an example of a system with uncontrollable modes. Specifically, in this case the even number vibration modes, which have a nodal point at the centerline are not excited and cannot, therefore, be identified.

2.2 Observability

While controllability is a reflection of the ability of the inputs to excite all modes observability is related to whether one can see the contribution of all modes in the output. Formally, a state x_p is observable at time step p if its knowledge, together with the past of the input and the output

$(0 < k \leq p)$, is sufficient to uniquely determine the initial condition[1]. The unobservable modes are the components (in the basis of the modes) of the initial conditions that cannot be resolved from analysis of the model and the data. A mathematical requirement for observability can be derived by substituting equation 2.28 into the output equation, namely

$$y_p = C_d A_d^p x_0 + C_d [B_d \quad A_d B_d \quad A_d^2 B_d \quad \dots \quad A_d^{p-1} B_d] \begin{bmatrix} u_{p-1} \\ u_{p-2} \\ \vdots \\ u_0 \end{bmatrix} + D_d u_p \tag{2.31}$$

from where

$$y_p - [D_d \quad C_d B_d \quad C_d A_d B_d \quad C_d A_d^2 B_d \quad \dots \quad C_d A_d^{p-1} B_d] \begin{bmatrix} u_{p-1} \\ u_{p-2} \\ \vdots \\ u_0 \end{bmatrix} = C_d A_d^p x_0 \tag{2.32}$$

or

$$z_p = C_d A_d^p x_0 \tag{2.33}$$

where the definition of z_p is evident. Writing equation 2.33 at all time stations from 0 to $p-1$ one has

$$Z_p = \begin{bmatrix} z_0 \\ z_0 \\ \vdots \\ z_{p-1} \end{bmatrix} = \begin{bmatrix} C_d \\ C_d A_d \\ \vdots \\ C_d A_d^{p-1} \end{bmatrix} x_0 \tag{2.34}$$

which allows one to conclude that x_0 is unique iff the observability block P_p $(pm \times n)$

$$P_p = \begin{bmatrix} C_d \\ C_d A_d \\ \vdots \\ C_d A_d^{p-1} \end{bmatrix} \tag{2.35}$$

[1] In a strict sense observability denotes the ability to determine a state from current and future values of outputs and inputs, the proper term for the ability to determine a state from current and past values is constructibility. In identification, however, it is common to use the term observability to refer to constructability and we've chosen to follow common usage.

is of rank n. Again, from the Cayley-Hamilton theorem one concludes that the rank of P_p does not increase beyond the value attained when $p = n$ so

$$\text{rank}\begin{bmatrix} C_d \\ C_d A_d \\ \vdots \\ C_d A_d^{p-1} \end{bmatrix} = n \tag{2.36}$$

is a necessary and sufficient condition for observability. The previous simple beam example can also be used to illustrate a non-observable situation. In particular, let there be an arbitrary number of shakers but only a single output sensor at the centerline. In this case, independently of what modes are excited the even number ones cannot be identified because they have no contribution in the response at the center of the beam.

2.3 Non-uniqueness of Realizations

The matrices of a state space realization are not unique. This well known result can be readily confirmed by noting that if one introduces a change in coordinates from the state vector x to a new state vector, z, namely

$$x = Tz \tag{2.37}$$

where T is an arbitrary non-singular matrix, the input to state and state to output map becomes

$$z_{k+1} = T^{-1}A_d T z_k + T^{-1}B_d u_k \tag{2.38}$$

$$y_k = C_d T z_k + D_d u_k \tag{2.39}$$

so the quadruple $\{T^{-1}A_d T\ ,T^{-1}B_d\ , C_d T$ and $D_d\}$ describes the same input-output relationship as $\{A_d,\ B_d,\ C_d,$ and $D_d\}$. An important point to note is that all the possible matrices A_d are related by a similarity transformation and thus have the same eigenvalues.

Oversized State-Vectors

Another reason why realizations are not unique is because state vectors with dimensions larger than the true order are feasible without changing the input-output map. To illustrate let a given system of order n be described by the quadruple $\{A_d\ B_d\ C_d\ D_d\}$ associated with the state vector x. Consider the discrete time equations;

$$w_{k+1} = Rw_k \tag{2.40}$$

where no attempt is made to assign any physical significance to R and w. Defining the expanded state vector $\eta = \{x\ w\}^T$ one can write

$$\eta_{k+1} = \hat{A}\eta_k + \hat{B}u_k \tag{2.41}$$

$$y_k = \hat{C}\eta_k + Du_k \tag{2.42}$$

where

$$\hat{A} = \begin{bmatrix} A_d & 0 \\ 0 & R \end{bmatrix} \tag{2.43}$$

$$\hat{B} = \begin{bmatrix} B_d \\ 0 \end{bmatrix} \tag{2.44}$$

and

$$\hat{C} = [C_d \quad 0] \tag{2.45}$$

As can be seen, the matrices in equations 2.43–2.45 have the same input-output map of the original quadruple and are thus a valid realization. Of all possible realizations with identical input-output maps the ones with the smallest state space dimension are said to be minimal. An important point to note is that only minimal realizations are observable and controllable.

2.4 Markov Parameters

Inspection of equation 2.31 shows that the output at time step k can be expressed as

$$y_k = C_d A_d^k x_0 + [D_d \quad C_d B_d \quad C_d A_d B_d \quad C_d A_d^2 B_d \quad \dots \quad C_d A_d^{k-1} B_d] \begin{bmatrix} u_k \\ u_{k-1} \\ \vdots \\ u_0 \end{bmatrix} \tag{2.46}$$

or

$$y_k = C_d A_d^k x_0 + [Y_0 \quad Y_1 \quad Y_2 \quad \dots \quad Y_k] \begin{bmatrix} u_k \\ u_{k-1} \\ \vdots \\ u_0 \end{bmatrix} \tag{2.47}$$

where the matrices

$$\begin{aligned} Y_0 &= D_d \\ Y_k &= C_d A_d^{k-1} B_d \end{aligned} \tag{2.48}$$

are known as the Markov Parameters (MPs). A noteworthy point is the fact that the MPs are independent of the transformation matrix T, a fact that is easily confirmed by substituting the general expression for the realization matrices (which appear in the paragraph right after equation 2.39) in equation 2.48.

The MPs of the continuous time system have the same form given by equation 2.48, with the discrete time matrices replaced by the continuous time ones. In the literature it is common to refer to the discrete time MPs as pulse response functions, a terminology that is literally correct only if the input is delivered using a zero order hold, i.e., if in equation 2.15 $f_0 = 1$ and $f_1 = 0$.

3 Extracting Markov Parameters from Measurements

Identification of a state space model from input-output observations (as presented here) has two main computational blocks: 1) extraction of MPs from the data and 2) derivation of a quadruple $\{A_d\ B_d\ C_d\ D_d\}$ whose MP sequence coincides (in the ideal case) with the identified one. In this section we examine the first block, namely, the extraction of the MP sequence from the data.

3.1 Time Domain Deconvolution

We begin by writing equation 2.47 as

$$y_k = C_d A_d^k x_0 + \Upsilon_k \begin{bmatrix} u_k \\ u_{k-1} \\ \vdots \\ u_0 \end{bmatrix} \tag{3.1}$$

where Υ_k is the collection of MPs (in a block row format) from zero to k. Writing equation 3.1 for $k = 0, 1, 2, 3..\ell.$, one gets

$$L_\ell = P_{\ell+1}^T X_\ell + \Upsilon_\ell U_\ell \tag{3.2}$$

where

$$L_\ell = [y_0 \quad y_1 \quad \cdot \quad y_\ell] \tag{3.3}$$

$$X_\ell = \begin{bmatrix} x_0 & & \\ & \cdot & \\ & & x_0 \end{bmatrix} \tag{3.4}$$

$$X_\ell = \begin{bmatrix} u_0 & u_1 & \dots & u_\ell \\ 0 & u_0 & \dots & u_{\ell-1} \\ & & \ddots & \vdots \\ & & & u_0 \end{bmatrix} \tag{3.5}$$

and P_ℓ is the controllability matrix defined by equation 2.35.

Solution for Asymptotically Stable Systems

A count of unknown parameters and available equations shows that equation 3.2 gives a unique solution for the MP sequence only when the initial condition is zero and there is a single input. One can take advantage, however, of the fact that for stable systems the MPs decay (because the eigenvalues of A_d are all insides the unit circle) and thus, with negligible error, can be taken as identically zero beyond a certain index. Equation 3.6, shown next, is the same as equation 3.2 written out with the subscript p introduced at some point in the MP sequence for notational convenience

$$[y_0 \;\cdot\; y_p \; y_{p+1} \; y_\ell] = [C_d \;\cdot\; C_d A_d^{p-1} \; C_d A_d^p \; C_d A_d^{\ell-1}] \begin{bmatrix} x_0 & & \\ & \cdot & \\ & & x_0 \end{bmatrix} + \\ [Y_0 \;\cdot\; Y_p \; Y_{p+1} \; Y_\ell] \begin{bmatrix} u_0 & u_1 & \dots & u_\ell \\ 0 & u_0 & \dots & u_{\ell-1} \\ & & \ddots & \vdots \\ & & & u_0 \end{bmatrix} \tag{3.6}$$

Assuming that Y_p is the last MP to be taken as non-zero and noting that this also implies that $C_d A_d^k = 0$ for $k \geq p$ one can write

$$[y_0 \;\cdot\; y_p \; y_{p+1} \; y_\ell] = [C_d \;\cdot\; C_d A_d^{p-1}] \begin{bmatrix} x_0 & 0 & \cdot & 0 \\ 0 & x_0 & \cdot & 0 \\ 0 & 0 & x_0 & 0 \end{bmatrix} + \\ [Y_0 \;\cdot\; Y_p] \begin{bmatrix} u_0 & u_1 & \dots & u_\ell \\ 0 & u_0 & \dots & u_{\ell-1} \\ 0 & 0 & \dots & u_{\ell-p} \end{bmatrix} \tag{3.7}$$

where only the first p columns of the initial condition matrix are non-zero. Considering now only time stations having time indices h or larger, where $h \geq p+1$, the initial condition effect disappears and one has

$$[y_h \;\cdot\; y_\ell] = [Y_0 \;\cdot\; Y_p] \begin{bmatrix} u_h & u_{h+1} & \dots & u_\ell \\ \cdot & \cdot & \dots & u_{\ell-1} \\ u_{h-p-1} & u_{h-p} & \dots & u_{\ell-p-1} \end{bmatrix} \tag{3.8}$$

which, with obvious definitions can be written as

$$L_{h\ell} = \Upsilon_p U_{h\ell p} \tag{3.9}$$

In equation 3.9 $U_{h\ell p} \in \mathbb{R}^{r\times(p+1)\times(\ell-h)}$ so, by taking $(\ell - h) > r(p+1)$ one gets over-determination and (assuming that the excitation is such that $U_{h\ell p}$ is full row rank) one has

$$\Upsilon_p = L_{h\ell} U^{\dagger}_{h\ell p} \tag{3.10}$$

where † stands for pseudo-inversion. Equation 3.10 gives the desired sequence of non-zero MPs but often proves impractical because the size of the matrix $U_{h\ell p}$ needed to allow for sufficient decay when damping is low can be large. Consider, for example, a system where the first vibration mode has 2% of critical damping and a natural period of two seconds. For this level of dissipation one needs around 36 cycles for the response to decay to about 1% of the original, or a total of 72 seconds. If the time step is 0.01 sec (which is reasonable if one is interested in a bandwidth of up to 30 or 40Hz) then the smallest dimension of the matrix $U_{h\ell p}$ is around 7000. A more computationally efficient alternative to compute the MP sequence is described next.

3.2 Observer Markov Parameters

Although the MP sequence is infinite, it is evident — since these parameters are related to the finite dimensional triple $\{A_d\ B_d\ C_d\}$ by equation 2.48 — that there is only a finite number of independent terms. The key to deriving compact representations of the input-output relationship is recursion, i.e., the use of previous model predictions in the prediction of the current output. There are many equivalent recursive descriptions of the input-output map and the one that has the smallest number of terms is known in identification as the one-step-ahead predictor. The one-step-ahead predictor has the form

$$y_k = \sum_{i=1}^{n_\alpha} \alpha_i y_{k-i} + \sum_{j=0}^{n_\beta} \beta_j u_{k-j} \tag{3.11}$$

where the α's and β's are matrices of appropriate dimensions. Equation 3.11 is known as the ARX model in many fields and, in digital signal processing (in its scalar form) as the Infinite Impulse Response (IIR) filter. The α's and β's can be computed given a set of measurements and it is not difficult to show that the MPs of the underlying system can be easily obtained once these matrices are known. In the state-space framework the matrices are known as the Observer Markov Parameters (OMP).

To clarify the connection between the OMPs and the state space form we add and subtract the term Gy_k to the right-hand side of the state space recurrence in discrete time and get

$$x_{k+1} = Ax_k + Bu_k + Gy_k - Gy_k \tag{3.12}$$

$$y_k = Cx_k + Du_k \tag{3.13}$$

where G is at this point arbitrary. Combining equations 3.12 and 3.13 gives

$$x_{k+1} = (A + GC)x_k + (B + GD)u_k - Gy_k \tag{3.14}$$

or

$$x_{k+1} = \bar{A}x_k + \bar{B}\nu_k \tag{3.15}$$

where

$$\bar{A} = A + GC \tag{3.16}$$

$$\bar{B} = [(B + GD), -G] \tag{3.17}$$

and

$$\nu_k = \left\{ \begin{array}{c} u_k \\ y_k \end{array} \right\} \tag{3.18}$$

Equations 3.15 and 3.13 provide an input to state and state to output map that has the same form as in the standard case, except for the fact that the input vector is enlarged in the state recurrence. Following exactly the same approach of the previous section one finds that the MPs of this system (known as an observer) are given by

$$\bar{Y}_0 = D \tag{3.19}$$

$$\bar{Y}_k = C\bar{A}^{k-1}\bar{B} \tag{3.20}$$

and that they can be obtained from the data as

$$\bar{\Upsilon}_p = L_{h\ell}V^{\dagger}_{h\ell p} \tag{3.21}$$

where $V_{h\ell p}$ is given by

$$V_{h\ell p} = \begin{bmatrix} u_h & u_{h+1} & \cdots & u_\ell \\ \nu_{h-1} & \nu_h & \cdots & \nu_{\ell-1} \\ & & \ddots & \cdots \\ \nu_{h-p-1} & & & \nu_{\ell-p-1} \end{bmatrix} \tag{3.22}$$

The supra bar in the previous expressions is used to indicate reference to the observer system. Although $V_{h\ell p}$ has more rows than $U_{h\ell p}$ the advantage of the observer structure resides in the fact that G is free to be selected so the eigenvalues of $\bar{A}$ can be moved towards the origin and, as a consequence, the index p beyond which the OMPs can be discarded (i.e, the value of k in equation 3.20 above which one can take $\bar{Y}_k = 0$) can be much smaller than the value that applies when computing the MPs. Once the OMPs are obtained the MPs havc to be extracted from the results but this operation, as shall be shown, is computationally inexpensive.

Dead Beat Observer

For ideal conditions (noiseless data, finite dimensional system and finite dimensional inputs) the matrix G can be selected so that all the eigenvalues of $\bar{A}$ are placed at the origin. A matrix where all the eigenvalues are zero is nilpotent, which means that $\bar{A}^k = 0$ for some $k \geq p$. When G is selected to attain a nilpotent $\bar{A}$ the resulting system is known as a dead beat observer. As shall be shown G does not need not be computed explicitly but is indirectly specified when one selects the number of non-zero entries in the OMP sequence.

On the Number of Non-Zero OMPs

The number of non-zero OMPs is a free-parameter in the identification process. The minimum value depends on the true order and the number of output sensors and is typically given by the ratio between the two, namely, n/m. Although it may appear that there is a need to know the true order to specify the number of OMPs this is not the case. Specifically, in the ideal case it suffices to specify a number of OMPs that is larger than the minimum for which a nill potent condition can be realized.In practice the number of non-zero OMPs is selected so the value does not impose a limitation on the effective model order one hopes to identify (since the real order is likely infinite).

To expand on the specifics assume the excitation that can be delivered has little energy outside the bandwidth $\{f_{min} f_{max}\}$ (in Hz) and that one estimates that the number of (complex) modes within this bandwidth is q then, to ensure that the non-zero OMP specification does not restrict the order one takes $p > q/m$ where m is the number of outputs. To accommodate the noise without distorting the system modes it is common to take p significantly larger than the lower limit noted. A guidance that the writer has found useful is

$$p = \frac{10 \cdot q}{m} \geq \frac{\gamma}{\Delta t \cdot f_{min}} \tag{3.23}$$

where γ depends on the noise level. Recommendations for γ as large as one have been put forth (Schoulkens *et al.* 1994) but $\gamma = 0.25$ appears to be generally sufficient. Examination of how the identification results vary with the selection of p is advisable.

3.3 Extracting System Markov Parameters from Observer Markov Parameters

The OMPs have two parts, namely;

$$\bar{Y}_k = C\bar{A}^{k-1}\bar{B} = C\bar{A}^{k-1}[(B + GD), -G] = [\bar{Y}_k^{(1)} \quad \bar{Y}_k^{(2)}] \tag{3.24}$$

therefore

$$\bar{Y}_k^{(1)} = C\bar{A}^{k-1}(B + GD) \tag{3.25}$$

and

$$\bar{Y}_k^{(2)} = -C\bar{A}^{k-1}G \tag{3.26}$$

Recalling that the MPs are given by

$$Y_k = CA^{k-1}B \tag{3.27}$$

and that

$$Y_0 = \bar{Y}_0 = D \tag{3.28}$$

one has, taking $k = 1$

$$\bar{Y}_1^{(1)} = C[B + GD] \tag{3.29}$$

and

$$\bar{Y}_1^{(2)} = -CG \tag{3.30}$$

Post multiplying equation 3.30 by Y_0 and adding the result to equation 3.29 gives

$$\bar{Y}_1^{(1)} + \bar{Y}_1^{(2)}Y_0 = CG \tag{3.31}$$

and since

$$Y_1 = CB \tag{3.32}$$

It follows that

$$Y_1 = \bar{Y}_1^{(1)} + \bar{Y}_1^{(2)}Y_0 \tag{3.33}$$

Continuing in analogous fashion for $k = 2$, $k = 3$ etc, one concludes (by induction) that the MPs can be extracted from the OMPs using the recurrence

$$Y_k = \bar{Y}_k^{(1)} + \bar{Y}_k^{(2)}Y_0 + \sum_{j=1}^{k-1} Y_j\bar{Y}_{k-j}^{(2)} \tag{3.34}$$

4 From Markov Parameters to a System Realization

4.1 The Eigensystem Realization Algorithm

The Eigensystem Realization Algorithm (ERA) is a technique for extracting a realization (Ho and Kalman 1965) from a MP sequence (Juang and Pappa 1985). The first step in the (standard) algorithm is to form a $\alpha m \times \beta r$ block Hankel matrix as shown in equation 4.1. The block Hankel matrix H_k is taken as the collection of MPs starting at time step $k+1$ and having α blocks in the row direction and β blocks in the column direction (where α and β are arbitrary, except for lower bounds are needed to ensure that the rank of the Hankel Matrix stabilizes). One has, therefore

$$H_k = \begin{bmatrix} Y_{k+1} & Y_{k+2} & \cdots & Y_{k+\beta} \\ Y_{k+2} & Y_{k+3} & \cdots & Y_{k+\beta} \\ \cdots & \cdots & \cdots & \cdots \\ Y_{k+\alpha} & & & Y_{k+\beta+\alpha-1} \end{bmatrix} \tag{4.1}$$

The reason for adding the word standard in the parenthesis of the previous paragraph is because in the most general form of the ERA algorithm the MPs need not be ordered sequentially as in equation 4.1. A discussion on the possible advantages of a non-sequential organization is not, however, essential for the objectives in this chapter so we restrict consideration to the form in equation 4.1. The key point in this section is the fact that the block Hankel matrix for the k index can be expressed in terms of the system matrix A_d raised to the power k and the observability and controllability blocks computed for α and β steps, namely recalling that $P_\alpha = [C_d \quad C_d A_d \quad \ldots \quad C_d A_d^{\alpha-1}]^T$ and $Q_\beta = [B_d \quad A_d B_d \quad A_d^2 B_d \quad \ldots \quad A_d^{\beta-1} B_d]$, one can see that

$$P_\alpha A_d^k Q_\beta = \begin{bmatrix} C_d A_d^k B_d & C_d A_d^{k+1} B_d & \cdots \\ C_d A_d^{k+1} B_d & C_d A_d^{k+1} B_d & \cdots \\ \vdots & \vdots & \vdots \end{bmatrix} \tag{4.2}$$

or

$$H_k = P_\alpha A_d^k Q_\beta = \begin{bmatrix} Y_{k+1} & Y_{k+2} & \cdot & Y_{k+\beta} \\ Y_{k+2} & Y_{k+3} & \cdot & Y_{k+\beta} \\ \cdot & \cdot & \cdot & \cdot \\ Y_{k+\alpha} & \cdot & \cdot & Y_{k+\beta+\alpha-1} \end{bmatrix} \tag{4.3}$$

therefore

$$H_0 = P_\alpha Q_\beta \tag{4.4}$$

and

$$H_k = P_\alpha A_d^k Q_\beta \tag{4.5}$$

Since the rank of a matrix product is not larger than the smallest rank of the intervening matrices one concludes that the rank of H_k is no more than n. To ensure that this rank is not limited by the matrix dimensions the constants α and β must, therefore, satisfy

$$\alpha \geq \frac{n}{m} \tag{4.6}$$

and

$$\beta \geq \frac{n}{r} \tag{4.7}$$

The realization is extracted as follows: performing a singular value factorization of equation 4.4 gives

$$H_0 = R\Sigma S^T \tag{4.8}$$

In the ideal case the order of the system is the rank of H_0 which is simply the number of non-zero singular values in Σ. In practice the singular values in Σ are never exactly zero and an effective order is selected by taking a given set of singular values as non-zero and rounding the others down to zero. With n as the effective order of the observable and controllable part of the system one has

$$\Sigma = \begin{bmatrix} \Sigma_n & 0 \\ 0 & 0 \end{bmatrix} \tag{4.9}$$

from where one gathers that equation 4.8 can be written as

$$H_0 = R_n \Sigma_n S_n^T \tag{4.10}$$

where R_n contains the first n columns of R and S_n the first n columns of S. Splitting the diagonal singular value matrix into the product of two matrices (E_1 and E_2)

$$E_1 E_2 = \Sigma_n \tag{4.11}$$

one has

$$H_0 = (R_n E_1)(E_2 S_n^T) \tag{4.12}$$

Finally, taking

$$P_\alpha = R_n E_1 \tag{4.13}$$

and

$$Q_\beta = E_2 S_n^T \tag{4.14}$$

and recalling the definition of the observability and controllability matrices one concludes that

- The first m rows of P_α provide a realization for C_d.

- The first r columns of Q_β provide a realization for B_d.

The system matrix A_d can be obtained in a number of different ways, one is from the block Hankel matrix for $k = 1$, namely, given that

$$H_1 = P_\alpha A_d Q_\beta = R_n E_1 A_d E_2 S_n^T \tag{4.15}$$

and the fact that R_n and S_n are orthonormal one gets

$$A_d = E_1^{-1} R_n^T H_1 S_n E_2^{-1} \tag{4.16}$$

Balanced Realization

A widely utilized selection for the E_1 and E_2 matrices is

$$E_1 = E_2 = \Sigma^{1/2} \tag{4.17}$$

the resulting realization being referred to as balanced because it is as observable as it is controllable.

5 Summary of Deterministic State- Space Identification Using ERA

- Collect input output data
- Select an initial time station h, a final time station ℓ, and a number of non-zero Observer Markov Parameters p; taking $h \geq p$ assures that initial conditions are not relevant in the identification. A recommendation on the number of non-zero OMPs is given in equation 3.23. The number of time steps ℓ should be such that the time period$(\ell - h)\Delta t$ is no less than a number of times (10 or more if possible) the longest period to be identified.
- Form the matrices $L_{h\ell}$ and $V_{h\ell p}$ (see equations 3.8, 3.9 and 3.22)
- Solve for the OMPs using equation 3.21
- Separate the OMPs into their two components (the first r columns and the remainder) using 3.25 and 3.26.
- Solve for the MPs using equation 3.34.
- Select α and β and form the Hankel matrices H_0 and H_1 using equation 4.1. Equations 4.6 and 4.7 provide guidance.
- Perform the factorization of equation 4.8 and decide on the effective order by looking at the singular values of Σ.
- Evaluate E_1 and E_2 using equation 4.17 (for a balanced realization)
- Take C_d as the first m rows or equation 4.13
- Take B_d as the first r columns of equation 4.14
- Compute A_d from equation 4.16

- D_d is the first OMP.
- Compute the continuous time state-space matrices using equations 2.18, 2.20, 2.24 and 2.25.

6 Mode Shapes, Frequencies and Damping Ratios

The eigenproperties in the identified bandwidth can be easily extracted from the continuous time realization results. Specifically, a) the eigenvalues of the system matrix contain the information on the natural frequencies and the damping and b) the (complex) mode shapes in the coordinates of the sensors are given by the product of the C_c matrix time the eigenvectors of A_c. To illustrate how these results come about consider the spectral decomposition of A_c, namely

$$A_c = \Phi \Lambda \Phi^{-1} \tag{6.1}$$

Introducing a change of basis from the state vector x to the amplitudes Y given by

$$x = \Phi Y \tag{6.2}$$

one has

$$\dot{Y} = \Lambda Y + \Phi^{-1} B_c u \tag{6.3}$$

and

$$y = C_c \Phi Y + D_c u \tag{6.4}$$

Equationss.6.3 and 6.4 gives the state space realization in Jordan canonical (or modal) form - on this basis, therefore

$$A_{c,\text{mod}} = \Lambda \quad B_{c,\text{mod}} = \Phi^{-1} B_c \quad C_{c,\text{mod}} = C_c \Phi \quad D_{c,\text{mod}} = D_c \tag{6.5}$$

To clarify the connection of the Jordan state space representation with the conventional modal parameters in structural dynamics lets examine the unforced response to an initial state x_0. From equation 6.3 one has

$$Y_{j,t} = \mathrm{e}^{\lambda_j t} Y_{0,j} \tag{6.6}$$

where $Y_{0,j}$ is the jth entry of the initial condition vector Y_0 which, from equation 6.2, is related to x_0 by

$$Y_0 = \Phi^{-1} x_0 \tag{6.7}$$

The eigenvalue λ_j can be expressed as

$$\lambda_j = a_j + b_j i \tag{6.8}$$

From equations 6.8 and 6.6 one has

$$Y_{j,t} = \mathrm{e}^{a_j t}\mathrm{e}^{i b_j t} Y_{0,j} \tag{6.9}$$

or, given Eulers identity,

$$Y_{j,t} = \mathrm{e}^{a_j t}(\cos(b_j t) + i\sin(b_j t))Y_{0,j} \tag{6.10}$$

As can be seen, the solution, therefore, is a harmonic with circular frequency b_j modulated by the exponential function $\mathrm{e}^{a_j t}$.

SDOF Analogy

In a SDOF under free vibration the exponent a_j is equal to the product of the undamped frequency times the fraction of critical damping and b_j is the damped circular frequency, using this analogy one has

$$a_j + b_j i = -\omega_j \xi_j + i\omega_j\sqrt{1-\xi_j^2} \tag{6.11}$$

which provides two equations from where to solve for ω and ξ. We note that in an arbitrarily damped MDOF system the frequency obtained from equation 6.11 is not the undamped natural frequency (although the result is usually very close) except in the special case where the damping is classical. What holds without approximation, however is the fact that the real part of the eigenvalue gives the exponential rate of decay and the imaginary part the damped natural frequency.

Latent Vectors (Complex Mode Shapes)

The transformation from the modal amplitudes to the physical displacements is

$$\Psi = C_c \Phi \tag{6.12}$$

The shapes obtained from equation 6.12 (same as equation. 6.5) are typically referred to as complex modes, although it is more proper to call them latent vectors (Lancaster 1966) since they are not a basis that diagonalizes any of the system matrices. It is perhaps worth noting explicitly that the MDOF system cannot vibrate freely in a single complex mode because the associated response is complex. What the system can do, however, is to execute motion at a single frequency in a shape that is a combination of one latent vector and its complex conjugate. When the damping is zero, or classical, the vectors in equation 6.12 can be normalized to purely real so the conjugate of the mode is the same as the mode and one can speak of vibration in a single mode.

6.1 Undamped Mode Approximation

In practice there is often interest in extracting undamped modes from identification results because only the stiffness and the mass matrices (K and M) of Finite Element Models can be estimated from the known physics. More specifically, if one wishes to adjust model parameters to bring the model as close as possible to the identified eigensolution the undamped result is ideal because it decouples the damping mechanism from the K and M update. While a number of procedures have been proposed for approximating the undamped modes from the latent vectors, the most common approach simply normalizes the latent vectors so that the largest entry (in absolute value) is real and then discards the imaginary parts. It can be shown that this approximation is reasonable for modes where the eigenvalue gap is not too small (where the gap is the distance to the closest eigenvalue).

6.2 Discriminating Between Computational Modes and System Modes

Because the singular values of the Hankel matrix are not necessarily ordered so that all the system modes appear first it is often necessary, to capture some weakly represented system modes, to select an order that includes singular values associated with computational modes. Many indicators designed to help discriminate between computational and system modes have been proposed (Papa and Elliot 1993) but the problem is one for which there is no entirely satisfactory solution at the time of writing. Perhaps the most successful approach, albeit tedious, consists of repeating the identification using different selected orders and formulating what is known as a stabilization diagram i.e., a chart that keeps track of when a certain mode appears in the identification and when it does not. The heuristic operating assumption being that modes that appear consistently are system modes and those that come and go are computational. There are, of course, some indicators based on a priori knowledge that can also be used to discriminate system modes from computational ones. The simplest one is the level of damping. Specifically, if a structure is lightly damped and a mode in the identification shows a damping that is notably outside the expected range then it is very likely computational. In many structural systems values of damping larger than 10% are unlikely and this threshold can, in many instances be set significantly lower.

Another widely used indicator is the relative phase angles of the associated eigenvectors. The idea here is that if an eigenvalue is not too close to another then the associated eigenvector entries, provided the damping is not too large (say less than 10%) have phase angles that are nearly zero or pi. If a mode displays strong complexity and is not associated with an eigenvalue for which the gap is small one can, therefore, confidently discard it as computational. A metric that measures how close a certain mode is to being nearly real can be defined by noting that when a vector can be normalized to real the real and the imaginary parts are linearly related. The two singular values of the matrix obtained by taking the real and the imaginary parts of the complex mode as columns can, therefore, be used to define a metric to judge modal complexity. With s_2 as the smaller singular value and s_1 as the largest one can define (among many possible forms)

$$coef = \left(1 - \left(\frac{s_2}{s_1}\right)^{0.5}\right) \times 100 \tag{6.13}$$

so a perfectly real mode gives $coef = 100$ and a mode where the imaginary part is orthogonal to the real, $coef = 0$. The weighted modal phase colinearity coefficient ($mpcw$) advanced by Pappa and Elliot (1993) is a similar metric.

6.3 Realization Matrices in the Real Domain after Discarding Computational Modes

Let the matrices in continuous time in modal canonical form be A_{cm}, B_{cm} and C_{cm} as given by equation 6.5. Assume that an examination of quality indicators suggests keeping only the n_1 set of the originally n modes whose indices are collected in the vector $nm = [p \quad j \quad q \quad \text{etc}]$ (n_1 terms). The realization in modal coordinates for the reduced system is then

$$A_{reduced} = A_{cm}(nm, nm) \tag{6.14}$$

$$B_{reduced} = B_{cm}(nm, :) \tag{6.15}$$

$$C_{reduced} = C_{cm}(:, nm) \tag{6.16}$$

where we have used the notation (:) to imply all (as in MATLAB). A point to note is that the reduced realization is complex and that in some cases it may be of interest to transfer it to the real domain. This can be done by change of basis for which

$$T^{-1} A_{reduced} T = \text{real} \tag{6.17}$$

$$T^{-1} B_{reduced} = \text{real} \tag{6.18}$$

$$C_{reduced} T = \text{real} \tag{6.19}$$

Examination shows that if one satisfies the requirement for $A_{reduced}$ then equations 6.18 and 6.19 are automatically satisfied. One possibility is to look for a T that is unitary, i.e., a matrix whose conjugate transpose is equal to its inverse. If such a T exists then we would have

$$T^* A_{reduced} T = \text{real} \tag{6.20}$$

where the superscript on the T matrix indicates complex conjugation. Taking, without loss of generality, the eigenvalues in $A_{reduced}$ ordered such that the complex conjugate pairs appear next to each other one finds, after some analysis, that a block diagonal T having the building block shown in equation 6.21 achieves the objective

$$T_b = \frac{1}{\sqrt{2}} \begin{bmatrix} 1 & -i \\ 1 & i \end{bmatrix} \tag{6.21}$$

We conclude by noting that the direct transmission matrix was left out of the discussion because this matrix is best computed once the triple $\{A_c, B_c, C_c\}$ is available using equation 2.25.

6.4 Coordinate Expansion and Transfer Matrices

Transfer matrices are the linear transformation between inputs and outputs in the Laplace domain computed on the assumption of zero initial conditions. Focusing attention on the relation between forces and displacements one has, taking a Laplace transform and combining the state and the output equations

$$y_s = C_c^{\text{dis}} [I \cdot s - A_c]^{-1} B_c u_s \tag{6.22}$$

therefore

$$G(s) = C_c^{\text{dis}} [I \cdot s - A_c]^{-1} B_c \tag{6.23}$$

Equation 6.23 can be generalized to velocity and acceleration measurements by noting that (Bernal 2006)

$$C_c^{\text{dis}} = C_c A_c^{-b} \tag{6.24}$$

where, as noted in connection with equation2.25, $b = 0, 1, 2$ for displacements velocities or accelerations. The general expression for the transfer matrix $G(s)$ (from force to displacement) is thus

$$G(s) = C_c A_c^{-b} [I \cdot s - A_c]^{-1} B_c \tag{6.25}$$

Pole Residue Form

Expressing the system matrix in spectral form

$$A_c = \Phi\Lambda\Phi^{-1} \tag{6.26}$$

where $\Lambda = \text{diag}(\lambda_1, \lambda_2, ..., \lambda_n)$ and taking

$$C_c A_c^{-b}\Phi = [\phi_1 \phi_2 \cdot \phi_n] \tag{6.27}$$

and

$$\Phi^{-1}B_c = \begin{bmatrix} \varphi_1^T \\ \varphi_2^T \\ \cdot \\ \varphi_n^T \end{bmatrix} \tag{6.28}$$

gives

$$G(s) = \sum_{j=1}^{N} \frac{R_j}{s - \lambda_j} \tag{6.29}$$

where R_j is the residue for mode j given by

$$R_j = \phi_j \varphi_j^T \tag{6.30}$$

Equation 6.29 is known as the pole-residue for of the transfer matrix.

Analytical Expansion

Equations 6.25 or 6.29 give rows and columns of the transfer matrix associated with the measured outputs and inputs respectively. For systems with symmetric mass, damping and stiffness matrices, however, the transfer matrix is symmetric and this fact can be used, provided there is at least one collocated input-output coordinate, to expand the experimental partition to the union of the input-output coordinates. To illustrate how this is done consider the situation where all the coordinates are collocated. Given that s can be taken arbitrarily close to each λ_j one concludes that each of the modal contributions in equation 6.29 must be symmetric and from this it follows that

$$\chi_j \phi_j = \varphi_j \tag{6.31}$$

so one can write

$$G_s = \sum_{j=1}^{N} \frac{\phi_j \phi_j^T \chi_j}{s - \lambda_j} \tag{6.32}$$

where χ_j is known as the modal constant (Heylen et.al. 1998, Ewins 1984) . Because the ratio of equation 6.31 holds at every entry in the mode it is apparent that to compute χ_j one does not actually need to have full collocation but is suffices that there be at least one collocated coordinate. We conclude by noting that when there is more than one collocation the ratio in equation 6.31 differs in each entry due to noise and other inevitable approximations. In these cases accuracy is generally promoted by computing χ_j using the collocated coordinate where ϕ_j has the largest absolute value.

7 Stochastic Identification

The stochastic identification problem is typically solved on the premise that the excitation is a realization of a stationary broad band random process. In a state-space framework the problem is defined as the task of extracting the matrices A_d and C_d and (sometimes) the covariance of the process noise and the output noise from analysis of the output signals. The unknown stochastic system is represented by

$$x_{k+1} = A_d x_k + w_k \tag{7.1}$$

$$y_k = C_d x_k + v_k \tag{7.2}$$

where w_k and v_k are (assumed to be) white noise vectors so one has

$$E(x_k, v_k^T) = 0, \quad E(x_k, w_k^T) = 0 \tag{7.3}$$

Algorithms to obtain stochastic realizations using subspace techniques can be found in the text by Van Overschee and Moore (1996) and Katayama (2005). The approach presented here is elementary and is intended to illustrate the basic ideas involved. We begin by noting that since the state is stationary one can write

$$E(x_{k+1}, x_{k+1}^T) = \Sigma \tag{7.4}$$

Performing the products and imposing the conditions in equation 7.3

$$\Sigma = A_d \Sigma A_d^T + Q \tag{7.5}$$

where Q is the covariance of the process noise, namely

$$Q = E(w_k, w_k^T) \tag{7.6}$$

The output covariance matrix Λ_i is

$$\Lambda_i = E(y_{k+i}, y_k^T) \tag{7.7}$$

Substituting equation 7.2 into equation 7.7 gives

$$\Lambda_i = E[(C_d x_{k+i} + v_{k+i})(C_d x_k + v_k)^T)] \tag{7.8}$$

which, given that the measurement noise is (assumed) white simplifies to

$$\Lambda_i = E(C_d x_{k+i} x_k^T C_d^T) \quad (\text{for } i \neq 0) \tag{7.9}$$

and

$$\Lambda_0 = C_d \Sigma C_d^T + R \tag{7.10}$$

where R is

$$R = E(v_k v_k^T) \tag{7.11}$$

From equation 7.1 one has

$$x_{k+i} = A_d^i x_k + A_d^{i-1} w_k + A_d^{i-2} w_{k+1} + \ldots + w_{k+i+1} \tag{7.12}$$

which, after post-multiplying by x_k^T and taking expectation gives

$$E(x_{k+i} x_k^T) = A_d^i E(x_k x_k^T) = A_d^i \Sigma \tag{7.13}$$

Substituting equation 7.13 into 7.9 one gets

$$\Lambda_i = C_d A_d^{i-1} (A_d \Sigma C_d^T) \tag{7.14}$$

Comparison of equation 7.14 and equation 2.48 shows that the output covariance can be treated as MPs of a system having the triplet $\{A_d,\ A_d \Sigma C_d^T,\ C_d\}$. The ERA algorithm can, therefore, be used in standard fashion to identify this triplet. We close by recalling that the eigenvalues and eigenvectors of the system depend only on A_d and C_d.

8 Appendix. Localization of Damage from Changes in a State Space Realization

This appendix presents a technique to determine the position of damage from an examination of changes in transfer matrices. The theoretical support and the implementation details can be found in Bernal (2002) and Bernal (2007b).

Example of a Damage Locating Vector (DLV) Localization

Consider a beam where the left side is fixed and the right side is simply supported with 4 sensors equally spaced along the length. We model damage as the appearance of rotational springs at the positions $x = 0.35L$ and $x = 0.75L$ where x is measured from the left side. Since the DLV theorem is exact and we're not including approximations from noise and truncation the actual rotational stiffness used to represent the damage is immaterial. For specificity, however, we take the springs as $k_r = 5EI/L$ where EI is the flexural stiffness of the cross section and L is the total length of the beam. With $E = I = L = 1$ in some consistent units the numerical results are:

$$F_u = 10^{-3} \begin{bmatrix} 1.621 & 2.784 & 2.603 & 1.525 \\ & 6.912 & 7.381 & 4.523 \\ & & 9.792 & 6.624 \\ sym & & & 5.461 \end{bmatrix},$$

$$F_d = 10^{-3} \begin{bmatrix} 1.857 & 3.377 & 3.011 & 1.842 \\ & 8.470 & 8.731 & 5.731 \\ & & 12.109 & 9.233 \\ sym & & & 8.525 \end{bmatrix}$$

So the change in the flexibility is

$$F_d - F_u = DF = 10^{-3} \begin{bmatrix} 0.236 & 0.593 & 0.408 & 0.316 \\ & 1.558 & 1.349 & 1.208 \\ & & 2.317 & 2.609 \\ sym & & & 3.063 \end{bmatrix}$$

A singular value decomposition of DF shows that the null space is of dimension two and a basis can be easily shown to be

$$DLV = \begin{bmatrix} 0.195 & -0.896 \\ 0.145 & 0.425 \\ 0.775 & -0.124 \\ 0.583 & 0.030 \end{bmatrix}$$

To apply the DLV theorem one takes the vectors in the null space, treats them as loads in sensor coordinates and computes the stress field. As can be seen from Figure 2, which plots the moment diagram for both of the DLV loads acting on a model of the undamaged beam, the aggregate information shows the damage to be located at $x = 0.35L$ and $x = 0.75L$.

The Dynamic Damage Locating Vector Approach (DDLV)

The DDLV approach (Bernal 2007b) is supported on a theorem which shows that the null space of the change in the transfer matrix at any location in the complex plane (provided the matrix is defective) is a basis for the

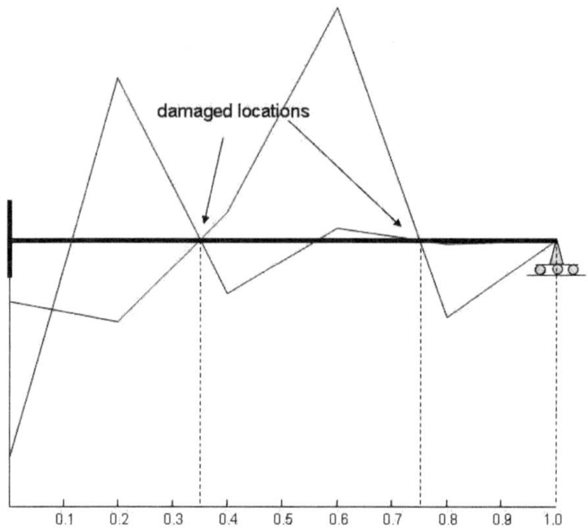

Figure 2. Moment diagrams (relevant stress field in this case) due to the DLV loads.

Laplace transform of excitations that lead to zero stress over the damage. The DDLV theorem is most easily applied in the Laplace domain, where the computations take a form entirely analogous to those in the static DLV technique; flexibility on the selection of where (and how many locations) in the complex plane to evaluate the transfer matrices allows for redundancy that can be taken advantage to promote robustness. The basic operational steps are illustrated next in the context of an example.

Example of a DDLV Localization

To illustrate a situation where the advantages in the spatial resolution offered by the DDLV theorem is evident we consider a 4 story uniform shear building with 2 sensors, one in level # 1 and one on the roof (level # 4). Damage is simulated as a reduction of 25% in the stiffness of level # 2. For generality we consider a non-classical damping distribution defined by a single dashpot located between mass number 1 and mass number 2. All springs in the undamaged state have a constant equal to 100 and the masses and the dashpot are equal to one in appropriate consistent units. In this academic example the selection of the values of s selected to realize the transfer matrices is immaterial so we illustrate the computations for two arbitrary selections $s = 2i$ and $s = -0.5 + 12i$.

Point # 1 ($s = 2i$)

Assume that realizations have been obtained from vibration measurements in the reference and the damaged state. The transfer matrices evaluated using equation 6.32 at $s = 2i$ are

$$G_u = 10^{-3} \begin{bmatrix} 12.315 - 0.0066i & 16.036 - 0.0561i \\ sym & 57.87 - 0.474i \end{bmatrix},$$

$$G_d = 10^{-3} \begin{bmatrix} 12.433 - 0.0133i & 17.027 - 0.1124i \\ sym & 66.245 - 0.949i \end{bmatrix}$$

So the change in the transfer matrix (at $s = 2i$) is

$$G_u - G_d = DG = 10^{-3} \begin{bmatrix} 0.117 - 0.0067i & 0.991 - 0.0536i \\ sym & 8.370 - 0.476i \end{bmatrix}$$

A singular value decomposition shows that the null space is of dimension one and a basis is

$$\ell = \begin{bmatrix} -0.9931 \\ 0.1176 \end{bmatrix}$$

which, only coincidentally, happens to be real. In the DDLV approach the vector that is used to load the structure is given by

$$L_{st} = K_m G_m f$$

where K_m and G_m are the stiffness and the transfer matrices of the model in the undamaged state (the later at the select s value) and $f = l$ with zeros added at the unmeasured coordinates. Performing the operations one gets

$$L_{st} = \begin{bmatrix} -1.0344 \\ -0.0414 \\ -0.0397 \\ 0.0811 \end{bmatrix}$$

for which the story shear distribution (the relevant stress resultant) is

$$V = \begin{bmatrix} 0.0811 \\ 0.0414 \\ 0 \\ -1.0344 \end{bmatrix}$$

which shows the damage is on the second level.

Point # 2 ($s = -0.5 + 12i$)

In this case the results are

$$G_u = 10^{-3} \begin{bmatrix} 1.868 - 1.69i & 8.565 - 1.889i \\ sym & -8.088 + 2.328i \end{bmatrix}$$

$$G_d = 10^{-3} \begin{bmatrix} 1.185 - 1.203i & 9.914 - 2.113i \\ sym & 7.492 + 2.225i \end{bmatrix}$$

So the change in the transfer matrix (at $s = -0.5 + 12i$) is

$$G_u - G_d = DG = 10^{-3} \begin{bmatrix} 3.053 - 0.487i & 1.349 - 0.224i \\ sym & 0.596 - 0.103i \end{bmatrix}$$

A singular value decomposition shows that the null space is of dimension one and a basis is

$$\ell = \begin{bmatrix} -0.4046 \\ 0.9145 + 0.0061i \end{bmatrix}$$

Substituting in the formula for L_{st}

$$L_{st} = \begin{bmatrix} 0.8600 - 0.2359i \\ 1.2646 - 0.2359i \\ -0.5815 - 0.0485i \\ -0.6830 + 0.2844i \end{bmatrix}$$

for which the story shear distribution (the relevant stress resultant) is

$$V = \begin{bmatrix} -0.683 + 0.285i \\ -1.265 + 0.236i \\ 0 \\ 0.860 - 0.236i \end{bmatrix}$$

which again shows that the damage is on level two.

Bibliography

P.J. Antsaklis and A.N. Michel *Linear systems*, Birkhuser, Boston 2006.

D. Bernal. Discrete to continuous time transfer for band limited inputs. *Journal of Engineering Mechanics, ASCE*, 133(12): 1370-1377, 2007a.

D. Bernal. Damage localization from the null space of changes in transfer matrices. *AIAA Journal*, 45(2): 374-381, 2007b.

D. Bernal Load vectors for damage localization. *Journal of Engineering Mechanics, ASCE*, 128(1):7-14, 2002.

D. Bernal. Flexibility-based damage localization from stochastic realization results. *Journal of Engineering Mechanics, ASCE*, 132(6):651-658, 2006.

D.J. Ewins. *Modal testing: theory and practice*, 1984. Research Studies Press Ltd, England.

G. Golub and C. Van Loan. *Matrix computation*, 1989. The Jonhs Hopkins University Press, Baltimore, Maryland.

H. Hanselmann. Implementation of digital controllers a survey. *Automatica*, 23(1): 7-32, 1987.

W. Heylen, S. Lammens, and P. Sas. *Modal analysis theory and testing*, 1998. Katholieke Universiteit Leuven.

B.L. Ho and R.E. Kalman. Effective construction of linear state-variable models from input output functions. *Regelungstechnik* 12: 545548, 1965.

J. Juang and R. Pappa. An eigensystem realization algorithm for modal parameter identification and model reduction. *Journal of Guidance Control and Dynamics*, 8(5):610-627, 1985.

J.N. Juang. *Applied system identification*, 1994. Prentice Hall, Inc., Englewood Cliffs, New Jersey.

T. Kailath. *Linear systems*, 1990. Englewood Cliffs, Prentice-Hall, N.J.

T. Katayama. *Subspace methods for system identification*, 2005. Springer.

P. Lancaster. *Lambda-matrices and Vibrating Systems*,1966. Pergamon Press.

L. Junjg. *System identification: theory for the user*, 1987. Prentice Hall, Englewood-Cliffs.

R. S. Pappa and K.B. Elliot. Consistent mode indicator for the eigensystem realization algorithm. *Journal of guidance and Control*, 16(5):832-838, 1993.

J. Schoukens, R. Pintelon and H. Van Hamme. Identification of linear dynamic systems using piecewise constant excitations: use, misuse and alternatives. *Automatica*, 30(7):1153-1169, 1994.

T. Södertröm, and P Stoica. *System identification*, 2001. Prentice Hall International, Series in Systems and Control Engineering, UK.

H. Ubenhauen and G.P. Rao *Identification of continuous systems*, 1987. North-Holland, Amsterdam.

P. Van Overschee and B.L De Moor. *Subspace identification for linear systems: theory, implementation, applications*, 1996. Kluwer Academic Publishers, Boston.

A Review of Ground Motion Record Selection Strategies for Dynamic Structural Analysis

Iunio Iervolino[†] and Gaetano Manfredi[†]

[†] Dipartimento di Ingegneria Strutturale, Università degli Studi di Napoli Federico II

Via Claudio 21, 80125, Naples, Italy.

Abstract Non-linear dynamic analysis is recognized as the more accurate tool for seismic evaluation of structures in the case of both probabilistic assessment and design. The key issue in performing this kind of analysis is the selection of appropriate seismic input (e.g. ground motion signals), which should allow for a correct and accurate estimation of the seismic performance on the basis of the hazard at the site where the structure is located. To this aim several procedures have been proposed, they require specific characterization of real ground motion records via the so called ground motion intensity measures (mainly related with elastic spectral features of the record) proven to be generally efficient in the estimation of the structural performance. This kind of approach requires specific skills as well as detailed probabilistic evaluation of the seismic threat to be available to the engineers. For this and other reasons codes worldwide, in many cases, try to acknowledge these procedures in an approximate fashion.

In this paper recent and advanced literature on the topic is presented and discussed. The current best practice in record selection is reviewed for the case of probabilistic seismic risk analysis and for code- based seismic assessment and design with special attention to the prescriptions of Eurocode 8 for both buildings and bridges. Finally, some light is briefly shaded on the effects of time scaling of records and its use in shake-table structural testing.

1 Introduction

The issue of selecting the seismic input is seen to be one of the most critical in the seismic assessment of structures via non-linear dynamic analysis. It is sometimes considered more important even than structural modeling. Therefore, this problem has been the subject of large research recently. In general, the signals that can be used for the seismic structural analysis are of three types: (1) artificial waveforms; (2) simulated accelerograms; and (3) natural records (Bommer and Acevedo, 2004).

Spectrum-compatible signals of type (1) are obtained, for example, generating a power spectral density function from a code-specified response spectrum, and deriving signals compatible to that. However, this approach may lead to accelerograms not reflecting the real phasing of seismic waves and cycles of motion, and therefore energy. Simulation records (2) are obtained via modeling of the seismological source and may account for path and site effects. These methods range from stochastic simulation (Boore, 2003) of point or finite sources to dynamic models of rupture. However, they often require setting of some rupture parameters, such as the rise-time, which are hard to determine. Some state-of-the-art simulation methods seem to overcome these shortcomings, but they are not yet readily available to engineers.

Finally, of type (3) are ground-motion records from real events. The availability of on-line, user-friendly, databases of strong-motion recordings, and the rapid development of digital seismic networks worldwide, have increased the accessibility to recorded accelerograms, which, therefore, have become the most promising candidates for the seismic assessment of structures, both for code-related purposes and probabilistic risk analysis. However, due to the large record-to-record variability in records representing a specific scenario (i.e. a magnitude-distance pair), a number of points arise regarding the criteria for appropriate selection and manipulation of such records. The aim is to obtain a correct and accurate estimation of the structural performance in a way that it does not require a undesirably large number of non-linear dynamic analyses to run. Therefore, earthquake engineering research has focused lately on the selection of real ground-motions for non-linear structural analysis and effective procedures, with various degrees of simplicity in application, have been developed to link real ground motion records to the hazard at the site and to estimate probabilistically the structural seismic performance.

Although the record selection for probabilistic seismic risk assessment should be regarded as a reference case, codes only try to approximate such an approach for assessment and design purposes. This is because all the information needed to perform state-of-the art procedures for record selection are, nowadays, seldom available to engineers. Moreover, codes also try to warrant standard and conservative procedures, at least until probabilistic seismic hazard data are not made broadly available and/or certified by authorities.

In the case of code-based seismic structural assessment, another issue regarding the use of real recordings, whose spectra are generally non-smooth, is the selection of a set compatible with a code-specified spectrum. Several approaches have been developed to manipulate real records in order to match a target spectral shape, either by frequency-domain or by time-domain modification methods such as the wavelet transform. The wavelet transform basically consists of using modulating functions, selectively located in time to modify the spectrum of the signal, where and when it is needed, in order to match the target spectrum, see Hancock et al. (2006) for details. Although these methods produce records perfectly compatible with code's prescriptions, and have the additional advantage of reducing the dispersion in the response and hence the required sample size, some studies show that they may lead to a non-conservative estimation of the seismic response (Carballo and Cornell, 2000; Bazzurro and Luco, 2003), and therefore natural ground motion records still seem to be the most appropriate option as input signals.

In the following the recent issues in record selection for probabilistic assessment of structures are reviewed first, subsequently the code procedures are discussed with specific reference to the European regulation (Eurocode) in the case of both one-component and multi-dimensional record selection for bridges and building structures. Finally, the effects of time scaling of real records and its use in shake-table structural testing are briefly discussed.

2 Record Selection for probabilistic assessment of structures

The objective of the probabilistic assessment of structures is defined as the estimation of the probability of the structure reaching a limit state (e.g. collapse or failure) over a period of time (i.e. one year or the design life of the structure). The limit state is often formulated as a function of the required level of seismic performance for the structure, or the seismic demand, and the supply of such performance intrinsic to the structure, or the seismic capacity. Both capacity and demand have to be expressed in terms of a global structural performance indicator as, for example, the maximum interstory drift ratio (MIDR) as it is related to the rotation of elements in moment resisting frame structures, for which semi-empirical capacity model are available. Cornell (2004) discusses the problem of estimating such a probability via non-linear dynamic analysis. In such a case, in fact, the failure rate in one year for example, may be formulated separating the estimation of the structural response from the probabilistic characterization of the seismic threat it is subjected to because of the seismicity of the site where it is located. By the total probability theorem one can write:

$$P_f = \int_M \int_R P[F|M,R]f(M,R)\, dm\, dr \qquad (2.1)$$

where $P[F|M,R]$ is the probability of the structure getting the limit state conditioned to some earthquake characteristic, herein represented by magnitude (M) and source-to-site distance[1] (R); while $f(M,R)$ is the joint probability density function related, for example, to the annual occurrence of any (M,R) pair.

In principle, $P[F|M,R]$ should be estimated by non-linear dynamic analyses using and adequate number of ground motion records consistent with the (M,R) domain of interest. In particular, if relationships expressing the likelihood of observing a certain magnitude and source-to-site distance (e.g. a Gutenberg-Richter relationship and a distribution of earthquake location to be provided by seismologists) are available, the engineer should get a sample of ground motion records for each (M,R) pair and run non-linear dynamic analysis for each record in the sample. The observed frequency of collapses over the size of the sample is an estimation of the failure probability conditional to that specific (M,R) pair.

The issue of establishing the adequate sample size for each (M,R) pair raises. It depends on the desired standard error in estimating the response quantity of interest (i.e. MIDR) and the dispersion of the variable to measure itself. Cornell (2004) clarifies that, arbitrarily assuming as 0.1 the acceptable value for the latter and 0.8 as the coefficient of variation of the failure-level MIDR for a moment-resisting frame, the necessary sample size is about $(0.8/0.1)^2$, or 64. Assuming that the domain of threatening magnitude and distance pairs can be partitioned in 20 bins, then the number of required records is in the order of 10^3 which is clearly impractical also because easily accessible record databases hardly would have, in each (M,R) bin, enough records which also match the appropriate faulting stile or, more importantly, soil conditions. Therefore alternative (more efficient) strategies based on a more refined selection of ground motion records have been developed and are reviewed in the following sections.

[1]Other earthquakes characteristics components may be faulting style and others. It is currently under discussion which of these parameters is statistically significant for the structural response, however it is appropriate here to focus on M and R considering them exhaustive.

3 Smart strategies to estimate P_f: ground motion intensity measures

Smarter approaches aimed at reducing as much as possible the sample size required to correctly estimate the structural response have been developed. These approaches are based on the concept of ground motion intensity measure or IM, and are quite easy to apply at least in their basic representation, and therefore are starting to be a widespread practice.

An IM is a ground motion feature, which should be a proxy for the earthquake potential (e.g. it should be a relatively good predictor of the structural response), which therefore, allows to estimate the response with only "few" analyses given a level of accuracy. Typical ground motion IMs are the peaks of the acceleration, velocity, and displacement signals (PGA, PGV and PGD) respectively. This is also because the seismic hazard is often represented in terms of probability of exceedance of these quantities. Linear spectral ordinates, especially accelerations, at the fundamental period of the structure, $\boldsymbol{S}_a(\boldsymbol{T}_1)$, are also often used as IMs for probabilistic assessment of structures. This is mainly because $\boldsymbol{S}_a(\boldsymbol{T}_1)$ is the response of a linear single degree of freedom system (SDOF) and therefore it should be, in principle, more correlated with the structural global non-linear performance in respect to peaks of ground motion. Integral signal's parameters, i.e. the Arias intensity, are possible IMs, although they are considered more related to the energy dissipation rather than to displacement-related structural responses as MIDR (Iervolino et al., 2006). More sophisticated IMs as non-linear spectral ordinates or vector valued IMs are also the subject of current investigation and will be discussed in the following.

Regardless of its definition, introducing an IM implies to have an additional variable to condition $\boldsymbol{P}_f$ on, and Eq.(2.1) may be rewritten as follows:

$$\boldsymbol{P}_f = \int_{IM}\int_{M}\int_{R} \boldsymbol{P[F|IM,M,R]f(IM|M,R)f(M,R)}\, dm\, dr\, d(im) \tag{3.1}$$

where **f(IM|M,R)** is the conditional distribution of the chosen IM to magnitude and distance (e.g. an attenuation law). ***P[F|IM,M,R]*** , is the failure probability of the structure as a function of magnitude, distance and IM. If ***P[F|IM,M,R]*** $=$ ***P[F|IM]***, then the IM is said sufficient since its ability to predict the structural response given IM is the same as given the whole set of IM, M and R (e.g. the response is independent of M and R given IM).

The sufficiency of the IM is a desirable property, which has an intuitive interpretation: since the IM is a ground motion feature which not only reflects the source features of the event, as magnitude, including also the path to the site and the site response, it represents a lager level of information about the shaking, which overwhelms the dependency of the structural response on M and R. In such a case, recognizing that

$$\boldsymbol{P(IM)} = \int_M \int_R \boldsymbol{f(IM|M,R)f(M,R)}\, dm\, dr \tag{3.2}$$

is the probability density function related to the frequency of occurrence of the IM in the time period of interest, Eq. (3.1) may be rewritten as:

$$\boldsymbol{P}_f = \int_{IM} \boldsymbol{P[F|IM]f(IM)}\, d(im) \tag{3.3}$$

If the IM is the PGA or another spectral quantity as the spectral acceleration at the fundamental period of the structure, $\boldsymbol{f(IM)}$ is simply the result of a common probabilistic seismic hazard analysis or PSHA (Cornell, 1968). This is why several studies focused on the estimation of structural response via IMs for which hazard curves are readily available.

Advantages of introducing an IM are easily recognized. In fact, if $\boldsymbol{S}_a(\boldsymbol{T}_1)$ is concerned, assuming that the COV of MIDR is about 0.4 (Shome et al., 1998), this will require about 15 records at each IM level. Therefore, if 10 levels are needed to cover the IM range of interest (i.e. from 0.1 g to 1.0 g) then the total number of runs is only about 150 which is an order of magnitude smaller than applying Eq. (2.1) to estimate P_f with the same accuracy. If only 15 records are selected and their intensity is progressively increased to cover the IM range of interest, then $\boldsymbol{P[F|IM]}$ (which is also called the fragility function of the structure) may be estimated with a very limited number of records via the so called incremental dynamic analysis, although this requires that such IM scaling do not bias the estimation of the seismic response (to follow).

4 Fundamentals of Incremental Dynamic Analysis

Incremental Dynamic Analysis (IDA) is a numerical simulation method to estimate the structural seismic performance. As extensively discussed by Vamvatsikos and Cornell (2002), it requires subjecting a structural model to a set of ground motion records, each of those scaled to multiple levels of intensity, represented by an IM. It is possible to recognize in IDA an analogy with the static pushover; in the IDA the seismic loading, in the form of ground motion record, is scaled rather than the horizontal forces distribution is incremented.

The main targets the IDA allows to reach are: (1) the description of the structural seismic response or "demand" versus the IM (e.g. a range of potential levels of a ground motion record); (2) the estimation of the seismic capacity of the structure; (3) the determination of a probabilistic characterization of the variability of capacity and demand from record-to-record, and therefore the fragility function for the structural model considered.

IDA requires that a single acceleration time-history is scaled (e.g. all its values are amplified by a common scaling factor, SF), to several values of the ground motion's destructive potential (e.g. the intensity). The scaling of the records by SF is basically the modification of the linear IM used (e.g., PGA, PGV or $\boldsymbol{S}_a(\boldsymbol{T}_1)$). SF is computed simply by Eq. (4.1), where $\boldsymbol{IM}_T$ is the target (desired) intensity level, and is the original value of the intensity, $\boldsymbol{IM}_U$ (e.g. of the record as unscaled). Multiplying all the values of the signal by the scaling factor, the resulting modified record, will have the IM coincident with the target value.

$$\boldsymbol{SF} = \frac{\boldsymbol{IM}_T}{\boldsymbol{IM}_U} \tag{4.1}$$

While the scaled IM monitors the level of the seismic input or loading on the structure, the parameter used to monitor the response of the structure to that ground motion (i.e. the MIDR) will be referred to as engineering demand parameter or EDP.

To cover a specific range of interest of IM, the record may be scaled several times to get, each time, a specific target level. If a nonlinear dynamic analysis for the structure is run for each of these IM levels and the EDP value is recorded, then the result of IDA for a single record is a plot of a response variable (EDP) versus the IM; this is called IDA curve. The IDA curve is different from record-to-record. In fact, since the IM is only a proxy for the potential of the ground motion, the structural response to two records scaled to the same IM, will be different as the response is also influenced by other features of the signal not captured by the IM the two accelerograms share. For example, if two records are scaled to have the same $\boldsymbol{S}_a(\boldsymbol{T}_1)$, the response may still differ if the two spectra are not similar in the range of periods beyond T_1, because the period lengthening during the shaking renders the structure sensitive to such portion of the records' frequency content.

In Figure 1 an example of IDA results for a reinforced concrete frame subjected to a set of about 20 real records is shown, in that case the EDP is MIDR and the IM is $\boldsymbol{S}_a(\boldsymbol{T}_1)$, where the fundamental period of the building is about 0.7 sec (De Risi, 2007). Each line illustrates the response of the structure to a specific accelerogram scaled several times to different intensities. Note that IDA curves are often represented with IM (the independent variable) on the vertical axis rather than on the horizontal one, this is to have a plot somewhat analogue to the push-over curves where the total base shear is the ordinate and the displacement is the abscissa.

If the structural model allows to account for some failure mode, which can be monitored via the EDP, a softening of the IDA curve may occur at high IM levels, signaling the onset of dynamic instability. The curve then goes flat at the maximum value of IM analogously to static instability, as it is the point where deformations increase indefinitely for small increments in the IM.

It is finally important to note that IDA curves are not necessarily monotonic (see the lower IDA curve in Figure 1 for example) because as the accelerogram is scaled up, weak response cycles in the early part of the response time-history become strong enough to "yield" the structure, the properties of the structure are altered for the subsequent stronger cycles (e.g. period elongation). Therefore, it may happen that a structure that showed high response at a given intensity level, may exhibit equal or even lower response when subjected to higher seismic intensities.

IDA obtained by a sample of ground motion records may be used to get a probabilistic representation of the seismic demand conditioned to IM. See for example Figure 1, where median, and 16 %, 84 % IDA curves are also plotted. For each (given) IM level those values read on the thick curves help to characterize the variability of the response.

IDA curves can also be used to determine the capacity of the structure. However, may be not obvious how to determine the capacity on an IDA curve. There are two possible approaches to solve this problem: (1) the capacity point is defined as that corresponding to $EPD \geq EDP_C$ where EDP_C is the response parameter's value that reflects the onset of the limit-state, this is consistent with static analysis methods and may be coincident with the reaching of limit rotations in an element, but it may ambiguous to find this capacity point on an IDA curve (see upper panel of Figure 2 for example); (2) the capacity may be defined in terms of a specific IM level (see bottom panel of Figure 2) that corresponds to the plateau in the curve, If $IM = IM_C$ then the limit-state is exceeded, a disadvantage of this latter method is that the IM_C level has to be determined curve by curve.

Clearly it is possible to have, therefore, a probabilistic characterization of the capacity when multiple IDA curves are available. In Figure 3 an example of the inferred distribution of collapse drift is given; each star in the plot represent the point where capacity is reached in each IDA curve (Jalayer, 2003).

IDA may also directly help to compute the seismic risk, by Eq. (3.3).

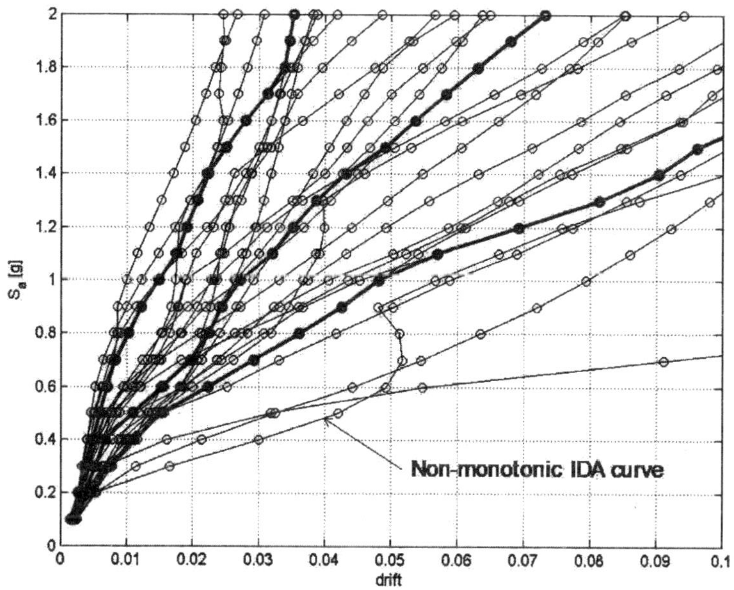

Figure 1. IDA curves in terms of MIDR vs. $S_a(T_1)$ for an underdesigned RC frame (De Risi, 2007). Thick curves represent medians and 16-84 percentiles of MIDR conditioned to IM.

In fact, the fraction of records causing collapse at each IM level provides an estimation of the $P[F|IM]$ term, which is the seismic fragility of the structure. In Figure 4 the determination of the fragility is represented for a case when the capacity of the structure is deterministic; the solid and dashed curves in the left panel represent the median and 18-84 percentiles of the distribution of the demand at each specific IM level. The probability that the capacity is exceeded according that distribution, is the failure probability for that IM.

5 The use of first-mode spectral acceleration as an IM

In describing the fundamentals of IDA it was assumed that the IM is the spectral acceleration at the fundamental mode of the structure, this is because not only hazard curves are readily available for $S_a(T_1)$, but also because it has been shown to be a sufficient and, to some extent, an efficient IM in many seismic risk engineering applications. It is to recall here

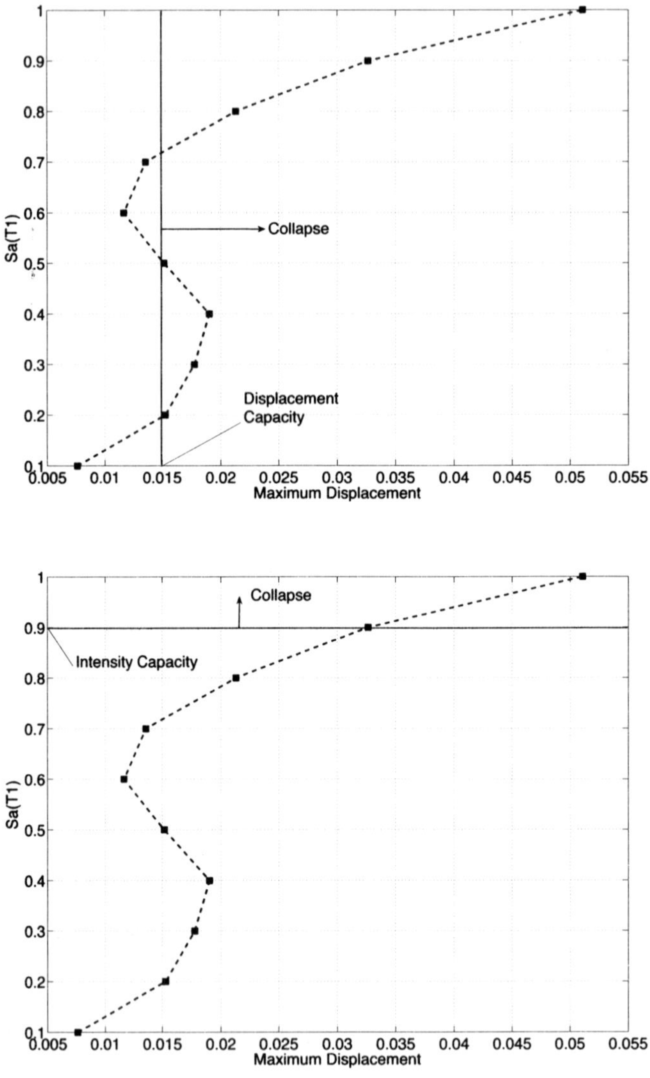

Figure 2. Measuring capacity on IDA curves.

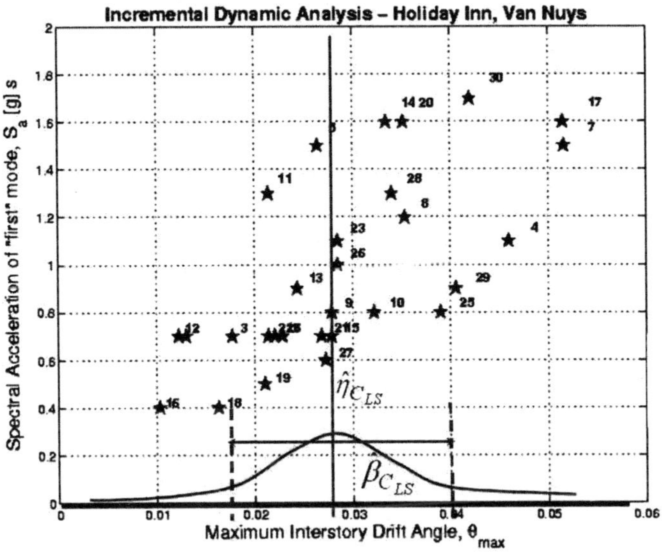

Figure 3. Probability density function of capacity from multiple IDA curves. Courtesy of Dr. Fatemeh Jalayer (2003).

that a sufficient IM makes the structural response variable conditionally independent of earthquake magnitude and distance, e.g. allow to use Eq. (3.3) to compute the seismic risk. At the same time a certain IM is defined efficient if the structural response, given IM, has a relatively small dispersion showing explanatory power of the seismic demand (e.g. the 16%, 84% IDA curves are "close" to the median curve).

Several studies have demonstrated that the displacement-based response measures, given $S_a(T_1)$, are independent of magnitude and distance (i.e. Iervolino and Cornell, 2005) and also that for drift response $S_a(T_1)$ is better (efficiency-wise) than PGA. Sufficiency of $S_a(T_1)$ has an important reflection on the record selection for seismic structural assessment. In fact, when real recordings are concerned, the current state of best practice is based on the determination of magnitude and distance most contributing to determine the $S_a(T_1)$ value corresponding to a specified probability[2]

[2]This procedure is called disaggregation of seismic hazard and allows one to determine the M and R scenario causative for the acceleration corresponding to the probability of interest in the hazard curve, see Bazzurro and Cornell (1999) for details.

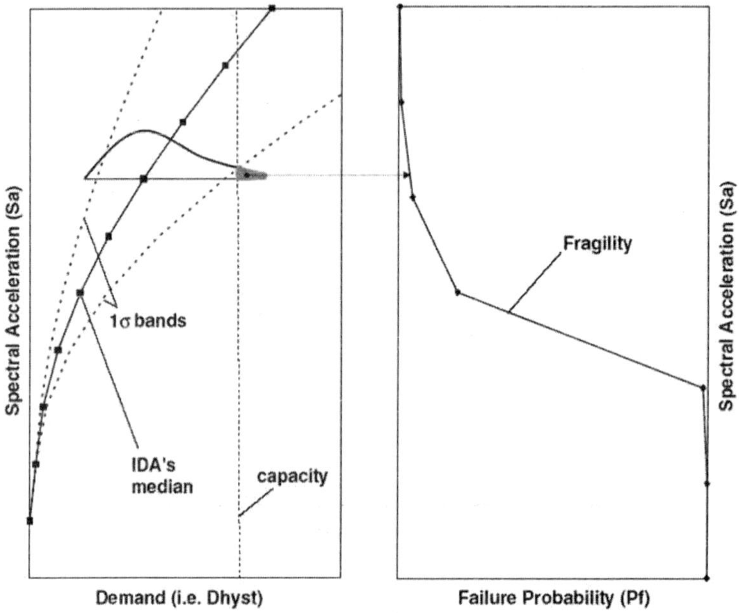

Figure 4. Example of determination of fragility form IDA curves (Iervolino et al., 2006).

(i.e. a 10% chance of exceedance in 50 years) in the hazard curve. This provides a scenario earthquake and therefore, the records are then chosen to match within tolerable limits these values of M and R. Sufficiency of the IM significantly simplifies this procedure as all this care about the selected records' may be avoided, and records may be, at least in principle, randomly chosen and then modified to have the $\boldsymbol{S}_a(\boldsymbol{T}_1)$ value of interest just by the linear amplitude scaling described above.

It is also worth to note that it has also been recently demonstrated that $\boldsymbol{S}_a(\boldsymbol{T}_1)$ may not be particularly efficient, nor "sufficient", for some structures. If long periods of oscillation are called into question (i.e. tall structures), the higher modes typically play a larger role into the seismic response and $\boldsymbol{S}_a(\boldsymbol{T}_1)$ has less prediction power than for first-mode dominated structures; it may be also insufficient because it is not able to capture the spectral shape in a range of frequencies where it depends on the magnitude. For soft-soil $\boldsymbol{S}_a(\boldsymbol{T}_1)$ may also be insufficient. There are also cases for which PGA may be a more suitable IM than $\boldsymbol{S}_a(\boldsymbol{T}_1)$, this happens for example, when the EDP to estimate is the peak floor acceleration, which is an important response variable correlated to inertia forces and therefore with non-structural damage (Taghavi and Miranda, 2003).

5.1 Using consistent Sa

In performing seismic assessment of structures via dynamic analysis it is important to bear in mind that structural engineers and seismologists sometimes intend Sa differently. This mismatch is due to the decomposition of ground-motion along two directions (Baker and Cornell, 2006a). For the aims of non-linear seismic assessment of structures, $\boldsymbol{S}_a(\boldsymbol{T}_1)$ is considered as the one along a single axis. Conversely seismologists may compute ground-motion prediction equations using the geometric mean of the spectral accelerations in the two horizontal directions; this is because using one arbitrary component would lead to a larger dispersion of hazard curves. Both uses of $\boldsymbol{S}_a(\boldsymbol{T}_1)$ are legitimate, but inconsistent if combined for the probabilistic seismic assessment of structures. Therefore, it is preferable to define the same Sa in both the hazard and response. This means either that in the seismic risk analysis of structures one should use hazard curves that use one-component $\boldsymbol{S}_a(\boldsymbol{T}_1)$, or estimating structural response using the geometric mean of the two components as an IM. This latter method has the advantage of not requiring new ground-motion prediction equations for hazard analysis. However, it will introduce additional dispersion into the response prediction and Sa will result less efficient. Alternatively, if the structural response is estimated using a single axis $\boldsymbol{S}_a(\boldsymbol{T}_1)$, while hazard refers to the mean of the two components, the dispersion of the response may be inflated, as proposed by the cited authors, to reflect that which would have been seen if the mean $\boldsymbol{S}_a(\boldsymbol{T}_1)$ had been used as the intensity measure.

5.2 Sa and records' duration

First-mode spectral acceleration has also been proven to be sufficient in respect of duration, at least for SDOF structures and if displacement-related response measures are of concern (Iervolino et al., 2006). It was shown, in the mentioned study, choosing three different record samples representing different duration scenarios. For each of these samples the non-linear response has been evaluated via incremental dynamic analysis, considering several EPDs, from kinematic ductility to equivalent number of cycles. This allowed to compare median responses to different duration bins.

In this framework duration has been found to be statistically insignificant to displacement ductility demand assessment, regardless of oscillation period and backbone curve. Conversely duration has been found to strongly affect, as expected, other demand parameters accounting for cyclic behavior such as hysteretic ductility or equivalent number of cycles (Cosenza et al., 1993). This is shown in Figure 5 where median IDA curves are given, for a bilinear SDOF, for the three record sets representing the duration scenarios (rhombuses, squares and triangles for short, moderate and long duration respectively).

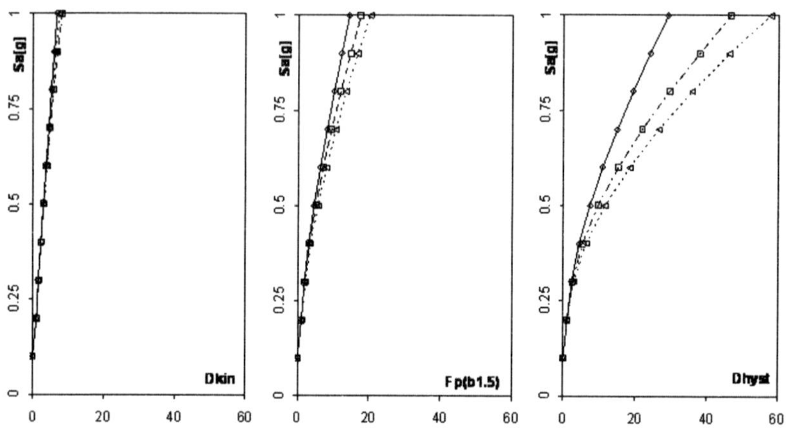

Figure 5. Comparison of median IDA curves, for an elastic-perfectly-plastic SDOF with a 1 sec oscillation period, in terms of kinematic ductility (left), plastic fatigue (center) and hysteretic ductility (right), (Iervolino et al., 2006).

It may be seen that curves are perfectly superimposed in the case of kinematic ductility, while for plastic fatigue the energy content of ground motion starts to play a role and median demand is proportional to the duration of the set that generated the curve. This effect, as expected is magnified in the case of hysteretic ductility.

Therefore, at least for the purposes of displacement-related demand assessment, it seems that one should also not take too much care in selecting records to match a specific duration value given they have (or are scaled to) a common spectral acceleration level.

5.3 Sa-based scaling of records and epsilon

Amplitude scaling of different records to get a target (common) intensity is useful to estimate the fragility of the structure; this do not bias the estimation of the response and also reduces its variability, achieving larger efficiency of the IM (Iervolino and Cornell, 2005). In fact, Figure 6 shows a record set used for non-linear dynamic analysis of a bilinear SDOF. The records are first used as unscaled (circles), and this is called cloud analysis; after the records in the set are scaled to a common Sa (which is the median of the set unscaled) and the response is computed. Results shows that the median response is similar in the two cases indicating no bias induced by scaling (this property is called robustness, e.g. the structural response estimated with scaled records is virtually the same as that from unscaled records featuring the same IM level). Moreover, the dispersion of the response (proportional to the horizontal bars in the figure) is significantly reduced with scaling.

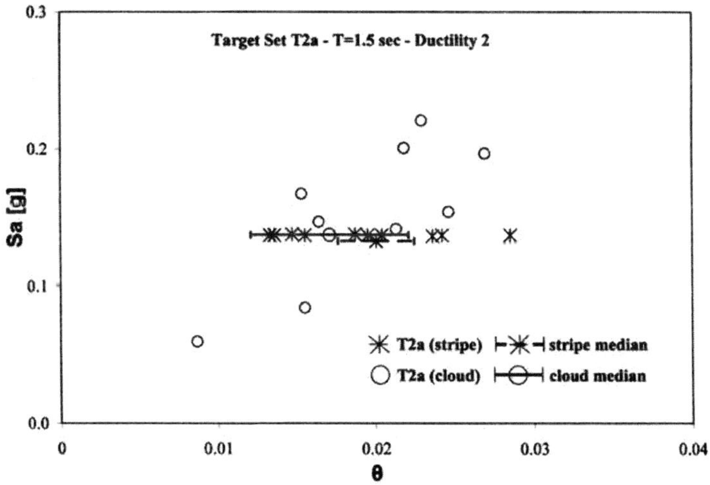

Figure 6. Variability of drift before and after scaling a set of records to the same IM level for a bilinear SDOF system with a 1.5 sec oscillation period (Iervolino and Cornell, 2005).

However, there are cases in which scaling of the spectral acceleration of the record would introduce a bias in the estimation of the structural response. It may be the case when the record shows a deviation of a record's Sa from the value predicted by the ground-motion prediction equation (Baker and Cornell, 2006b). This deviation is called *epsilon*[3] or "normalized residual". High (positive) epsilon values are associated with peaks in the spectrum at the fundamental mode of the structure, and hence with more benign nonlinear structural behavior. In fact, during the shaking the effective period of the structure lengthens descending the peak toward a less energetic portion of the frequency content. Therefore, scaling down a positive epsilon record (to match a specific spectral acceleration value for example) would introduce an un-conservative bias in the demand estimation because, due to the lengthening of the period during the shaking, the structure will be sensitive to a part of the spectrum which is away from the peak; conversely scaling up a negative epsilon record could lead to an overestimation of the seismic response (Figure 7).

Even though some researchers believe epsilon is not an intrinsic ground-motion feature, PSHA disaggregation of seismic hazard for epsilon often shows that high IM levels, contributing directly to rare maximum interstory drift ratio levels, are associated with high values of epsilon. Therefore, when selecting records for analyses at these high IM levels, one should consider to choose them among those having the right epsilon, in order to have the correctly deviating spectral shape around the period of interest, for a more efficient and unbiased estimation of structural response. This is more important than matching records with scenario M and R values.

A method has also been proposed (Baker and Cornell, 2006b) for developing a target spectrum which accounts for the effect of magnitude, distance and epsilon. This spectrum, discussed in the following of the paper, allows the selection of records that only have a spectral shape that matches the mean spectrum from the causal event, without taking care of appropriate magnitude, distance and specific epsilon.

5.4 Near-source

It has been briefly reviewed above why in seeking for characteristics to mirror in the record selection one should look to any systematic effect on spectral shape as epsilon is. Therefore, it is prudent to avoid selecting records from soft soil sites or from near-source records showing directivity effects. In fact, it should briefly be mentioned that a site located close to the source of a seismic event may be in a geometrical configuration, in respect to the propagating rupture, which may favor the constructive interference

[3]Epsilon (ε) is defined as the difference between the log of the spectral acceleration, at a given period, of a record and that predicted by an ordinary ground-motion prediction equation divided by the standard deviation of the residuals.

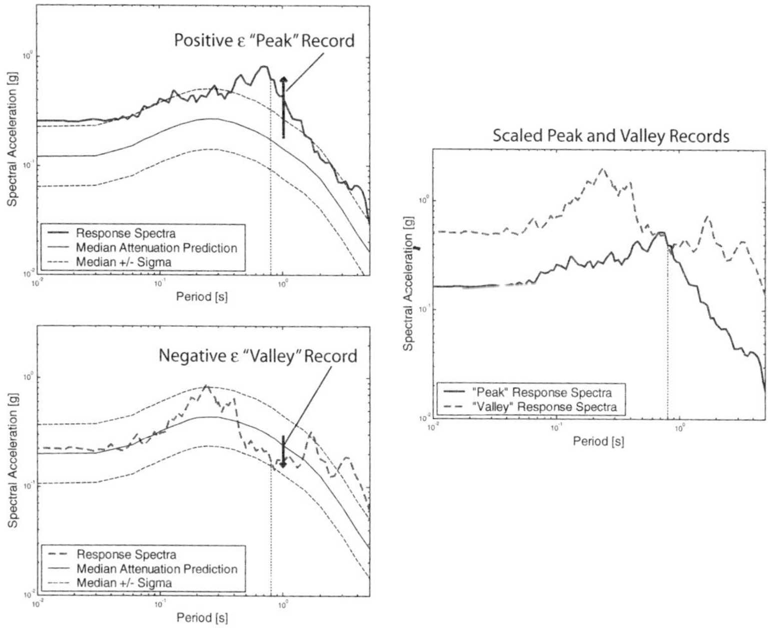

Figure 7. Scaling a negative ε record and a positive ε record to the same spectral acceleration at the period of 0.8 s. Courtesy of Jack W. Baker, see Baker and Cornell (2005) for details

of waves (synchronism of phases causing building up of energy) traveling to it, which may result in a large velocity pulse, Figure 8 (Tothong et al., 2007). This situation, for dip-slip faults, requires the rupture going toward the site and the alignment of the latter with the dip of the fault, whereas for strike-slip faults the site must be aligned with the strike; if these conditions are met the ground-motion at the site may show forward directivity effects (Somerville et al., 1997).

Parameters driving the amplitude of the pulses are related to the above-discussed rupture-to-site geometry, while empirical models positively correlating the earthquake's magnitude to the period of the pulse have been proposed by seismologists (Somerville, 2003). Pulse-type records are of interest for structural engineers because they: (1) may induce unexpected demand into structures having the fundamental period equal to a certain fraction of the pulse period; (2) such demand may not be adequately captured by ground-motion intensity measures such as first mode spectral acceleration (Tothong and Cornell, 2006). Therefore, common record selection practice

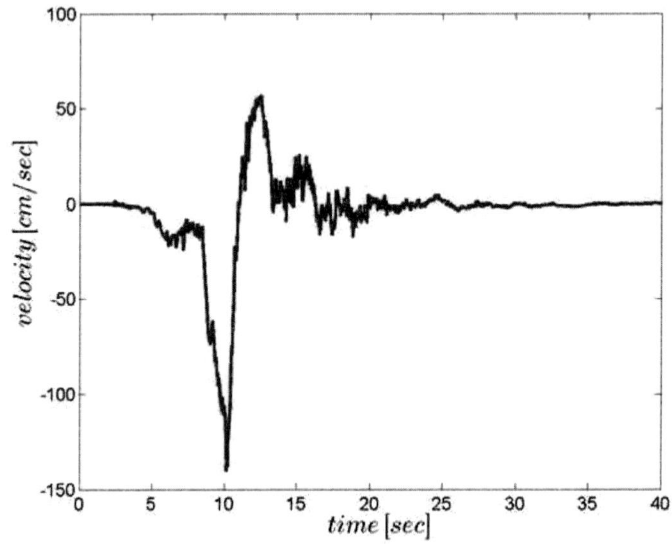

Figure 8. A directivity-related velocity pulse in the Lucerne record of the 1992 Landers earthquake (Tothong et al., 2007).

and classical PSHA do not apply in the near-source. In fact, the latter requires ground-motion prediction relationships able to capture the peculiar spectral shape driven by the pulses and an estimation of the likelihood of pulse occurrence, while the former should produce record sets reflecting the pulse features compatible with the near-source PSHA (Iervolino and Cornell, 2007).

6 Beyond first mode-spectral acceleration: advanced IMs

Several recent studies propose IMs alternative to the first-mode spectral acceleration with the main purpose of having synthetic parameters able to capture the spectral shape in a period range related to the non-linear behavior of the structure, and to overcome some of the shortcomings of $\boldsymbol{S}_a(\boldsymbol{T}_1)$. They include both scalar and vectors, linear and non-linear quantities. A detailed review is given by Tohong and Luco (2007); in their paper the authors compare the use of different scaling methods for vector-valued and advanced scalar ground motion IMs.

An example of vector-valued IM is given by $\boldsymbol{S}_a(\boldsymbol{T}_1)$ and ε, which for the reasons discussed in section 5 has been found to be an effective predictor of structural response because it is an implicit measure of spectral shape. Using the vector $\boldsymbol{S}_a(\boldsymbol{T}_1)$ and ε allows to properly take into account the effect of peak and valleys in the spectrum of the record for which $\boldsymbol{S}_a(\boldsymbol{T}_1)$ alone is insufficient (Baker, 2007). In general the main benefit provided by vector-valued IMs is increasing efficiency, which means that fewer nonlinear dynamic analyses will be needed to characterize the relationship between structural response and the IM. However, vector-valued IMs have the disadvantage that they are more complicated to use in respect to scalar IMs. Therefore, advanced scalar IMs are worth to be investigated. One successful attempt in this direction is that of Cordova et al. (2001) which propose an IM, called $\mathbf{S}^*$, which is the product of $\boldsymbol{S}_a(\boldsymbol{T}_1)$ and a non-dimensional parameter power function of the ratio of $\boldsymbol{S}_a(\boldsymbol{T}_f)$ over $\boldsymbol{S}_a(\boldsymbol{T}_1)$, Eq. (6.1), where $\boldsymbol{T}_f$ is a period accounting for the reduced stiffness of the structure because of non-linear behavior. This IM has proven to be more efficient than $\boldsymbol{S}_a(\boldsymbol{T}_1)$ as it somehow accounts for the spectral shape in a broader range.

$$\boldsymbol{S}^* = \boldsymbol{S}_a(\boldsymbol{T}_1) \cdot \left(\frac{\boldsymbol{S}_a(\boldsymbol{T}_f)}{\boldsymbol{S}_a(\boldsymbol{T}_1)} \right)^{\alpha} \tag{6.1}$$

An extension of this concept of capturing the spectral shape in a range is the so called Sa average, Eq. (6.2), which is the geometric mean of the spectral acceleration in an interval from T_1 until a final period of interest. Clearly, this IM is able to provide improved information about Sa values at a range of periods.

$$S_{a,avg}(T_1, \cdots, T_n) = \left(\prod_{i=1}^{n} \boldsymbol{S}_a(\boldsymbol{T}_i) \right)^{1/n} \tag{6.2}$$

Another advanced scalar IM for structural demands that are dominated by a first mode of vibration, is that based on inelastic spectral displacement, $\boldsymbol{S}_{di}$, of a bilinear SDOF. This IM which can be advantageous if compared to both elastic spectral acceleration and the vector consisting of $\boldsymbol{S}_a(\boldsymbol{T}_1)$ and ε. It has been show to be sufficient, 50% more efficient (50% reduction in variability of response) than $\boldsymbol{S}_a(\boldsymbol{T}_1)$, and robust in respect to scaling. Tohong and Luco (2007) shows, via IDA, for the cases investigated that, using $\boldsymbol{S}_{di}$, the dependence of the structural responses on ε is substantially reduced, which therefore avoids the need to include ε in a vector IM. $\boldsymbol{S}_{di}$ also allows to overcome the problems with near-source pulse-like records, this is because that IM *can distinguish the amount of period elongation induced by the record.*

On the other hand, $\boldsymbol{S}_{di}$ is less efficient, sufficient and robust for higher-mode-sensitive structures (i.e. long-period buildings). These shortcomings are largely due to the fact that using $\boldsymbol{S}_{di}$ alone does not capture the ground motion frequency content at higher-mode periods. However, also $\boldsymbol{S}_{di}$ can incorporate second-mode effects, for example, via a scalar modification function of the elastic spectral displacement of the second mode and the elastic participation factors of the first two modes of the structure of interest. Authors show that this advanced IM leads to an increase in the efficiency, because it reduces the conditional dispersion of the response spectra.

It is finally worth to mention that the choice of an IM should not be unrelated to the possibility of computing the probabilistic seismic hazard for that IM. In fact, computing the seismic risk via Eq. (3.3) requires the distribution of the IM over a period of time at the site of interest. Therefore, it may be said that the best candidates to be ground motion intensity measures are those for which attenuation relationships are easy to compute.

7 Code-based record selection

It has been described, so far, the main issues in real records selection and scaling for non-linear dynamic analysis with the purpose of seismic risk assessment, which requires integration of fragility and seismic hazard. To discuss the code-based procedures to select the seismic input for nonlinear time-history analysis of structures, the concept of Uniform Hazard Spectrum (UHS) has to be clarified first. The UHS is an elastic spectrum defined on the basis of the seismic hazard at the site where the structure is supposed to be located. In particular, if for such a site, curves representing the probability of exceedance of elastic spectral acceleration, for example in 50 years, are available for several oscillation periods, $\boldsymbol{T}$, in a range of engineering interest, it is possible to enter all this curves with a the same specified probability (i.e. 10%) and get the corresponding spectral ordinate, one for each $\boldsymbol{T}$ value. These uniform probability ordinates determine a spectrum if plotted versus the corresponding $\boldsymbol{T}$ values. Such a spectrum is called Uniform Hazard Spectrum and it is often used as a reference for structural assessment and design via non-linear dynamic analysis.

As, discussed above, advanced practice today (i.e. in U.S.) would find a seismologist responsible for providing input to an engineer who has to do a nonlinear dynamic assessment for design. The seismologist would provide (1) a probabilistic seismic hazard analysis, (2) for one probability level, a uniform hazard spectrum, and (3) for such level, a suite of n accelerograms for use in nonlinear dynamic analyses. Typically the seismograms have been selected to reflect the likely magnitudes, distances, and other earthquake parameters thought to dominate the hazard at the site; this choice is guided disaggregation of hazard (Bazzurro and Cornell, 1999). Finally, the records are usually scaled to match the UHS level at the period corresponding to the first mode of the structure.

In Figure 9 (top), the hazard curves for a site in southern Italy are reported for several oscillation periods in the range of 0-2sec (data provided by the Istituto Nazionale di Geofisica e Vulcanologia, http://essel.mi.ingv.it/, accessed November 2007). These curves are all entered at 10% probability and the corresponding acceleration values are plotted versus each period to form the UHS, which is displayed in the bottom panel along a set of records scaled, as an example, to the level of the UHS corresponding to T = 0.6 sec.

The engineer will subsequently run time history analyses for each of the n accelerograms in the chosen set, and observe for each one or more measures of structural performance, for example, MIDR. If the average of MIDR of the n records exceeds 7% (in a steel moment resisting frame) he may conclude that frame failure is likely given that ground motions.

Code-based procedures (i.e. in Europe) apparently try to approximate this procedure. In fact, to design using an UHS corresponding to a small probability of exceedance is, in principle, analogous to choose a conservative value of the action in the load-resistance factor design and therefore it is consistent with the common design philosophy of codes worldwide, which allows the practitioner to check the seismic structural performance in semi deterministic conditions where the action are amplified and the capacity is reduced on a probabilistic basis.

However, the use of UHS requires the seismic hazard at the site provided for all the national territory, national agencies often take care of this. In the U.S., for example, hazard curves and UHS for design may be downloaded by the USGS website (http://earthquake.usgs.gov/research/hazmaps/design/); recently also Italy has such a service by the mentioned website and it is likely that the next generation of Italian codes will allow for the use of site-specific UHS for design and assessment of structures. This is not the case for many other countries where seldom engineers are able to easily obtain PSHA data for the site of interest, therefore the record selection procedure described above is often only approximated as happens in the case of Eurocode 8 (EC8), record selection prescriptions of which are discussed below.

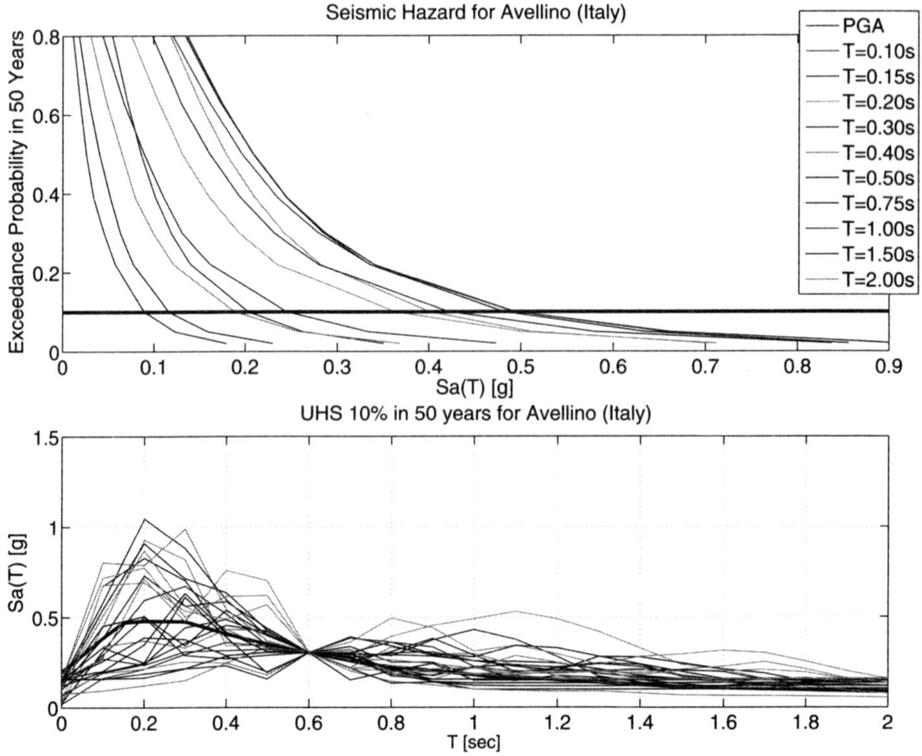

Figure 9. 10% in 50 years exceedance probability Uniform Hazard Spectrum (bottom) for the site of Avellino (Italy), from the INGV hazard data (top) available at http://esse1.mi.ingv.it/.

7.1 EC8 prescriptions for buildings and bridges

In EC8 the seismic input for time-history analysis is defined after the elastic response spectrum. In Part 1 (CEN, 2003), which applies for buildings, the spectral shapes are given for both horizontal and vertical components of motion. In Section 3.2.2 two spectral shapes, Type 1 and Type 2, are defined. The latter applies if the earthquake contributing most to the seismic hazard has surface waves magnitude not grater that 5.5, otherwise the former should be used. In Figure 10 the 5% damped elastic spectra for the five main soil classes are given as normalized in respect to a_g, which is the anchoring value of the spectral shape.

a_g is to be determined depending on the seismicity of the site of interest.

In Italy, for example, the seismic territory is divided into four zones representing different hazard levels, where seismic resistant design is mandatory only in the upper three zones. The ag values for the Zone 3, 2 and 1 are 0.15g, 0.25g and 0.35g respectively. These values are related to the probabilistic seismic hazard analysis (PSHA) for the site of interest. In fact, if the PGA (on rock) with a 10% exceeding probability in 50 years falls in one of the intervals]0.25g, 0.35g],]0.15g, 0.25g], or]0.05g, 0.15g], then the site is classified as Zone 1, 2 or 3 respectively (OPCM n.3519, 2006). It is clear, therefore, the indirect relationship between seismic hazard and the code spectrum, which may be considered a crude approximation of the

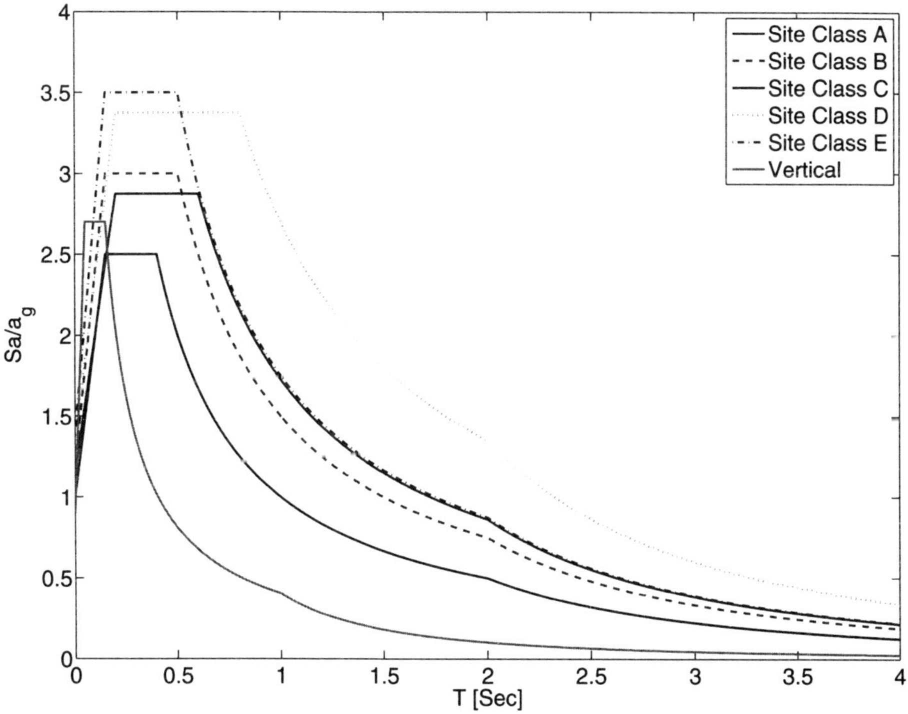

Figure 10. EC8 horizontal and vertical type 1 spectral shapes.

Eurocode 8 part 1 allows the use of any form of accelerograms for structural assessment. The main criterion the set of accelerograms should satisfy, regardless they are natural, artificial or simulated, is that in the range of periods between $0.2T_1$ and $2T_1$, where T_1 is the fundamental period of the structure in the direction where the accelerogram will be applied; no value of the mean 5% damping elastic spectrum, calculated from all time histories, should be less than 90% of the corresponding value of the 5% damping elastic response spectrum. The upper limit accounts for the lengthening of period due to the non-linear structural behavior, while the lower considers the contribution of higher modes to structural response.

Some duration prescriptions are given for artificial accelerograms, while recorded or simulated records should be adequately qualified with regard to the seismogenetic features of the sources and to the soil conditions appropriate to the site.

The set has to be made of at least 3 accelerograms, however, the code allows the consideration of the mean effect on the structure, rather than the maximum, if at least seven non-linear time-history analyses are performed. In the case of spatial structures, the seismic motion shall consist of three simultaneously acting accelerograms representing the three spatial components of the shaking; therefore 3 is the minimum number of triplets to be used. However, the vertical component of the seismic action should be taken into account only in special cases, as long span elements, not applying to most common structures. Therefore, sets for analysis of common spatial structures are made up of 14 records. The code also specifies that the same accelerogram may not be used simultaneously along both horizontal directions.

EC8 Part 2 (CEN, 2005) refers to the same spectral shapes of Part 1 in order to define the seismic input for time-history analysis of bridges. The requirements for the horizontal seismic input for dynamic analysis are somehow similar to those for building but not coincident. The main differences are: a) for each earthquake consisting of a pair of horizontal motions, the SRSS spectrum shall be established by taking the square root of the sum of squares of the 5%-damped spectra of each component; b) the spectrum of the ensemble of earthquakes shall be formed by taking the average value of the SRSS spectra of the individual earthquakes of the previous step; c) the ensemble spectrum shall be scaled so that it is not lower than 1.3 times the 5% damped elastic response spectrum of the design seismic action, in the period range between $0.2T_1$ and $1.5T_1$, where T1 is the natural period of the fundamental mode of the structure in the case of a ductile bridge. The prescriptions for the vertical component of motion in part 2 are the same as part 1, except the fundamental vertical mode has to be considered in place of T_1.

Note that EC8 Part 1 specifies that the chosen set should not underestimate the code spectrum more than 10%. The prescription of EC8 Part 2 seems to be equivalent because the SRSS, for equal spectra in the two directions, is exactly one of the two spectra times 1.4. Part 2 also allows the consideration of the mean effects on the structure, rather than the maximum, when non-linear dynamic analysis is performed for at least seven independent ground motions. The effects of the vertical seismic component on the piers may be omitted in cases of low and moderate seismicity, while in zones of high seismicity these effects need only be taken into account special cases. EC8 Part 2 has also specific prescriptions for near-source conditions. The code also prescribes when the spatial variability of ground motion has to be considered.

The EC8 criteria clearly try to relate the assigned spectrum with the hazard at the site (the code spectrum is related to the hazard for the site of interest only through the anchoring value). Moreover, prescribing that records chosen for time-history analysis should reflect magnitude and other scenario parameters, the code tries to reproduce record selection based on disaggregation of seismic hazard.

In Figure 11 an example of a horizontal natural records set matching the EC8 Part 1 spectrum for a site with ag = 0.25g in the range 0-2 sec, is given (Iervolino et al., 2008). In Figure 12 a set reflecting spectral matching criteria of EC8 Part 2 in terms of horizontal and vertical component of motion is given (Iervolino et al., 2007).

It emerges from the cited studies that the code prescriptions do not allow to control the record-to-record variability (important to estimate the response with a limited number of analyses) and also that may be unfeasible to satisfy the matching to any specific source parameter and at the same time having the required average spectral compatibility if real records are concerned. Therefore, spectrum matched or synthetic records seems to be favored by EC8 and by codes that prescribe the matching with a specific (smooth) spectral shape in general.

7.2 Beyond the uniform hazard spectrum: conditional mean spectrum including epsilon

It has been discussed that the current best practice (not yet taken in by many codes worldwide) in record selection for seismic structural assessment relies on the concept of uniform hazard spectrum. However the UHS has some limitations that should be bared in mind when using it. The UHS is built considering hazard curves independently, which do not allow to properly account for the actual probability of joint occurrence of spectral ordinates at different periods. Moreover, the low-period range of the UHS is often dominated by small earthquakes "close" to the site, while the high-period branch is dominated by far and high magnitude events (Baker

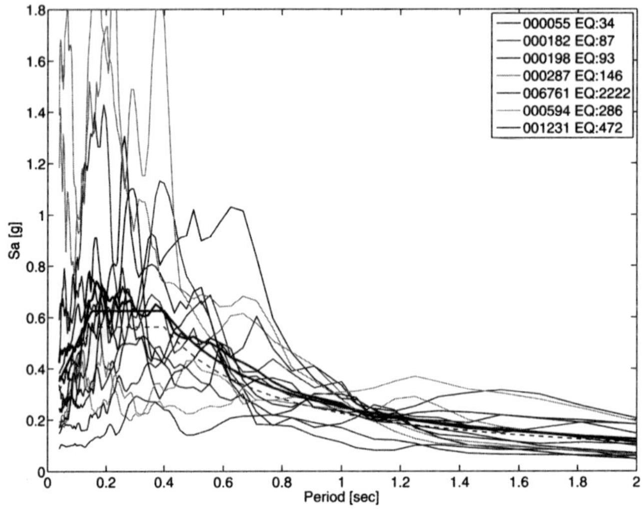

Figure 11. Example of a record set matching, in the average, the EC8 part 1 criteria for a site with ag = 0.25g (Iervolino et al., 2008).

and Cornell, 2006b). For these reasons the UHS hardly may represent the spectrum from a single threatening event, and the UHS is often quite conservative in probabilistic terms. In fact, even when a record has a spectral acceleration value as large as the UHS at a given period, it is unlikely to be as high as the uniform hazard spectrum at other periods.

A method has been proposed by (Baker and Cornell, 2006b) to develop a target spectrum which overcomes the limitations of the uniform hazard spectrum. Such a spectrum is an extension, to include epsilon, of that used in the U.S. nuclear industry and accounts for the effect of magnitude and distance from disaggregation. This spectrum is called conditional mean spectrum including epsilon (CMS-ε); the fundamental steps of the procedure to built the CMS-ε are:

1. Evaluate, from the spectral acceleration hazard curve corresponding to the fundamental mode of the structure, the value of $\boldsymbol{S}_a(\boldsymbol{T}_1)$ corresponding to the exceeding probability of interest;
2. Disaggregate the PSHA curve at the determined value of $\boldsymbol{S}_a(\boldsymbol{T}_1)$ and obtain the design (scenario) M, R, ε_{T_1} values;
3. Compute, for any period T_2 different from the fundamental period

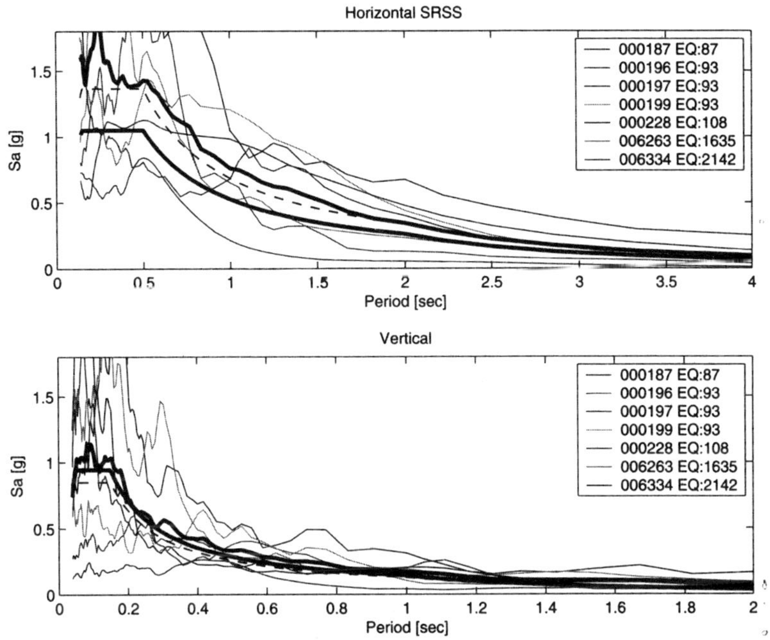

Figure 12. Example of a record set matching, in the average, the EC8 part 2 criteria for horizontal (top) and vertical (bottom) components of ground motion (Iervolino et al., 2007).

of the structure, the approximate median spectral ordinate under the assumption of jointly Gaussian distribution of the logarithms, e.g. via Eq. (7.1).

$$\mu_{ln\mathbf{Sa}(T_2)|ln\mathbf{Sa}(T_1)} = \mu_{ln\mathbf{Sa}(T_2)|\mathbf{M},\mathbf{R}} + \sigma_{ln\mathbf{Sa}(T_2)|\mathbf{M}}\rho_{ln\mathbf{Sa}(T_1),ln\mathbf{Sa}(T_2)}\varepsilon_{T_1} \tag{7.1}$$

where $\mu_{ln\mathbf{Sa}(T_2)|\mathbf{M},\mathbf{R}}$ is the mean of the log of Sa for the disaggregated magnitude and distance (from an attenuation relationship, which is usually expressed in logarithmic terms), $\sigma_{ln\mathbf{Sa}(T_2)|\mathbf{M}}$ is the standard deviation of the residuals of the attenuation at the same period and conditioned to the disaggregated magnitude; $\rho_{ln\mathbf{Sa}(T_1),ln\mathbf{Sa}(T_2)}$ is the correlation coefficient of the log of the spectral ordinates at the two periods (i.e. from Baker and Cornell, 2006c), and ε_{T_1} is the disaggregated epsilon value;

4. Repeat step 3 for all values of T_2 in the interval of interest and get the conditional mean spectrum accounting for epsilon.

In Figure 13 examples of conditional mean spectra are compared to the UHS. It is clear how the CMS-e is more probabilistically consistent than the UHS, and therefore allows to control more directly the likelihood of the performance one is designing for. Moreover, it has been shown that searching records with a shape similar to that of the conditional mean spectrum including epsilon allows to not take care in selection of appropriate magnitude, distance and specific epsilon coming from disaggregation. In fact, records with a spectral shape similar to that of the CMS-ε induce on the structure the same unbiased response than records chosen to reflect scenario M, R and ε_{T_1}. Therefore, CMS-ε seems to be a good candidate to overcome the shortcomings related to the UHS in respect of which is also less conservative.

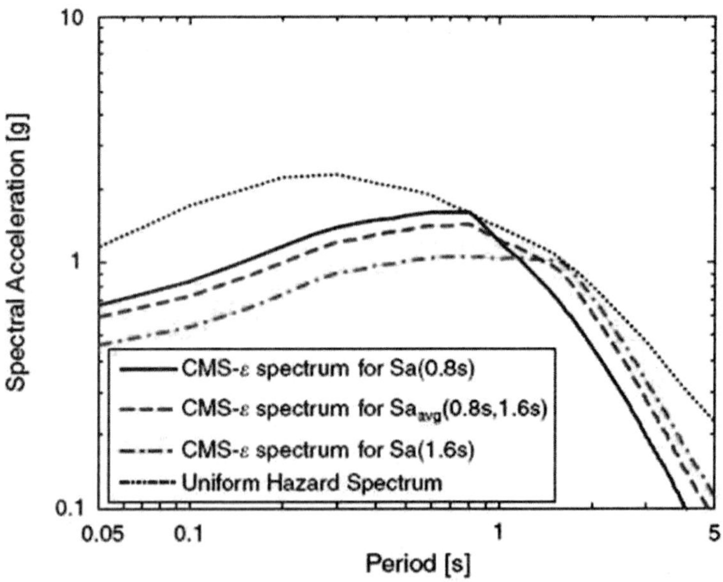

Figure 13. Examples of conditional mean spectra compared to the corresponding UHS. Courtesy of Jack W. Baker see Baker and Cornell (2005) for details.

7.3 Time scaling of records and shake table testing

It is finally to mention a manipulation procedure of real ground motion records that is nowadays not used frequently, if at all, for the computer-aided simulation of structures, while it is still popular in the shake table testing of structures if the specimen is reduced-scale. This procedure is the time-scaling of records; it consists of multiplying the time scale by a factor, similarly to what is done in amplitude scaling of records via the IM discussed above. However, while the records subjected to amplitude scaling show response spectra amplified by the scaling factor and keep the same shape of the unscaled record, the time scaling changes all properties involving time. In particular, scaling a record in time by a scaling factor $\boldsymbol{SF}$ implies that the ordinates of the original acceleration, velocity and displacement spectra have to be multiplied by 1, $\boldsymbol{SF}$, and $\boldsymbol{SF}^2$, while the abscissas have to be divided by to obtain the spectra of the modified signal (Kaldjian and Fan, 1968). When both time and amplitude scaling are performed the effects of the two procedures on the spectra should be superimposed.

Because time scaling changes the frequency content of the signal it was used when there was lack of recordings, since the earthquake engineering community has gotten more records (e.g. in the last 20 years) and some spectral matching scheme accomplishes the frequency content changes more intelligently this procedure has been almost abandoned. However, in the case of shake-table testing of structures time scaling of records may still be useful. In fact, usually shake-table infrastructures are limited in dimensions and load capacity so that it is not possible to test full size structures and reduced scale models have to be employed instead. In this case real ground motion records used as an input for tests are compressed in time by a factor equal to the square root of the scaling factor of the specimen to be tested in respect to the full-scale case. In this way the frequency content of the compressed record at the natural frequency of the undamaged reduced-scale test structure is the same as that of the un-compressed record at the natural frequency of the test structure without scaling (Hashemi and Mosalam, 2006).

8 Summary

Both the increase in easy available computing power and in the accessibility to natural ground motion recordings have enhanced the attractiveness of non-linear dynamic analysis for seismic structural assessment. In the paper the main issues and recent literature regarding seismic input preparation, when real ground motion records are concerned, have been reviewed. It emerges that, in the probabilistic structural assessment, the record selection is driven by the purpose of achieving small computational effort to estimate the response correctly and accurately. In the code-based design, in principle, the hazard is disaggregated to determine the design scenario and records are chosen to mach it.

In the probabilistic assessment, the number of records required to estimate the seismic risk may be extremely large and may be driven by many parameters. The introduction of a ground-motion intensity measure simplifies the assessment of seismic risk. To reduce the number of records to estimate the structural performance one should use a ground motion intensity measure able to capture the spectral shape in a range the structure is sensitive to, rather than try to match any event feature as magnitude and distance or faulting style. (This, procedure also takes advantage of amplitude scaling of records, while time scaling is usually used in experimental testing of reduced size specimens.)

If displacement response is of interest, it is widely accepted that first-mode spectral acceleration is a more efficient IM in respect to PGA, as it allows to select records without taking care of magnitude, distance and related features as duration. Moreover, it allows to estimate the response with a smaller uncertainty. However, $\boldsymbol{S}_a(\boldsymbol{T}_1)$ has also shown some limitations when employed in incremental dynamic analysis which may lead to an incorrect estimation of the performance in respect to unscaled records, especially for structures for which the response is dominated by a wider portion of the spectral shape, or if the hazard is contributed by near-source pulse-like events. Therefore, advanced vector-valued and scalar IMs have been proposed, an example of that is an IM which accounts for the spectral ordinates at the first and lengthened periods. Also the inelastic spectral displacement is an efficient scalar measure representing an sufficient and robust alternative to vectorial IMs. However, it requires a specifically modified hazard analysis as it happens for sites dominated by near source effects.

The more advanced code procedures for record selection and structural assessment are based on the concept of uniform hazard spectrum which has been shown to produce conservative estimation of seismic performances. A possible improvement is represented by the conditional mean spectrum which has a more sound probabilistic basis. However, often codes do not allow even the use of UHS for a number of reasons. First of all, the discussed procedures requires a large amount of information especially regarding the hazard at the site seldom available to engineers. Moreover, as they are not fully standardized, the record selection procedures involve a series of semi-arbitrary choices of the practitioner which is reasonable to implicitly control. The result is that, for example in Europe, the code-based record selection prescription are based on standard spectral shapes and are not easy to apply to real records favoring the use of spectrum matching records; although they have shown some, even significant, limitations in the estimation of the structural performance.

Bibliography

J.W. Baker. Probabilistic structural response assessment using vector-valued intensity measures. *Earthquake Engineering Structural Dynamics*, 36(13):1861–1883, 2007.

J.W. Baker and C.A. Cornell. *Vector-valued ground-motion intensity measures for probabilistic seismic demand analysis.* Technical report, Stanford University, Stanford, CA, 2005. Blume Earthquake Engineering Center Report No. 150.

J.W. Baker and C.A. Cornell. Which spectral acceleration are you using? *Earthquake Spectra*, 22(2):293–312, 2006a.

J.W. Baker and C.A. Cornell. Spectral shape, epsilon and record selection. *Earthquake Engineering and Structural Dynamics*, 35(9):1077–1095, 2006b.

J.W. Baker and C.A. Cornell. Correlation of response spectral values for multi-component ground motions. *Bulletin of the Seismological Society of America*, 96(1):215–227, 2006c.

P. Bazzurro and C.A. Cornell. Disaggregation of seismic hazard. *Bulletin of the Seismological Society of America*, 89:501–520, 1999.

P. Bazzurro and N. Luco. *Report for Pacific Earthquake Engineering Research (PEER)Center Lifelines Program Project 1G00.* Technical report, 2003.

J.J. Bommer and A.B. Acevedo. The use of real earthquake accelerograms as input to dynamic analysis. *Journal of Earthquake Engineering*, Vol.8: 43–91, 2004. Special Issue, 1.

D.M. Boore. Simulation of ground-motion using the stochastic method. *Pure and Applied Geophysics*, 160:635–676, 2003.

J.E. Carballo and C.A. Cornell. Probabilistic seismic demand analysis: spectrum matching and design. Technical report, Reliability of Marine Structures Program, Department of Civil and Environmental Engineering, Stanford University, 2000. Report No. RMS-41.

CEN. *Eurocode 8: Design of structures for earthquake resistance.* Brussels, Belgium, 2003. Part 1: General rules, seismic actions and rules for buildings, Final Draft prEN 1998.

CEN. *Eurocode 8: Design of structures for earthquake resistance.* Brussels, Belgium, 2005. Part 2: Bridges, Final Draft prEN 1998-2.

P.P. Cordova, S.S.F. Mehanny, and C.A. Deierlein, G.G.and Cornell. Development of new seismic intensity measure and probabilistic design procedure. *Proceedings of the 3rd U.S.-Japan Workshop on Performance-Based Seismic Design Methodology for Concrete Buildings*, 2001.

C.A. Cornell. Engineering seismic risk analysis. *Bulletin of the Seismological Society of America*, 58:1583–1606, 1968.

C.A. Cornell. *Hazard, ground-motions and probabilistic assessment for PBSD.* in performance based seismic design concepts and implementation. 2004. PEER Report 2004/05.

E. Cosenza, G. Manfredi, and R. Ramasco. The use of damage functionals in earthquake engineering: A comparison between different methods. *Earthquake Engineering and Structural Dynamics*, 22(10):855–868, 1993.

R. De Risi. *Valutazione probabilistica di una struttura in CA con input sismico EC8-compatibile.* Technical report, Laurea triennale in Ingegneria Civile, Universita degli Studi di Napoli Federico II, 2007.

J. Hancock, J. Watson-Lamprey, N.A. Abrahamson, J.J. Bommer, A. Markatis, E. McCoy, and R. Mendis. An improved method of matching response spectra of recorded earthquake ground-motion using wavelets. *Journal of Earthquake Engineering*, 10(special issue 1):67–89, 2006.

A. Hashemi and K.M. Mosalam. Shake-table experiment on reinforced concrete structure containing masonry infill wall. *Earthquake Engineering and Structural Dynamics*, 35:1827–1852, 2006.

I. Iervolino and C.A. Cornell. Record selection for nonlinear seismic analysis of structures. *Earthquake Spectra*, Vol. 21, No.3:685–713, 2005.

I. Iervolino and C.A. Cornell. Probability of occurrence of velocity pulses in near-source ground motions. *Bulletin of the Seismological Society of America*, 2007. (Submitted for publication).

I. Iervolino, G. Maddaloni, and E. Cosenza. Eurocode 8 compliant real record sets for seismic analysis of structures. *Journal of Earthquake Engineering*, 12(1):54–90, 2008.

I. Iervolino, G. Maddaloni, E. Cosenza, and G. Manfredi. Selection of time-histories for bridge design in eurocode 8. Pavia, Italy, 18 - 20 April 2007. Proc. of 1st US-Italy Seismic Bridge Workshop. EUCENTRE.

I. Iervolino, G. Manfredi, and E. Cosenza. Ground-motion duration effects on non-linear structural response. *Earthquake Engineering and Structural Dynamics*, 35:21–38, 2006.

F. Jalayer. *Direct Probabilistic Seismic Analysis: Implementing Non-Linear Dynamic Assessments.* PhD thesis, Department of Civil and Environmental Engineering, Stanford University, CA, 2003.

M.J. Kaldjian and W.R.S. Fan. Intensity and time scale effects of accelerogram. Technical report, Industry Program of The College of Engineering, University of Michigan, 1968.

OPCM n.3519. *Criteri per lindividuazione delle zone sismiche e la formazione e laggiornamento degli elenchi delle medesime zone.* Gazzetta Ufficiale della Repubblica Italiana, 108, 2006.

N. Shome, C.A. Cornell, P. Bazzurro, and J.E. Carballo. Earthquakes, records and nonlinear responses. *Earthquake Spectra*, 14(3):469–500, 1998.

P.G. Somerville. Magnitude scaling of near fault rupture directivity pulse. *Physics of the Earth and Planetary Interiors*, 137:201?12, 2003.

P.G. Somerville, N. Smith, R. Graves, and N. Abrahamson. Modification of empirical strong ground-motion attenuation results to include the amplitude and duration effects of rupture directivity. *Seismology Research Letters*, 68:199–222, 1997.

S. Taghavi and E. Miranda. *Response assessment of nonstructural building elements.* Technical report, Pacific Earthquake Engineering Research Center, Dept. of Civil and Environmental Engineering, Stanford University, Stanford, CA, 2003. Report No. PEER.

P. Tohong and N. Luco. Probabilistic seismic demand analysis using advanced ground motion intensity measures. *Earthquake Engineering and Structural Dynamics*, 36(13):1837–1860, 2007.

P. Tothong and C.A. Cornell. *Probabilistic seismic demand analysis using advanced ground motion intensity measures, attenuation relationships, and near-fault effects.* Technical report, Pacific Earthquake Engineering Research Center, University of California, Berkeley, CA, USA, 2006. PEER Report 2006/11.

P. Tothong, C.A. Cornell, and J.W. Baker. Explicit directivity-pulse inclusion in probabilistic seismic hazard analysis. *Earthquake Spectra*, 23(4): 867–891, 2007.

D. Vamvatsikos and C.A. Cornell. Incremental dynamic analysis. *Earthquake Engineering and Structural Dynamics*, 31(3):491–514, 2002.

Shaking Table Testing

Rogério Bairrão[†]
[†] LNEC / DE / NESDE
Av. do Brasil, 101
1700-066 LISBOA, PORTUGAL

Abstract This chapter reflects the first of the four lectures that the author has presented at Udine, Italy, at CISM–International Centre for Mechanical Sciences, in June 2007, in the aim of the course "New Approaches to Analysis and Testing of Mechanical and Structural Systems". He is being managing experimental activities in seismic engineering since 1993, when LNEC (the Portuguese National Laboratory for Civil Engineering) has started the construction of a large triaxial shaking table at the new earthquake testing hall of its Structures Department. Following a presentation of the main objectives of the Seismic Engineering experimentation, a brief state-of-the-art is made concerning its main laboratory methodologies, with a more clear focus on the shaking table tests. The leading problems related to the main testing methodologies (static, pseudo-dynamic, and shaking table) are identified and discussed. The advantages and disadvantages of each one are also highlighted. Moreover, a brief presentation of the centrifuge tests is introduced and, finally, the future developments needed are identified.

1 Objectives of the Seismic Experimentation

The paramount objective of Seismic Engineering is to contribute in deep for the policies of prevention, formation or development leading to the mitigation of the seismic risk concerning the structural damage or collapse. The safety verification of a seismic action involves several concepts (Borges, 1980): necessary conditions, performance criteria, probabilistic reliability, hazard scenarios and quality guarantee. In spite of several scientific areas being under constant development, it is still nowadays very common to get important knowledge following the occurrence of each new catastrophic earthquake in a developed zone of the globe. In fact, there is no better seismic testing "laboratory" than the real place where an earthquake takes place, because it is absolutely impossible to reproduce artificially, and simultaneously, all the concerned phenomena. Anyway, and despite the wide numerical methods and computer capability developments, the tests will be always indispensable because they are the only way to reproduce seismic actions allowing its systematic repetition and the necessary scientific observation of their effects. Consequently, the main objective of the Seismic Engineering experimentation is, basically, to allow the observation and analysis of the damage and behaviour of different types of structures, under different conditions, while submitted to a seismic action. In fact, the response of a structure to a seismic action is influenced by innumerable parameters whose implications must be widely analysed. For instance, not only the characteristics of the seismic action itself, but also the foundation and bordering structures interaction, the structural and material characteristics or the already existent damage induced by previous events.

Seismic engineering experimentation is also indispensable for the validation of analytical models and for the verification of new design methodologies or new strengthening or reinforcement structural techniques. The study of new materials behaviour and the analysis of new methodologies still in a pre-normative phase need also the use of experimental tests. It should be noticed too that, apart the multiple aspects already referred, experimentation with earthquake simulators provides the certification of several types of equipment in order to ensure their proper way of functioning during the occurrence of a major earthquake. In fact, shaking table tests of important equipment are quite common; mechanical, electrical and electronic equipment of various types as main-stations and sub-stations material, nuclear power plants control units or hospital vital devices, just to quote a few of the more relevant examples. Moreover the already referred activities, of seismic engineering itself or equipment certification, civil protection authorities can also use earthquake simulators for didactic campaigns. This utility is a role very disseminated in countries, like Japan, where the occurrence of earthquakes felt by all the population is part of their common life. Without the purpose of exhausting, there is however another aspect of the seismic engineering experimentation that should be mentioned; that is its use in order to evaluate the seismic vulnerability of different types of the building stock, with obvious applications in the aim of the insurance companies activities. Consequently, it is of paramount importance the improvement of the seismic engineering experimental conditions in order to raise new techniques to pay a better and clearly more important role in the seismic risk mitigation. In the following, the main methodologies currently used and their main limitations will be presented and discussed.

2 Laboratory Tests

In this chapter are successively presented, and commented, the more important laboratory methodologies in seismic engineering: quasi-static tests, pseudo-dynamic, centrifuge and, finally and with a clear accent on them, the dynamic simulator tests or, as they are more commonly referred, the shaking table tests. Other types of tests are mentioned very briefly because they are far from the context of the present text: the usual in-situ tests and a new type of experimental installation still under development phase in the USA that will simulate in-situ real strong ground motions (Armer, 1998). Concerning the common in-situ tests, the excitation can be induced by actuators or by eccentric mass shakers. In general the main purpose of these tests is the determination of the dynamic characteristics of the structures under analysis, for a linear behaviour, being usually obtained a correct identification of their main vibration modes (frequency, damping and shape).

This type of tests is basically limited by the maximal forces or by the hydraulic characteristics of the available equipment. So, trying to avoid that difficulty, another methodology has appeared to induce dynamic forces in a real structure of large mass and dimension: the use of explosive devices. When compared with the laboratory methodologies, presented in this chapter, the in-situ use of explosives has the advantage of considering a real soil-structure interaction but it does not allow the reproduction of real earthquake inputs at the level of the structure footing. In fact, it is not possible to generate with a single explosion a previously defined displacement history. It should be also mentioned that those maximal displacements are too small, the duration is too short and the characteristics of the vertical component of such an input is very different when compared with a typical real earthquake one. Consequently, this methodology can not be a good alternative to the laboratory tests when it is wished to analyse the response of a structure to a very well defined seismic input. Anyway, it should be noticed that in-situ tests are extremely useful because they allow the evaluation of the modal characteristics of real large and important structures like, for instance, entire buildings, bridges or dams. However, the analysis of the continuum behaviour of those structures allows an important evaluation of their response to solicitations more frequent than an earthquake, as for example wind forces or vibration induced by road or train traffic.

Concerning the test of very large structures, there is a new type of experimental installation still under development phase in the USA, as mentioned before. It was designed by Bechtel Corporation and the Nevada Testing Institute (NeTI) and it is an in-situ earthquake ground motion simulator capable of measuring soil-structure interaction as well as applying loads to large scale structures through a soil foundation. This very important equipment is basically a "ground shaking table" of absolutely uncommon dimensions: a "soil box" with a squared surface of 46x46m being 23m deep. Real structures will be erected over foundations built in the "soil box". Real strong ground motions will be generated using the RESCUE technique "Repeatable Earth Shaking by Controlled Underground Expansion" (Mote et al., 1998) developed by SRI International. It employs a complex explosive technique based on a matrix of controlled explosions that induce the soil movement through the lateral expansion of a grid of steel membranes. The main objective of this new technique is to allow a better seismic simulation of very large structures, real ones or at a very large scale, taking into account real soil-structure interaction. The future will show if that desideratum will be reached. An obvious handicap is the huge cost to be spent, in the equipment itself and also for the performance of each test.

2.1 Static Tests

The static tests, or quasi-static, are developed through a history of input displacements, or forces, previously defined. This kind of tests may be applied to a full structure, to structural elements or to a sub-structure. For the first case it is usual the use of a reduced scale, whereas when it is not a prototype under analysis it is more common to test specimens at real scale or very close to it. One of the advantages of testing full scale specimens, beyond the absence of all the problems concerning the scale effects, important issue that will be later fully discussed while presenting the shaking table tests, is the fact that it is very easy to get all the constitutive materials. In such cases the specimens can be built with, for instance, real concrete, steel or masonry easily available. Usually the deformation is the parameter under control for this type of tests, because to define a displacement history appropriate for the study under development, like it will be exemplified later, is easier than to foresee with exactitude an equivalent history of force. In fact, and often under non-linear conditions, it is very difficult the definition of a correct and convenient force history.

For safety reasons it is also preferable to use a displacement history, to avoid a possible accident if a brittle, and unexpected, rupture occurs. In such a case, and for a force-controlled test, the natural consequence is an actuator looking for an inappropriate position to get the necessary resistance for the application of the demanded force. There are innumerable possible configurations to perform static tests. A very important aspect to be considered for the static, and quasi-static, tests, is the stiffness of all the accessory structures. Particular attention must be paid to the reaction structures to be used to support the actuators that will impose the history of displacements. If the stiffness of these structures would not be adequate, it would introduce spurious errors of measurement because the effective history of displacement imposed to the specimen would be the one reproduced by the actuator deduced by the non neglectable one allowed by the reaction structure.

A possible example of static test is the behaviour analysis of a column with a constant vertical axial force, simulating the weight imposed to the real structure, simultaneously submitted to a horizontal displacement history. For this kind of tests, and to characterise correctly the P-Delta effects, it is important to provide a very good synchronisation between the history of displacements demanded at the top level of the column and the movement imposed to the reaction structure where the vertical constant force actuator is supported. The more basic static tests are monotonic and, in seismic engineering, they are used to obtain a standard for later cyclic quasi-static tests or even for dynamic tests. For instance, previously to a series of cyclic tests, it is extremely useful the accomplishment of monotonic tests to analyse the behaviour of a structural element under study and to define the more appropriate displacement histories to be applied during the cyclic tests. In fact, the behaviour of an element while submitted to a seismic input is mainly connected to its energy dissipation ability, up to the rupture. Its resistance and its capacity of deformation are the two main characteristics involved.

A stress-strain diagram of a monotonic test of a given element may be adopted as an involving curve of the diagram that would be obtained if the same element would be submitted to a succession of charges and discharges. Consequently, it is quite obvious the advantage of the performance of monotonic tests to define the values to be imposed for posterior cyclic tests. The results obtained through cyclic quasi-static tests pay, in turn, an important role in the preparation of pseudo-dynamic test campaigns that will be later referred. There are two main limitations of the quasi-static tests, concerning the behaviour of an element under a seismic solicitation; the fact of being accomplished under strain velocities clearly slower than the imposed by a real earthquake and the fact of not being used displacement histories close to real earthquake ones. Concerning the reduced strain velocities, being an indeed very relevant aspect for the pseudo-dynamic tests, the raising questions will be later discussed.

In the following we will see the more important aspects to be weighed while defining a displacement history to be adopted for a cyclic test. This history of displacements should be representative, as close as possible, of the response of the structural element to a seismic action and so it should be composed by a series comprising large cycles of charge, discharge, charge in the opposed direction and new discharge. It is however clearly known that the response of a specimen is absolutely influenced by the shape of the adopted cycles. So, in the absence of an ideal solution, the choice of the history of displacements to be adopted must be in the dependency of the precise objectives of the specific case under study.

The standard values for the displacements are usually associated to the ones previously defined by monotonic tests, as for instance to reach very precise yielding or rupture deformations such as spalling, longitudinal steels traction rupture, steel-concrete slip rupture, etc. Based on accurate values, relevant for each test under consideration, several examples of deformation histories effectively adopted for cyclic quasi-static tests can be found. The more common ones show simple displacement symmetrical cycles of progressive amplitudes, or with the repetition of two or three cycles for each amplitude value. In the following is shown an example used at the seismic engineering laboratory of the Structures Department of LNEC. It was the case of a test campaign of reinforced concrete plane frames, with masonry infilled panels, submitted simultaneously to vertical axial forces on both columns and to a history of imposed displacements at the frame plane and at the top beam level. It was decided to apply two very different types of history of displacements.

For the first series (Pires et al., 1995, and Combescure et al., 1996) it was considered a symmetrical displacements history with a constant maximum amplitude, but with an alternation of the velocity values and considering also an intermediary cycle of discontinuous increase. For the second series (Pires et al., 1998) successive histories of constant amplitude sinusoidal displacements were adopted. For both series, increasing values of amplitude were applied for the successive tests phases. For the first series, constant frequencies were kept throughout the tests (1 Hz for the fast cycles and 0.03 Hz for the slow ones) with the consequent increasing in velocity while the amplitudes were intensifying (from about 1 to 100 mm). For the second series, it was decided to keep a constant velocity (about 5 mm/s) what, for an amplitude increasing similar to the previous series, induced the need of a progressive frequency decreasing (from 1.326 Hz to 0.008 Hz). For the first series a conventional setup was used (Figure 1). For the second series it was decided to adopt a new setup, adapting a large shaking table facility as an actuator to perform quasi-static tests (Figure 2). For the first case, the setup can be basically described as a specimen, together with a set of reaction structures, dully fixed to a rigid floor. One actuator is attached to each column to impose constant vertical axial forces and a third actuator, at the top beam level, applies a history of horizontal displacements.

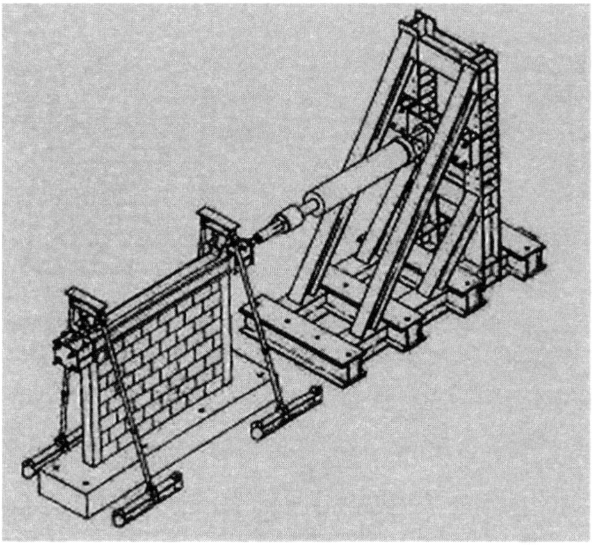

Figure 1. Conventional setup for static tests.

For the second case, the specimen and a reaction structure to ensure the constant vertical forces at the columns were fixed to the platform of the shaking table. View the existence of reaction walls close to the earthquake simulator, it was built a specific reaction structure to be fixed to one of these walls and a connecting rod to link it to the top beam level of the frame to be tested. With such setup, and applying the history of displacements to the footing of the specimen, it was possible to perform tests equivalent to the described for the first case. For that it was however compulsory to control the triaxial shaking table just in one transverse direction and under an indeed slow velocity when compared with an usually fitted for seismic tests.

In conclusion, we can declare that, although its limitations, the quasi-static cyclic tests are quite used in seismic engineering, and successfully, for the validation of analytical models of structural elements and for the evaluation of parameters to be used during posterior and more complex experimental activities.

2.2 Pseudo-Dynamic Tests

During the course where this text is inserted, a colleague has presented his lectures exactly on finite-elements based simulations and continuous

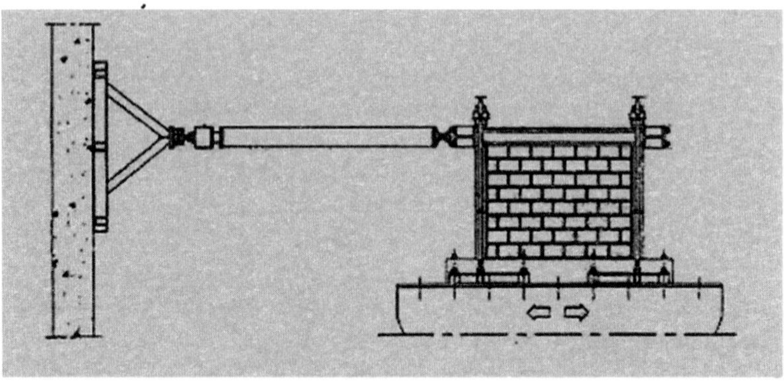

Figure 2. Using a shaking table to perform static tests.

pseudo-dynamic testing (Pegon and Queval, 2004). Consequently no details are now given related to the algorithms used in this methodology but just a brief description of this kind of tests and on their main advantages and disadvantages. The pseudo-dynamic tests methodology was initially introduced in Japan (Takanashi, 1975) at the Institute of Industrial Science, of the University of Tokyo and at the Building Research Institute of the Japanese Ministry of Public Works. At that time it was known as "On-line Computer Controlled Method of Dynamic Testing of Structural Systems".

This kind of tests is performed quasi-statically, computational methods and experimental measures being simultaneously used to generate a realistic simulation of a dynamic response of a specimen (Figure 3). The innovation of this methodology consisted in the way of solving the dynamic equilibrium equations; a step-by-step numerical integration with the following characteristics:

1. The terms relative to the inertial forces and to the viscous damping forces are computed numerically by any of the ways usually accepted.
2. The terms relative to the restitution forces and to the imposed forces are obtained experimentally, through appropriate measurements at the specimen under test.

This way of getting the terms relative to the restitution forces and to the imposed forces is very advantageous because it allows an efficient solution for the main problem inherent to solve the system of equations by an exclusive numerical methodology. In particular when the structural behaviour is directly related with strong phenomena of physical non-linearity, it is of paramount importance the evaluation of the stiffness characteristics of the specimen.

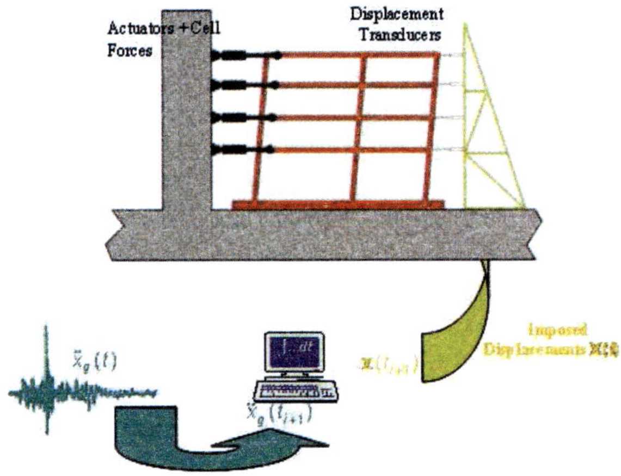

Figure 3. Pseudo-dynamic tests methodology.

In parallel, the meaningful time spent with the predictor-corrector algorithms used when this evaluation is obtained by numerical methods is very considerably reduced because these algorithms operate on parameters effectively measured at the physical model and not on their estimates. Moreover, with this methodology, the hysteretic damping, as a consequence of the inelastic deformations and the damage accumulation, is automatically taken into account thus resulting in realistic values of dissipated energy. So, a pseudo-dynamic test can be basically considered as a static test under an imposed displacement. However, it should be noticed that the use of control algorithms allowing a redefinition of the displacement history during the test itself, is extremely useful because it allows, for instance, considering the inappropriate response of a structure motivated by undesirable characteristics of the physical model.

Another very important aspect is the fact of the inertial forces being obtained numerically and not from the action of the test itself, as occurs with the dynamic experiments. Such procedure allows the adoption of a test duration that, for instance while intending to evaluate structure behaviour during a seismic action, has nothing to do with a typical less than one minute input usually used for the shaking table tests. A characteristic pseudo-dynamic test duration is several hours and that is why the "pseudo-dynamic" designation. Not being performed in real time allows the accomplishment of tests of indeed very heavy structures, and under severe inputs, without the necessity of a large hydraulic system. In fact, a relatively modest hydraulic pack is enough, when compared with the one that would be necessary to perform an equivalent dynamic test.

Another significant characteristic of this kind of tests is the possibility to pause them, at any time, waiting for an order to input the next step of displacement to be imposed to the specimen. This is an extremely advantageous feature because it allows a stopover during a test whenever it is useful, for reasons directly related to the test itself or for external ones. Therefore pauses are frequent during pseudo-dynamic tests, thus allowing all the common experimental activities checking and also, for instance, the change of some sensor gains or even their substitution if deteriorate or when their saturation levels are dangerously close. As a consequence, an additional big advantage of the pseudo-dynamic tests is the possibility to conclude them exactly before a predictable occurrence of an undesirable event, such as, for instance, an important rupture or collapse.

The most important experimental facility in Europe, able to perform large pseudo-dynamic tests, is the European Laboratory for Structural Assessment (ELSA) at the Ispra Joint Research Centre, in northern Italy. This facility has a very large reaction wall 4m thick, 21m wide and 16m height. The main side of its strong floor has the same width and is 25m long. Its load capacity allows a 200MNm bending moment and a 20MN horizontal base shear. Additionally, the strong floor grants a 240MNm bending moment. Both, the reaction wall and the strong floor, have a regular 1m square grid allowing anchor loads of 500kN axial force. Concerning its main hydraulic characteristics, the facility has a power pack with a 1500l/min flow at 21Mpa. The set of available actuators can provide loads up to 1MN and strokes up to 1m.

Although its success, the pseudo-dynamic methodology has also its drawbacks. The main problems are errors induced by the adopted mesh to discretize and shape the structure and by stress relaxation effects caused by the tests being performed during a very long duration. Another problematic issue is its very high sensitivity to control and measurement errors. Concerning the errors introduced by the adopted finite-elements mesh, beyond a necessary care to insure that both models (numerical and test-specimen) describe identical phenomena, another aspect is also essential. The physical limitations created by the maximal number of points where it is possible to install an actuator, or other important piece of equipment, must allow a proper description of the test-specimen dynamic characteristics. Such a characteristic of the pseudo-dynamic tests makes them inapt to analyse structures presenting a too uniform or dispersed mass distribution, because a mass matrix considering just their degrees of freedom experimentally reachable does not describe them adequately.

Another drawback of this kind of experiments is the impracticability of taking into account the vertical component of the dynamic inputs. Being very slow tests, they are also inadvisable for the global dynamic analysis of any structure composed by materials whose behaviour should be influenced by the strain rate, because the obtained results are clearly different when compared with the real ones. As previously referred another important matter for this type of tests is the sensitivity to control and measurement errors. In contrast to a conventional analysis (where the step-by-step errors are just of numerical nature and directly associated with the integration of the equations of movement) for the pseudo-dynamic tests the data set of each step is also influenced by the response of the experimental equipment. Consequently they become very sensitive to the accumulation effects of the, despite very small, eventual experimental errors. This fact was immediately revealed by the pioneer (Shing and Mahin, 1983), but also in more recent works (Negro and Magonette, 1998).

These kind of errors can easily occur when induced by the performance of the hydraulic power pack, responsible for the imposition of a very well suited history of displacements. Indeed, an improper displacement input for a given step, its too inaccurate measurement, or an imprecise force actuator one, induces next step false initial conditions and a consequent quick error propagation, leading to absolutely erroneous global results, although a small step error. Another hypothesis are errors induced by the spurious energy introduced at the system through high frequencies that my improperly excite high vibration modes of the specimen. Consequently, in seismic engineering, the pseudo-dynamic tests are more useful to analyse the behaviour of structures with a mass mainly concentrated in very well defined zones. In Ispra, at the ELSA laboratory, several seismic tests of reinforced concrete building and bridge structures have been under continuous development since the beginning of the last decade (Donea et al., 1996).

For the buildings, the masses have been considered as concentrated at the floor levels and the slabs as rigid in-plane diaphragms, thus allowing controlling just the generated horizontal displacements. For the bridges, the masses have been considered as concentrated at the deck level being the imposed displacements also horizontal and applied at the top of the columns. On the contrary, when it is foreseen the analysis of a structure with a much more distributed mass as it is the case, for instance, of a masonry wall, it becomes almost impossible to use the convenient number of actuators. In fact it would be too large to enable a correct simulation of the response of a displacement imposed at the footing level and perpendicular to the wall plane.

View the hybrid characteristic of the pseudo-dynamic tests (a combination of experimental and numeric procedures) the intention to develop a substructure-testing technique has appeared since the very beginning (Dermitzakis and Mahin, 1985). This technique is particularly attractive for the case of structures in which the non-linear phenomena occur at very localised zones, thus allowing an a-priori prediction of their exact location. In such way, the most important part of the structure is the only zone to be experimentally tested, and just for what concerns the stiffness matrix, being all the rest of the structure modelled by numerical methods. It is important to mention that this procedure allows no experimental limitations for the substructure to be just analytically studied. Consequently, its number of degrees-of-freedom can be extremely higher than the number considered for the substructure whose displacements are in fact controlled during the test.

The use of substructure techniques for pseudo-dynamic tests makes essential a parallel processing allowing a step-by-step exchange of data. One process being responsible for the pseudo-dynamic algorithm, controlling the physical test of one substructure, and another being responsible for a finite-element software, computing the behaviour of the other substructure. The pseudo-dynamic tests, being particularly useful to analyse the behaviour of very large mass structures, if allied to substructure-testing techniques, can also be very advantageous to study very big structures because very significant parts of those structures do not need to be physically present at the laboratory. This is of meaningful significance because it stands for the adoption of bigger scales with a consequent closeness between the specimen and the real structure.

Bridge structures are a paradigmatic example of this technique usefulness by admitting that the decks behave essentially in linear regime and that the non-linear phenomena occur exclusively at the columns. Following that assumption, it is possible to model a deck in a way totally analytical being just necessary to test physically at the lab the real part of the bridge where the non-linear phenomena are supposed to occur, the columns. Figure 4 shows a scheme of such a test performed in Ispra (Pinto, 1998). Another important attribute of this type of tests is being very well suitable for the analysis of long structures while submitted to time-delayed actions. Input motion variability can happen, for instance, at the successive columns of a long bridge during an earthquake event.

2.3 Centrifuge Tests

Following the intention to present the main laboratory methodologies of seismic engineering research, a brief presentation of the centrifuge tests is now made. Using the centrifuge forces, this kind of equipment can simu-

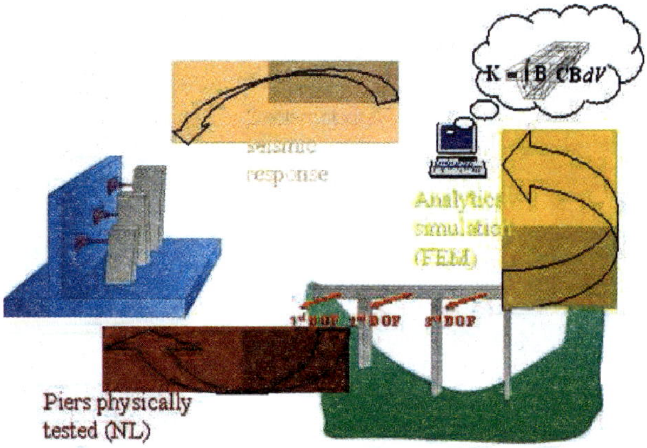

Figure 4. Example of a pseudo-dynamic test with sub-structuring.

late artificial gravitational fields very useful in geotechnical science research. Fitting the centrifuge velocity, and in accordance with the model/prototype similitude laws (to be fully discussed at the following chapter), it is possible to test a specimen at a scale $\underline{n}$ times smaller under an acceleration $\underline{n}$ times bigger than our common gravitational field. Consequently, rotating a physical model under an acceleration $\underline{ng}$ it is possible to reproduce the real stress-state of the prototype, and consequent behaviour, under good similitude conditions.

This technique allows the accomplishment of indeed very reduced scale tests, inducing however large and multiple difficulties to create the necessary structure specimens. The too small sizes of the constitutive materials are a very specific question. Therefore, the centrifuge tests are mainly used for research activities in the aim of the analysis of soil mechanics behaviour and foundation engineering.

At LNEC there is an illustrative centrifuge. This equipment comprises a rotating arm having at one extremity a dumping basket where the specimen-charge is introduced and, at the other extremity, an adjustable weight to counterbalance it. It presents a 1.55m radius measured from the centre of rotation to the basket centre and a 1.80m radius to the assembling platform, allowing a maximum 345rpm speed. The higher dimensions of the admissible charge are 500 x 700 x 500mm. The maximum weight of this charge is a function of the desired acceleration, decreasing linearly from 4kN, for a maximum acceleration of 100g, to 2kN, for a maximum acceleration of

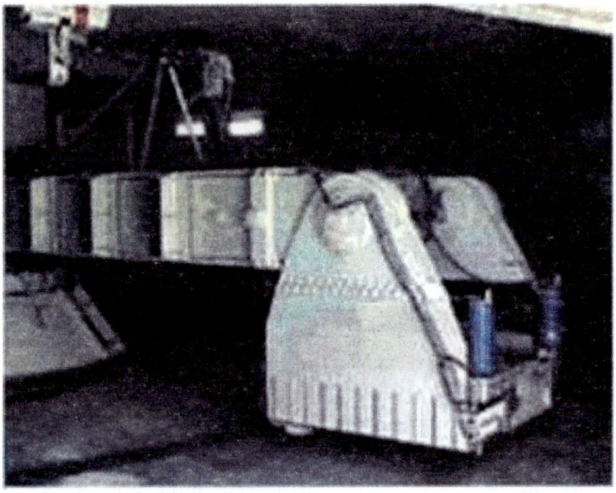

Figure 5. Detail of a centrifuge.

200g (Portugal et al., 1995). At the top of a pyramidal structure there are two rotating joins, a fluid and an oil-hydraulic one, allowing the access to the charge and the operation of the rotating arm automatic balance system. Additionally, it is very useful the presence of a video camera inside the dumping basket thus making possible to follow the behaviour of the specimens during centrifugation.

2.4 Shaking Table Tests

Introduction A control system and a testing platform, as rigid as possible and usually known as shaking table (where the specimens must be fixed) are the essential components of an earthquake simulator. During the tests, the testing platform must be moved by a set of servo-controlled hydraulic actuators. There is large number of possibilities for shaking table conceptions, mainly defined by the number of degrees of freedom under control. Consequently, there are shaking tables having from a single degree of freedom (typically, a horizontal translation) up to the more complex where the six degrees of freedom are simultaneously controlled (three translations and three rotations).

The uniaxial shaking tables are obviously the simulators of easiest conception. They have usually a very simple rail guiding system that annuls efficiently any undesirable motions and they are useful for the cases where a single direction of imposed displacements history is wished. For these cases it becomes possible to avoid the control of any spurious motions in the other directions that are never absolutely absent and may cause significant errors while interpreting the test results. At the biaxial shaking tables the two horizontal translations or the vertical and one horizontal are controlled. The actuators are usually placed in a position allowing also to control the yaw rotation, for the first case, and the pitch, or roll, for the second one.

The shaking tables with six degrees of freedom are the most versatile but also the more problematic because they need a good and simultaneous control system of the three translations, the three rotations and all their interactions. For the easiest configurations, these shaking tables have a system of eight actuators, being four of them vertical, one at each corner of the platform, and being the other four horizontal and in a star-shape position. The LNEC triaxial shaking table has a unique conception, different from the above mentioned, and is presented at the following chapter. Although the shaking tables be just one of the types of available equipment to be used for the seismic engineering analysis, they are very essential because they allow a repeatable behaviour investigation of complex structures, of relatively large dimensions, while submitted to real seismic inputs.

The more important limitations of these tests are clearly the ones generated by the maxima hydraulic and geometric characteristics of the facility. These constraints cause the need to accomplish the test of scale-reduced specimens with a consequent problematic that will be herein after discussed. The only way to reduce these limitations is to build bigger and of better performance shaking tables. To give an idea of the performance range of the equipment to be referred during the next chapters, it is now presented a synthesis of peak values recorded during relatively recent and important earthquake events, in Japan and in Portugal. Concerning Japan, Table 1 shows some values recorded during the Hyogo-Ken-Nanbu great Hanshin earthquake, Richter magnitude 7.2, occurred at Kobe area on January 17th 1995 (INCEDE newsletter, 1995). This event is chosen because it is simultaneously relatively recent and considering the fact that there is an indeed very large number of credible records because it took place at a very abundantly instrumented zone. It should be noticed the huge values that were recorded. Concerning Portugal, Table 2 shows the maxima acceleration values recorded at Horta, during the important Central Azores Islands earthquake, Richter magnitude 6.0, occurred on July 9th 1998 (Oliveira et al., 1998). This event is chosen for the same reasons as the Japanese one, but at national level. These records obtained at the Azores Islands were very significant. It should be referred that these acceleration values were more than the double of the prescribed by the Portuguese seismic design code, for the area and soil type where they were in fact measured

Table 1.

	Maritime Observatory	Higashi-Kobe Bridge	JR Takatori Station
Displacement H / V(mm)	210 / 110	490 / 150	420 / 60
Velocity H / V (cm/s)	90 / 40	91 / 35	138 / 20
Acceleration H / V (g)	0.82 / 0.33	0.33 / 0.40	0.67 / 0.29

Table 2.

	Prince of Monaco Observatory
Acceleration H / V (g)	0.39 / 0.32

The LNEC Shaking Tables Table 3 summarises the main characteristics of the three shaking tables of the seismic engineering laboratory of the Structures Department of LNEC. Moreover these basic figures, useful to allow an evaluation when later compared with other equipment, should also be referred some particular aspects with a more accurate attention concerning the triaxial shaking table. The LNEC uniaxial shaking tables have been successfully operating over thirty years with an adaptive control system developed in-house. It is a single simulator, with two platforms, that can be used, alternatively, to input horizontal or vertical displacement histories, being however compulsory to change the position of its single 200kN actuator. They are both steel platforms oblong shaped. The general design of

Table 3.

	Triaxial	Horizontal	Vertical
Platform Surface (m)	5.6 x 4.6	3.2 x 2.2	2.6 x 1.6
Platform Mass (ton)	40	3	1.5
Specimen Dead Weight (kN)	600	60	20
Specimen Height (m)	10+	8	7
Translations	X / Y / Z	X	Z
Rotations	0	0	0
Displacement X / Y / Z (mm)	±175	±100	± 20
Velocity H / V (cm/s)	70 / 42	21	21
Bare Harmonic Accel. H / V (g)	2.0 / 1.0	4.6	9.2
Frequency Range (Hz)	0-40	0-25	0-25

the LNEC triaxial shaking table was conceived in-house, as well as its platform and very special guiding system (Emilio et al., 1989). A new testing hall was built, the characterisation tests took place seven years later (Costa et al., 1996) and immediately after those certification experiments it started to operate successfully. This shaking table has a steel platform with about

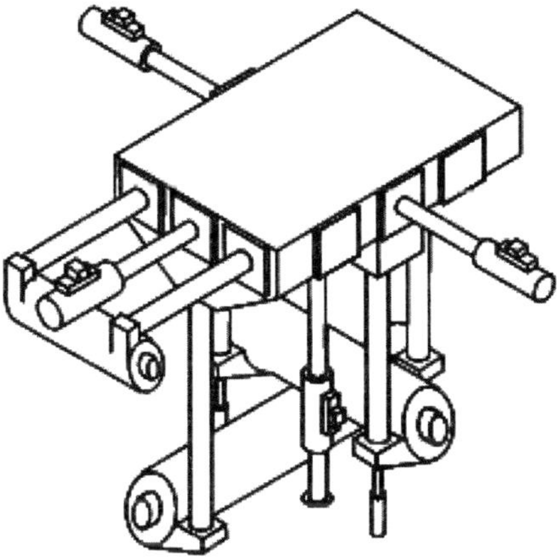

Figure 6. LNEC 3D shaking table scheme.

40 tons, triangular prismatic shaped and strongly rigidified inside. This piece of equipment is driven by a set of hydraulic actuators: two 300kN in the transverse direction, one 1000kN in the longitudinal and one 300kN in the vertical. There is also a pair of additional 500kN vertical passive gas actuators to ensure the static compensation of the platform/specimen dead weight.

These two additional actuators are controlled by a totally independent system, by means of nitrogen injection, for each specific case until the exact value of needed pressure be reached. In the hypothesis of testing an indeed very heavy specimen just submitted to horizontal displacement history components, it is foreseen to disconnect the dead weight compensation system and to absolutely fast vertically the platform using rigid blocks. A very unique, and dully patented, characteristic of this shaking table is to allow the simultaneous control of the three translations while mechanically preventing each of the three rotations. This feature is achieved through the action of a three torque-tube system of large dimensions, as schematised in Figure 6. These torque-tubes are structures 20mm thick and with large diameters: 1.2m to prevent the roll rotation, also 1.2m for the pitch and 0.8m for the yaw. This last one is a little smaller because the shaking table is expected to deal with higher overturning moments than torsional ones. This basic configuration can be easily changed to perform tests under non-

3D displacement histories. In those cases it is possible to block, for instance, the longitudinal translation, preventing the rotation of the yaw torque-tube by fixing it to the wall, as can be seen in Figure 6.

The advantages of such an immobilisation are basically two. It really annuls the spurious rotations, because no matter the high stiffness of the torque-tubes they have always some flexibility and, on the other hand, it allows the control of just the actual directions of the desired displacements. Concerning this last one, if the axes that should remain unmoved would not be blocked it would be necessary to control them simultaneously by imposing null displacement histories inputs. To compensate the flexibility of the torque-tubes, it could be designed three pairs of additional actuators (one pair for each rotation) to be inserted in each of the six connecting rods placed between the torque-tubes and the platform. However, these new actuators would also decrease the global system stiffness and would have to allow very high velocities. To avoid a large stiffness decrease they would need high diameter pistons that together with high velocities would involve an important hydraulic supply. Additionally to the digital control of the shaking table, there is a parallel safety mechanical system to limit physically the displacements of each axle, independently of the limits defined by the computer control system. In spite of the main characteristics summarised in Table 3, where the maxima available displacements for each direction are declared as $\pm 175mm$, it is now interesting to comment these values. These figures, like the ones of other characteristics, are never absolute but dependent of each very particular case. It is a team decision following the knowledge of each specific condition, to ensure that, even under a potential accident, it would be still possible to stop all the actuators before hitting their limits.

Just as an example, for these very easily measurable characteristics, the author of this text has decided to check, as carefully as possible, the real six maxima displacements of the LNEC triaxial shaking table, after disconnecting all the safety systems. The values found were far from the theoretical actuators strokes defined by the manufacturer ($\pm 200mm$). In fact, the measures obtained were +192mm; -186mm in the transverse direction, +191mm; -183mm in the vertical and +187mm; -199mm in the longitudinal. Concerning the power pack, it has three pump groups, each with 110kW, and an additional small one of 22kW, for boost purposes. This system allows a 690l/min oil flow under 210bar. A few years ago, an important upgrade of the hydraulic performance was accomplished to achieve the peak velocities mentioned in Table 3. To obtain these velocities with the same power pack it was decided to use an additional bank of nitrogen accumulators and to replace the previous servo-valves, piping and substations by higher capacity ones. In particular, to reach a velocity of about 40cm/s in the longitudinal direction, it was even necessary to drill the body of the 1000kN actuator working in that direction to install simultaneously two 770l/min servo-valves. In fact, having a 604.5cm^2 piston section it is needed an about 1450l/min oil flow to attain that velocity. With the new bank of nitrogen accumulators it is possible to reach those oil flows for short periods of time, in spite of keeping the old 690l/min power pack flow.

It should be also referred that an overhead crane with 400kN capacity and a very low speed control allows the construction, installation and removal of large specimens inside the laboratory. Another important feature is a floor to ceiling testing hall height of 10m enabling to analyse the seismic behaviour of relatively tall structures.

Shaking Tables in Europe The CASCADE Research Training Network was recently developed under the support of the European Commission, CASCADE being the acronym for Cooperative Advancements in Seismic and Dynamic Experiments. Several reports were published, being summarised in Table 4 the main features of the largest shaking tables in Europe then identified (Taucer and Franchioni, 2005) completed with additional information obtained directly from some of the researchers responsible for the facilities (Bairrao and Severn, 2007) and from the Internet institutional sites.

Table 4.

Institution	Location	Country	DF	Payload kN	Area m^2	Disp mm	Vel cm/s	Acc m/s^2	Freq Hz
BRISTOL Univ	Bristol	U.K.	6	150	9	150	60	75	100
CEA-Azalee	Saclay	France	6	1250	36	125	100	40	100
CEA-Vesuve	Saclay	France	1	240	12	100	100	12	100
CEDEX	Madrid	Spain	6	100	9	100	n.a.	50	60
CKTI	S.Petersburg	Russia	1	200	5	100	n.a.	10	50
IZIIS	Skopje	Maced.	2	720	25	125	75	10	80
ENEA	Rome	Italy	6	300	16	125	78	78	50
ESA - ESTEC	Noordwij	Netherl.	6	225	25	70	80	30	100
INCERC-A	Iasi	Romania	1	1400	100	130	80	6	20
ISMES-Master	Seriate	Italy	6	300	16	100	55	50	120
ISMES-Ariete	Seriate	Italy	1	250	10	80	70	25	50
KOERI	Istanbul	Turkey	1	100	9	120	120	30	50
LNEC	Lisbon	Portugal	3	1000	26	175	70	31	40
NTUA - LEE	Athens	Greece	6	200	16	100	100	40	100
PAVIA Univ	Pavia	Italy	1	1030	39	500	220	18	84
SOPEMEA	Villacoublay	France	2	300	9	210	60	30	100
TsNIISK	Moscow	Russia	1	1000	24	50	n.a.	20	n.a.

Shaking Tables Worldwide Again, from the same reference (Taucer and Franchioni, 2005) and complementing also with information obtained directly from some of the researchers responsible for the facilities and from the Internet institutional sites, Table 5 summarises the main features of the more important shaking tables operating in the World with a payload of 400kN or higher.

Table 5.

Institution	Location	Country	DF	Payload kN	Area m 2	Disp mm	Vel cm/s	Freq Hz
CEA	Saclay	France	6	1250	36	125	100	100
HAZAMA	Tsukuba	Japan	3	800	24	300	115	50
INCERC	Iasi	Romania	1	1400	100	130	80	20
IZIIS	Skopje	Maced.	2	720	25	125	75	80
KAJIMA	Tokyo	Japan	6	500	25	200	100	60
KUMAGAI-GUMI	Tsukuba	Japan	6	700	25	260	150	70
LNEC	Lisbon	Portugal	3	1000	26	175	70	40
MITSUBISHI Elect.	Tokyo	Japan	2	400	16	100	70	30
MITSUBISHI H.Ind.	Tokyo	Japan	3	1000	36	50	150	50
Nat. Inst. Dis. Prev.	Miki	Japan	6	12000	300	1000	200	15
Nat. Inst. Dis. Prev.	Tsukuba	Japan	1	5000	218	220	90	50
Nat. Inst. Rur. Eng.	Tsukuba	Japan	3	450	24	150	75	40
NISHIMATSU	Tokyo	Japan	6	650	30	500	150	50
Nuc. P. Eng. Corp.	Tadotsu	Japan	2	10000	225	200	75	30
OBAYASHI - GUMI	Tokyo	Japan	3	500	25	600	200	50
PENTA OCEAN	Yonkucho	Japan	6	600	27	300	100	70
Port Airp. Res. Inst.	Kanagawa	Japan	2	600	12	200	40	50
P. Works Res. Inst.	Tsukuba	Japan	6	3000	64	600	200	50
TsNIISK	Moscow	Russia	1	1000	24	50	n.a.	n.a.
United tates Army	Rese	USA	3	450	13	300	130	60
Univ. BERKELEY	Berkeley	USA	6	450	37	127	76	15
Univ. BUFFALO	Buffalo	USA	5	500	13	150	125	100
Univ. NEVADA	Reno	USA	2	450	19	300	100	30

Following the Hyogo-Ken-Nanbu great Hanshin earthquake, occurred at Kobe area on January 1995 the Japanese government has decided to build a new, and with huge performances, shaking table. This very important equipment briefly referred in the previous table was built in Miki, near Kobe, and was inaugurated in 2006 being nowadays the biggest shaking table in the world. It belongs to the Hyogo Earthquake Engineering Research Centre, of NIED –National Research Institute for Earth Science and Disaster Prevention.

Being a very unique piece of equipment, its more important characteristics deserve to be mentioned with special attention. Consequently, in addition to the already mentioned in the previous table, a few more details are now given (Ogawa et al., 2004, and Minowa, 2007). The cost of the facility was about 300,000,000 Euro. Its foundation mass is 200,000ton.

Figure 7. Miki shaking table during construction.

The maxima accelerations at maximum loading are $9m/s^2$ at the horizontal direction and $15m/s^2$ at the vertical. The highest allowable overturning moment is 150MNm for a 1g vertical acceleration and the highest admissible yawing moment is 40MNm for a 1.5g horizontal acceleration. Concerning the hydraulics, this shaking table has a set of 24 actuators (5 for each horizontal translation and 14 for the vertical movements). The total volume of oil needed is 800,000litres. Just to pilot the servo-valves a 650l/min flow is used. Finally, the total number of sensors is 896 acquisition channels for the specimen measurements, being available a tremendous additional number of 440 channels just for the control system.

Specimens and Input Definition One of the main shaking table tests problems is the fact that, usually, it becomes necessary to build reduced-scaled specimens because it is not possible to analyse the behaviour of full-scale prototypes in view of their dimensions or of the involved masses. Even in structural engineering analysis, very reduced scales are sometimes used (Jurukovski et al., 1994). However some limitations may be impossible to overcome as, for instance, the characteristics of the materials to be applied. A few examples can be mentioned. In the case of metallic structures, thermal effects at welded parts that are not important in a prototype may be critical in a very reduced scale specimen. In the case of brick masonry it is almost impossible to build a scaled wall without changing the global mechanical characteristics of the structure. In the case of reinforced concrete it becomes also impracticable a good modelling of the steel/concrete interaction phenomena. Without the purpose of a complete presentation of the similarity theory, it is however appropriate to refer its application to the particular case of the shaking table tests. To transfer to a prototype the results obtained with a reduced scale model, demands in fact an adequate dimensional management to ensure that both systems are similar concerning a very well defined set of physical quantities.

The fundamental theorem of the dimensional analysis is the $\underline{\pi}$ theorem (Buckingham, 1914) stating that *"if we have a physically meaningful equation involving a certain number,* ***n****, of physical variables, and these variables are expressible in terms of* ***p*** *independent fundamental physical quantities, then the original expression is equivalent to an equation involving a set of (****n-p****) dimensionless variables constructed from the original variables"*. Concerning the dynamic behaviour analysis of structural models, several combinations of three independent fundamental physical quantities have been adopted. For instance: length, mass and time; length, force and time; length, elasticity modulus and specific mass. Just to mention a few possible choices in the aim of seismic engineering. After choosing the three fundamental quantities, and according to the Buckingham theorem, it is possible to reduce the problem to a relationship among (**n-3**) dimensionless variables.

However, in shaking table experimentation, being necessary a large number n of physical variables, it is very important to select the dimensionless variable to be considered among several available possibilities. In this case, the two more important dimensionless variables are the Cauchy Number, if the restoring forces are derived from an elastic potential, and the Froude Number, if the restoring forces are derived from a gravity potential. As explained later, under particular situations, it is possible to respect simultaneously both. The Cauchy Number is defined as the ratio between inertial and elastic restoring forces:

$$Fi/Fe = (m.a)/(E.A) = (\rho L^3.v^2/L)/(E.L^2) = (\rho v^2)/E \tag{2.1}$$

The Froude Number is defined as the ratio between inertial and gravitational forces:

$$Fi/Fg = (m.a)/(m.g) = (\rho L^3.v^2/L)/(\rho L^3.g) = v^2/(Lg) \tag{2.2}$$

where $\underline{\rho}$ = specific mass, $\underline{\boldsymbol{L}}$ = length, $\underline{\boldsymbol{v}}$= velocity, $\underline{\boldsymbol{E}}$ = modulus of elasticity and $\underline{\boldsymbol{g}}$ = gravitational acceleration.

The scale ratios to be adopted to ensure Cauchy and Froude similitude are presented in Table 6. These ratios were obtained considering the relationship $\underline{\alpha_p / \alpha_m}$, where $\underline{\alpha_p}$ means a prototype quantity and $\underline{\alpha_m}$ means the same quantity but for the model. The three independent fundamental physical quantities adopted being length, elasticity modulus and specific mass. Thus, the symbols used are $\underline{\lambda} = \underline{L_p} \;/\; \underline{L_m}$ for the length ratio, $\underline{e} = \underline{E_p} \;/\; \underline{E_m}$for the modulus of elasticity ratio and $\underline{\rho} = \underline{\rho_p} \;/\; \underline{\rho_m}$ for the specific mass ratio. Under the condition $\underline{e} = \underline{\lambda \rho}$ Cauchy and Froude similitude are simultaneous.

Table 6.

Quantity	Symbol	Cauchy	Froude
Length	L	λ	λ
Modulus of Elasticity	E	e	e
Specific Mass	ρ	ρ	ρ
Area	A	λ^2	λ^2
Volume	V	λ^3	λ^3
Mass	m	$\rho\,\lambda^3$	$\rho\,\lambda^3$
Velocity	v	$e^{1/2}\,\rho^{1/2}$	$\lambda^{1/2}$
Acceleration	a	$e\,\rho^{-1}\,\lambda^{-1}$	1
Force	F	$e\,\lambda^2$	$\rho\,\lambda^3$
Moment	M	$e\,\lambda^3$	$\rho\,\lambda^4$
Stress	σ	e	$\rho\,\lambda$
Strain	ε	1	$e^{-1}\,\rho\,\lambda$
Time	t	$e^{-1/2}\,\rho^{1/2}\,\lambda$	$\lambda^{1/2}$
Frequency	f	$e^{1/2}\,\rho^{-1/2}\,\lambda^{-1}$	$\lambda^{-1/2}$

In Table 7 are presented those scale ratios, being the third column explicit for $\underline{e}$ and $\underline{\lambda}$ and the fourth column equivalent but explicit for $\underline{\rho}$ and $\underline{\lambda}$.

It is obvious that whatever the length ratio $\underline{\lambda} \neq \underline{1}$ implies the alteration of the elasticity modulus of the model, of its specific mass, or both. In order to deal with this fact, the solution usually adopted is to consider a very special condition of Cauchy and Froude simultaneous similitude: $\underline{e} = \underline{\rho}\,\underline{\lambda} = \underline{1}$. Under these conditions, the same material can be used because the prototype and the model materials have the same elasticity modulus. To avoid the necessary increase of specific mass ($\underline{\lambda}$ times), this quantity is artificially raised by adding dummy masses that must be spread in the best possible rigid and distributed way, to change minimally the dynamic characteristics of the model and to prevent any chocks during the tests. The scale ratios to be adopted to ensure simultaneously Cauchy and Froude similitude, under these very particular conditions are presented in

Table 7.

Quantity	**Symbol**	**Cauchy and Froude**	**Cauchy and Froude**
Length	L	λ	λ
Modulus of Elasticity	E	e	$\rho\,\lambda$
Specific Mass	ρ	$e\,\lambda^{-1}$	ρ
Area	A	λ^2	λ^2
Volume	V	λ^3	λ^3
Mass	m	$e\,\lambda^2$	$\rho\,\lambda^3$
Velocity	v	$\lambda^{1/2}$	$\lambda^{1/2}$
Acceleration	a	1	1
Force	F	$e\,\lambda^2$	$\rho\,\lambda^3$
Moment	M	$e\,\lambda^3$	$\rho\,\lambda^4$
Stress	σ	e	$\rho\,\lambda$
Strain	ε	1	1
Time	t	$\lambda^{1/2}$	$\lambda^{1/2}$
Frequency	f	$\lambda^{-1/2}$	$\lambda^{-1/2}$

Table 8.

Concerning the input definition, it is important to remark that the acceleration ratio is 1, but the scales of time and frequency must be considerably changed. As shown in the previous table, the accelerograms to use require to be compressed in time and expanded in frequency. Under this similarity, stress and strain have both unity ratios. This fact is useful because it allows a direct interpretation of stress/strain diagrams. That can be also found in another frequently used similarity condition. It is also a special case, but just under the Cauchy similitude and for $\underline{e} = \underline{\rho} = \underline{1}$. Its scale ratios are presented in Table 9.

In this case, concerning the construction of the specimen, as the prototype and the model have identical modulus of elasticity and specific mass, it is not necessary the use of dummy masses. Concerning the input definition, it is important to remark that in this case it is the velocity ratio that is equal to 1, and not the acceleration, as previously. The scales of time and frequency must be more considerably changed then for the previous situation, concerning the accelerograms. The choice of one of these very particular similitude set of characteristics, commonly known as acceleration similarity and velocity similarity, respectively, depends on the peculiarities of each specific experiment and facility.

Table 8.

Quantity	**Symbol**	**Cauchy and Froude** ($when\ \underline{e} = \underline{\rho}\,\underline{\lambda} = 1$)
Length	L	λ
Modulus of Elasticity	E	1
Specific Mass	ρ	λ^{-1}
Area	A	λ^2
Volumc	V	λ^3
Mass	m	λ^2
Velocity	v	$\lambda^{1/2}$
Acceleration	a	1
Force	F	λ^2
Moment	M	λ^3
Stress	σ	1
Strain	ε	1
Time	t	$\lambda^{1/2}$
Frequency	f	$\lambda^{-1/2}$

Table 9.

Quantity	**Symbol**	**Cauchy** ($when\ \underline{e} = \underline{\rho} = 1$)
Length	L	λ
Modulus of Elasticity	E	1
Specific Mass	ρ	1
Area	A	λ^2
Volume	V	λ^3
Mass	m	λ^3
Velocity	v	1
Acceleration	a	λ^{-1}
Force	F	λ^2
Moment	M	λ^3
Stress	σ	1
Strain	ε	1
Time	t	λ
Frcquency	f	λ^{-1}

Control The application of Control Systems Engineering for the dynamic simulation of seismic actions is a very complex problem consisting essentially in the conception of a system allowing an automatic adaptation to the specimen to be tested and to its dynamic characteristics changes during the tests. Three key aspects must be considered:

1. The transfer function of the specimen is unknown;
2. It can change during the test;
3. The system should be universal.

Concerning the first condition, in fact, view the complexity of the usual specimens to analyse, it is not possible to obtain analytically its initial transfer function, not even in elastic regime. Respecting the second condition, and a large majority of shaking table tests being conducted up to the rupture, the dynamic characteristics of the specimens are of course changed because of the damage accumulation caused during significant inelastic excursions. Finally, regarding the third condition, it is obvious the advantage of a unique control system usable for any test within those particular simulator capabilities.

Control algorithms will not be presented in the following, but just the methodology currently in use at LNEC. First of all, characterisation tests of the facility were performed. For that purpose, with a bare shaking table, it was decided to inject successive signals until finding a very particular one that produced the observation of a pink noise in the working frequency range of the simulator. It was based on an initial pink noise tentatively adapted to the dynamic characteristics of the equipment. In the following this signal will be designed as "the pseudo pink noise". A pink noise, or 1/f noise, is a signal with a frequency spectrum such that the power spectral density is proportional to the reciprocal of the frequency. Its name derives from being intermediate between a white noise ($1/f^0$) and a Brownian noise ($1/f^2$), frequently named as red noise. It was decided to avoid the use of a white noise (with a constant power spectral density) because it is important to prevent the high-frequency components and to privilege the low-frequency ones, considering that they are very significant for the more common civil engineering structural tests usually performed at LNEC. Before the adaptation process of a given input signal, commonly an accelerogram, a pre-processing procedure is needed to transform it in a file understandable by the command system and reproducible at the shaking table, taking into account their intrinsic limitations. This pre-processing consists basically in the following actions:

1. Resampling.

   The sampling frequency of the LNEC control system being 200Hz, it is necessary a resampling of the initial signal to get a 5ms sampling interval.

2. Zeros adding or truncation.
   This action is needed for computing the FFT of a multiple of 1024-sample blocks of input data.
3. Tapering.
   The tapering is performed using sine functions to smooth the extremities of the input signal.
4. Filtering.
   The filtering is accomplished to exclude the frequencies over the maximum of the simulator frequency range (not reproducible) and below 0.1Hz (hardly controllable and that would raise multiple difficulties).
5. Limits imposition.
   Concerning the imposition of limits, the main limitation at the LNEC triaxial shaking table is usually a maximum velocity, in a given direction. However, this can be avoided by using a more convenient similarity, as presented in the previous chapter.
6. File-type record.
   Finally, a very well defined file-type record is compulsory in order to be correctly identified by the command system.

Once concluded the pre-processing of the initial accelerogram the signal adaptation procedure can then start (a flow-chart of that procedure is presented in Figure 8.). To avoid the damage of the specimen to be analysed, a first adaptation procedure is made with a set of rigid masses, with identical weight. Before starting the adaptation of the accelerogram (target signal), the previously mentioned pseudo pink noise is injected to the shaking table and a traditional control approach is developed. The FFT of the pseudo pink noise and the FFT of the consequently measured signal are computed. From them, the system transfer function and its reciprocal

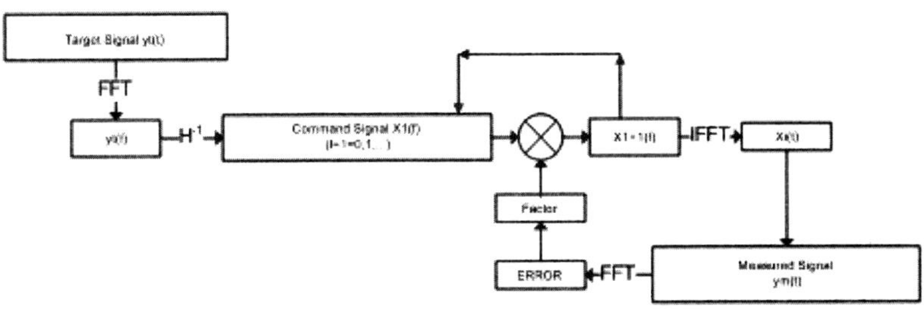

Figure 8. Traditional shaking table adaptation procedure.

are also obtained. These functions are estimated taking simultaneously in account displacement and acceleration measurements, in particular for the frequency ranges where a better coherence is found.

An iterative procedure is then developed until the demanded shaking table motion is reached, within an acceptable error range. The steps are the following:

1. It is injected a signal obtained from the application to the target signal (in the frequency domain) of the reciprocal of the system transfer function above mentioned.
2. This first signal derives a result still too far from the wanted target signal. This measured signal is then deducted from the target signal and an error signal is computed.
3. In the next step the signal injected is obtained from the signal injected in the previous step considering an additional signal to correct the observed error. This correction is made by applying to the error signal (in the frequency domain) the already mentioned reciprocal of the system transfer function.
4. The new signal then observed is, as previously, deducted from the target signal, being computed a new error signal and giving raise to an iteration loop.

Using this procedure, the number of iterations needed to reach the desired accelerogram, with an acceptable error, is deeply influenced by the characteristics of the exact target signal under analysis but that goal is typically reached after a maximum of five or six steps.

The control of shaking tables being a major concern in earthquake engineering experimental research, it was recently developed a performance benchmark of three major European shaking tables (Bairrao et al., 2006). Three laboratories were chosen (CEA, LNEC and NTUA), running important earthquake simulators, from a maximum specimen dead weight of 100kN to 1000kN, thus comprising a wide spread of seismic experimental activities. The characteristics of one single test specimen were defined and it was built in France and successively tested on each of the three shaking tables under uniaxial input signals, being transported by truck and ship among the laboratories. It was designed to simulate the global non-linear behaviour of a concrete building while being damaged during an earthquake, concerning a typical global stiffness reduction. It is important to highlight that it was conceived in order to allow consecutive similar tests in spite of its non-linear behaviour.

This system, being able to simulate a sudden collapse or a progressive degradation, could reproduce, in a repetitive and accurate way, the same non-linear conditions. It was possible to change its stiffness during the seismic tests whenever wanted and for an adjustable duration. Consequently, it has also permitted a performance comparison concerning the change of the control conditions and the influence of the intrinsic characteristics of each specific facility. The following main parts have constituted this specimen (Figure 9):

1. A complex footing for perfect connection to the three grid patterns;
2. A rigid steel frame;
3. A variable oscillation mass attached to the steel frame with a controlled rigidity;
4. A four controlled pressurized airbags system, to change the frequency during tests.

Figure 9. The benchmark specimen on the LNEC 3D shaking table.

The mass was guided transversally in order not to disturb the longitudinal motion during the tests. It was kept vertically by a bearing device designed to have a vertical frequency higher than 20Hz. The pressure of the air bags system has allowed the adjusting of the frequency of the movable part and the damping was ensured by a set of shock absorbers. The decreasing of the frequency, from about 7Hz to 3Hz during the tests, was obtained by a partial depressurization through specific gauged diaphragms. Two diaphragms of larger (60mm) or smaller (10mm) openings were used for the control of the discharge duration. A real-time adaptive control technology, initially developed for robotics, and based on the MCS –Minimal

Control Synthesis algorithms is currently under development (Stoten and Gomez, 2001). This new methodology, when it will be dully adapted to the control of large shaking tables, will be very useful. In fact, it does not demand a previous identification of the dynamic characteristics of the system to be controlled and it will take into account, on-line, the changes of those characteristics during the tests.

3 Future Developments

The improvement of shaking table technologies is clearly of paramount importance to reduce the seismic vulnerability of the building stock and also to mitigate the consequences of future, and inevitable, seismic events by contributing significantly to the amelioration of new construction techniques. Much advancement is on progress among the members of the European consortia of earthquake engineering laboratories and the more important needs of the experimental facilities were recently identified in detail (Taucer, 2005). Concerning specifically the shaking tables facilities, the improvement of two technologies has priority: Real-time Control and Sub-structuring. The current advantages of a real-time control were already previously referred, but that technology will be absolutely necessary to follow the sub structuring technique developments.

In fact, as it is clear in Table 5, the global capacities of the biggest European shaking tables are very moderate when compared with the ones of the more recent Japanese laboratories. Therefore the shaking table test of real and very large structures is not foreseen in Europe for the near future. Consequently, the use of sub-structuring techniques must be developed in order to fix to the simulator platform just a part of the structure while all the remainder will be tested on computer, like in the pseudo-dynamic testing procedure. To use this technique in shaking table tests it will be need a real-time control and an indeed very high speed of the data transfer between the control system of the shaking table and the computer system modelling the numerical part of the experiment.

Bibliography

G. Armer. *Final report under consulting agreement between the Nevada Testing Institute and G. A. Consultants.* April 1998.

R. Bairrao, M.J. Falcao, P. Carydis, H. Mouzakis, L. Karapitta, and J.C. Queval. NEFOREEE *-Performance benchmark of three major European shaking tables.* 1st European Conference on Earthquake Engineering and Seismology, Geneva, Switzerland, 2006. paper 110.

R. Bairrao and R. Severn. Experiences under *ECOLEADER* -European Consortium of Laboratories for Earthquake and Dynamic Experimental Research. ISBN 972-49-1972-2, 2007. CASCADE report n.8.

F. Borges. *Harmonisation of rules for seismic design.* Lisbon, Portugal, 1980. CSNI specialist meeting on Probabilistic Methods in Seismic Risk Assessment for Nuclear Power Plants.

E. Buckingham. *On physically similar systems.* illustrations of the use of dimensional analysis. *Physics Review*, (4):354–377, 1914.

D. Combescure, F. Pires, P. Cerqueira, and P. Pegon. *Test on masonry infilled reinforced concrete frames and its numerical interpretation.* Acapulco, Mexico, June 1996. 11th World Conference on Earthquake Engineering. paper 1731.

A. Costa, P. Morais, B. Wainwright, and A. Martins. *Characterisation of the new LNEC shaking table.* 1996. LNEC/C3ES report 148/96.

S. Dermitzakis and S. Mahin. *Development of sub-structuring techniques for on-line computer controlled seismic performance testing.* California, USA, 1985. UCB/EERC report 85/04.

J. Donea, G. Magonette, P. Negro, P. Pegon, A. Pinto, and G. Verzeletti. *Pseudo-dynamic capabilities of the ELSA laboratory for the earthquake testing of large structure. Earthquake Spectra*, 12(1), 1996.

F. Emilio, R. Duarte, F. Carvalhal, C. Costa, C. Vaz, and M. Correa. *The new LNEC shaking table for earthquake resistance testing. LNEC memory 757.* 1989.

D. Jurukovski, L. Tashkov, M. Bojadziev, and F. Zarri. *Shaking table test of 1:30 scale model of the Sports Palace Bologna with a continuous shell covering structure.* volume 3, pages 2461–2465, Vienna, Austria, August 1994. 10th European Conference on Earthquake Engineering.

C. Minowa. *Miki shaking table presentation during visit at the LNEC/DE Seismic Engineering and Structural Dynamics Division.* March 2007.

P. Mote, G. Armer, P. Gefken, D. Curran, D. McCallen, I. Arango, and O. Gurbuz. *Strong motion generator for full-scale structural testing at the Nevada test site.* Paris, France, September 1998. 11th European Conference on Earthquake Engineering.

P. Negro and G. Magonette. *Experimental methods in structural dynamics. European Earthquake Engineering*, pages 29–39, 1998. year XII, 1.98.

INCEDE newsletter. *University of Tokyo, Institute of Industrial Science, International Centre for Disaster Mitigation Engineering, special Great Hanshin Earthquake newsletter issue.* January 1995.

N. Ogawa, M. Sato, K. Ohtani, and T. Katayama. *Construction of a 3D full-scale earthquake testing facility. Journal of Japan Association for Earthquake Engineering*, 4(3), 2004. (special issue).

C. Oliveira, J. Guedes, L. Sousa, and A. Costa. *The Faial/Pico/S.Joao seismic crises following the July 9th earthquake recorded at the Azores stations (in Portuguese).* Faro, Portugal, November 1998. 1st APMG Meteorology and Geophysics Symposium and 2nd APMG/AME Portuguese/Spanish Meteorology Encounter.

P. Pegon and J.C. Queval. *Structures under seismic actions: Laboratory experimentation (in French).* In "Comportement dynamique des betons et genie parasismique"), Paris, France, 2004. Lavoisier editions.

A. Pinto. *Earthquake performance of structures. Behavioural, safety and economical aspects.* PhD thesis, Instituto Superior Tecnico,Lisbon, Portugal, 1998.

F. Pires, R. Bairrao, and S. Silva. *Earthquake behaviour study of masonry infilled reinforced concrete frames. Test of model MD6.* 1995. LNEC/C3ES report 137/95.

F. Pires, J. Rodrigues, A. Costa, and R. Bairrao. Repair of reinforced concrete frames infilled with masonry walls. Behaviour study under cyclic horizontal actions. Test of model I9R (in Portuguese). 1998. LNEC/C3ES report 146/98.

J. Portugal, L. Madeira, and C. Santos. *Geotechnical Department centrifuge. Reception and operation starting (in Portuguese).* 1995. LNEC/DG/CPCE report 290/95.

P. Shing and S. Mahin. Experimental error propagation in pseudo-dynamic testing. California, USA, 1983. UCB/EERC report 83/12.

D. Stoten and E. Gomez. *Adaptive control of shaking tables using the MCS -Minimal Control Synthesis algorithm. Phil. Trans. Royal Society of London,* (259), 2001.

K. Takanashi. *Non-linear earthquake response analysis of structures by a computer-actuator on-line system. Bulletin of the Earthquake Resistant Structures Research Centre,* (8), 1975. Institute of Industrial Science, University of Tokyo, Japan.

F. Taucer. Recent advances and future needs in experimental earthquake engineering. CASCADE report n.7. ISBN 972-49-1971-4, January 2005.

F. Taucer and G. Franchioni. Directory of European facilities for seismic and dynamic tests in support of industry. CASCADE report n.6. ISBN 972-49-1970-6, January 2005.

Continuous PsD Testing With Substructuring

P. Pegon*
European Laboratory for Structural Assessment, Institute for the Protection and Security of the Citizen, Joint Research Centre, European Commission, Ispra, Italy

Abstract This work presents the development and implementation of a domain decomposition approach for solving dynamics problems involving continuous testing with analytical substructuring in the non-linear range. This approach implements an inter-field parallel time integration procedure suitable to work with an explicit experimental synchronous process using small time steps, such as the one running continuous PsD testing, and an implicit, possibly iterative, analytical process using large time steps. The characteristics and the robustness of the proposed schemes are illustrated by means of a parametric numerical study. The results of the first validation under test conditions are also presented. For completness, the presentation includes a short presentation of the continuous PsD testing and its implementation at the ELSA laboratory and some general comments regarding substructuring.

1 Introduction

The Pseudo-Dynamic (PsD) test method combines the numerical time integration of the equations of motion of a structure, properly condensed on a limited number of degrees-of-freedom (DoF), and the experimental measurement of the reaction forces resulting from this motion, applied by means of actuators. This hybrid analytical/experimental character is taken into account when introducing the subtructuring technique (Dermitzakis and Mahin, 1985), allowing to obtain the dynamic response of a structure with only a part, usually the most vulnerable one, present in the laboratory. This technique has been implemented at the European Laboratory for Structural Assessment (ELSA) using two processes running in parallel and

*Many of the developments presented here are the results of the work of the ELSA team. The author would like to thank in particular Philippe Buchet, Georges Magonette, Francisco Javier Molina and Artur Viera Pinto for all these years of fruitful collaboration.

dialoguing with each other: one is responsible for the part of the structure present in the laboratory (the experimental structure) and the other one for the part of the structure (the analytical structure) which is only modelled (Buchet and Pegon, 1994, and Pegon, 1996). This distributed approach had been adopted because the implementation of the PsD method itself was already decentralized (Donea et al., 1996): one master PsD process piloting a network of controllers imposing displacements to the structure by means of ramps followed by hold periods. This approach lead to a campaign of tests on large scale bridges (Pegon and Pinto, 2000). Although these tests were performed in 1997, they were already distributed over the internet on heterogeneous hardware. In fact the master PsD process was running on PC/DOS Real Time system whereas the modelling process was running on a UNIX workstation, the connection between both being realized by means of Berkeley sockets. Finally, even if this point may have been insufficiently noted (Pan et al., 2006), these test also provided a convincing example of an on-line collaboration between a PsD oriented program running on the PC and holding the experimental DoFs, and a large multi purpose Finite Element software (Millard, 1993) running on the workstation and used for the modelling and the time integration of the DoFs of the analytical structure.

The ELSA laboratory has been involved from some years in the development of the new continuous PsD testing method. The continuous PsD testing (PsD testing without hold period, Magonette, 2001, and Magonette et al., 1998) avoids the load relaxation problem and consequently improves the quality of the results. It allows also a considerable reduction of the test duration and provides a much cleaner signal for force measurement. The continuous PsD testing has been implemented by means of a synchronous process with short control period (2ms) and small time step for the time integration.

If the continuity of the motion of the experimental structure provided by the continuous PsD method and the use of two parallel processes are to be preserved, this introduces some challenging difficulties for the implementation of substructuring, namely:

1. *Different time steps* are likely to be used by the two processes: small time steps for the experimental process, and large ones for the analytical process. This is a consequence of the very nature of the continuous PsD method. It is virtually impossible to ensure that, within the effective clock duration (2ms) of one laboratory time step, the analytical process is able to successively perform all the tasks required for advancing its solution, in particular when non-linearities are present. Moreover, it is foreseen that the experimental part could be distributed among various testing facilities located in different and

eventually distant places. In this case and in absence for the moment of any high-speed dedicated hardware for the communication, the overhead to transmit and receive information could be much larger than the effective duration of one step of the analytical process. It is thus highly desirable that the experimental process could be able to perform as many time steps as needed (typically from 10 to 10000) during one analytical time step, while preserving a smooth continuous trajectory.

2. *Different time integration schemes* are likely to be used by the two processes: explicit time integration for the experimental process and implicit time integration for the analytical one. In fact it is assumed that only few DoFs are controlled in the laboratory(ies) and that eventually many of them are taken into account in the analytical process. Since it is expected that the experimental structures will be advanced in time through small time steps, the conditional stability of the explicit scheme will be easily fulfilled. This is no longer possible to ensure for the analytical process, possibly working with far larger time steps: it is thus mandatory to rely on implicit, unconditionally stable scheme for handling the analytical part of the structure.

Although some implementations of the continuous PsD method with non-linear substructuring (Pegon and Magonette, 1999) along this line had been successfully performed (Magonette, 2001, and Magonette et al., 1998), some parts of the coupling algorithm were not working satisfactory, in particular when introducing very different time steps among the analytical and numerical processes.

Then, as no practical convincing solution to this problem was available, the continuous PsD testing with substructuring (also called continuous hybrid simulation) has been implemented somehow sacrifying the continuity of the experimental trajectory. In fact, the motion of all the DoFs are determined by the analytical process with a large time step and the continuous motion of the actuators is no longer the result of time integration but takes a predefined shape.

1. During the VAB project in 2001 (Pinto et al., 2004), a bridge with a linear deck and 6 non-linear piers was studied. Two piers where present in the laboratory and the rest of the structure numerically modelled. The continuous control of the horizontal pier displacement within the large time step follows an S-shape ramp followed by hold period.
2. This scheme has been more or less reproduced during the first geographically distributed (The MOST Experiment, 2003). Here a linear ramp followed by a hold period was introduced.

3. For providing a smoother trajectory of the actuators, an interpolation/extrapolation procedure has been proposed (Nakashima and Masaoka, 1999). Basically, during the computation of the next target displacement, the actuators are following a trajectory obtained from the extrapolation of the previous one. As soon as the effective target displacement is known, an interpolation is introduced to reach it. Use of third order polynomial shape allows to provide smooth command signal to the actuators. This procedure is also proposed for distributed testing (Mosqueda, 2003).

In the mean time, the research for an implementation featuring parallel and different time integration schemes working with different time step proceeded successfully as reported in this chapter. In particular, starting from a domain decomposition scheme presented in (Gravouil and Combescure, 2001), it is shown that it is possible to transform this essentially staggered asynchronous procedure in an inter-field parallel procedure suitable to work with one synchronous process, keeping most of the original characteristics of the scheme. The non-linear case is also investigated and a combination of a non-iterative Operator Splitting strategy (Nakashima et al., 1993, and Combescure and Pegon, 1997) and of a non-linear condensation technique is proposed to handle the analytical structure.

This chapter is organized as follows. The main features of the continuous PsD method are first introduced (section 2) and compared with the classical PsD method used previously. Some details are also given on the hardware and software architectures available at ELSA. Then a section introducing subtructuring and detailing the algorithms developed for continuous testing is proposed (section 3). The results of some numerical experiments are then presented (section 4) in order to illustrate the main features of the proposed methods. The results of the first experimental validation are then presented and commented (section 5), followed by concluding remarks (section 6) and references.

2 The Classical and the Continuous PsD Methods

2.1 Introduction

PsD testing consists of the step-by-step integration of the discrete-DoF equation of motion:

$$\mathbf{Ma} + \mathbf{r}(\mathbf{d}) = \mathbf{f}(t) \tag{2.1}$$

where $\mathbf{M}$ is the theoretical matrix of mass, $\mathbf{a}$ and $\mathbf{d}$ represent the unknown vectors of acceleration and displacement and $\mathbf{f}$ are known external forces that, in the case of a seismic excitation, are the inertial forces obtained by multiplying the specified ground acceleration a_g by the theoretical masses and the incidence vector of the earthquake $\mathbf{I}_g$.

$$\mathbf{f}(t) = -\mathbf{M}\mathbf{I}_g a_g \tag{2.2}$$

The unknown restoring forces $\mathbf{r}(\mathbf{d})$ are experimentally obtained at every time integration step by quasistatically imposing, generally by means of a hydraulic control system, the computed displacements (see Figure 1).

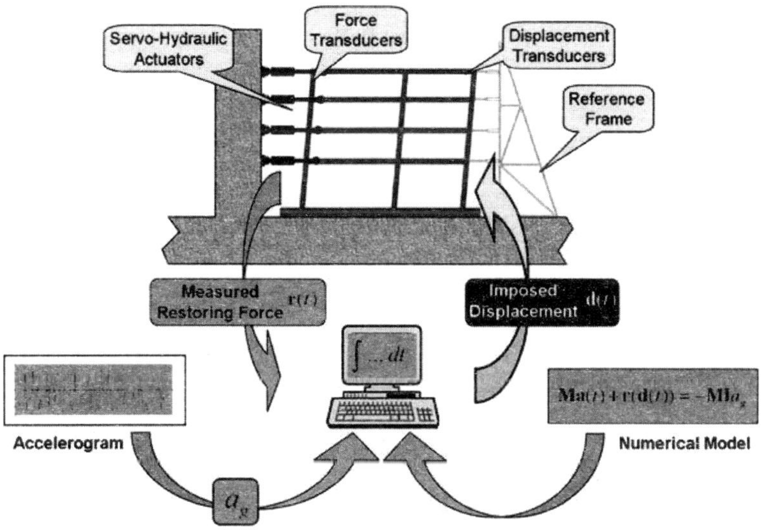

Figure 1. The PsD method.

Note that Eq.(2.1) does not involve a damping matrix. This is to underline that, in most of the cases, the damping matrix is useless at the laboratory since it is just a straightforward artefact which is used in pure analytical simulation to introduce an equivalent model for micro-hysteresis dissipation.

For every PsD test, two types of time can be introduced: the prototype or accelerogram time t, which corresponds to the one of the original problem with an earthquake excitation that may last for a few tens of seconds, and the experimental time T, which may extend to several hours for the execution of the test in the laboratory. Then, λ, the time scale factor, is defined as the proportion between experimental and prototype times, i.e

$$\lambda = \frac{T}{t} \tag{2.3}$$

Even when both PsD methods, classical and continuous, use the same integration scheme (typically explicit Newmark), there are some major differences. These differences ought to be very carefully explained, since they radically change the philosophy of substructuring.

2.2 The Classical PsD Method

In the classical PsD method, the time increment Δt for the integration of the equation of motion is chosen small enough to satisfy the stability and accuracy criteria of the integration scheme. The ground accelerogram a_{g} must be discretized with the prototype record increment a_{g}^{n}. The smaller this time increment is chosen, the larger the number of integration steps will be to cover the duration of the earthquake. The execution of every step will take place in the corresponding experimental time lapse, which uses to be in the order of several seconds. In fact, an increment ΔT of the experimental time T is split into four phases (see Figure 2):

1. A stabilising hold period ΔT_{h1} of the system motion after the ramp of the reference signal at the controller. In practise, this period allows the specimen to reach the computed displacement. If the computed displacement has not been accurately achieved, the measured force will not correspond to the computed displacement.
2. A measuring hold period ΔT_{h2} that allows reducing the load cell signal noise by averaging a number of measures. This could be important to reduce some random errors in the solution.
3. A computation hold period ΔT_{h3} for solving the next step of the time integration algorithm. This period includes also the transmission time if the system equations is solved in a different CPU from the controller itself.
4. A period of ramp ΔT_{ram} at the reference signal in order to smoothly change to the new computed displacement $\mathbf{d}^{n+1}$. During the three previous hold periods, the reference was maintained constant at the previous value of the computed displacement $\mathbf{d}^{n}$.

The accuracy in the imposed displacement and the measured force depends, apart from the characteristics of the experimental set-up, on the selected periods for stabilising, measuring and ramp, whereas the computation period is determined by the system equations and processor/network characteristics. It can thus vary from one step to the following one due

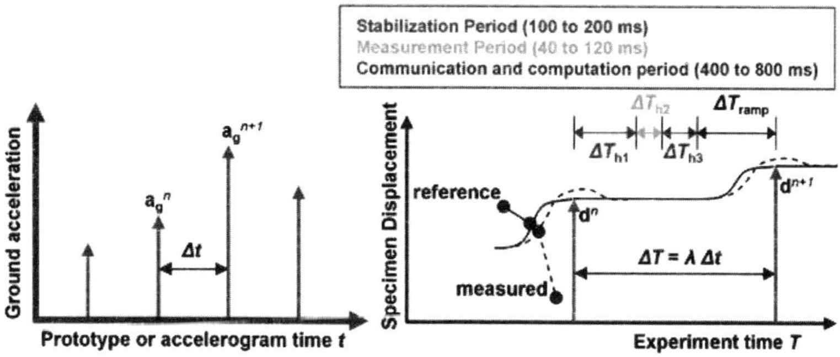

Figure 2. The classical PsD method.

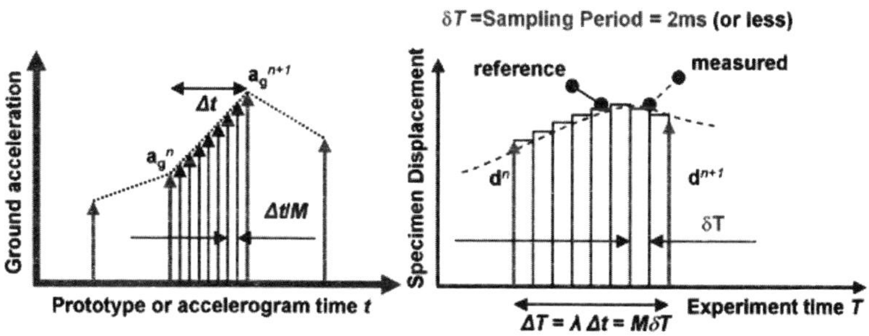

Figure 3. The continuous PsD method.

to e.g. irregularities in the network. Consequently the time expansion parameter λ of Eq.(2.3) is not constant. The duration of the ramp and the stabilisation period may also depend on the amplitude of the displacement increments and on the number of DoFs controlled during the test. In any case ΔT is much larger than 1s and since the test is *effectively* stopped at each time ΔT, there is a priori no practical difficulty (in terms of convergence speed, data transfer and bandwidth) for coupling the PsD process with other PsD processes running on other experimental platforms and/or handling an analytical structure. This can be done for instance using the α-Operator Splitting scheme, as demonstrated for a large substructured bridge test (VAB project in 2001, Pinto et al., 2004) or more recently during the first geographically distributed test (The MOST Experiment, 2003) followed by numerous distributed tests around the world.

The asynchronous nature of the classical PsD method is also a particular advantage for tests that require complex transformations of coordinates at every step, as for bi-directional testing (Molina et al., 1999).

2.3 The continuous PsD method

Within the continuous PsD testing, the motion of the actuators is no longer stopped at every step. On the contrary, the servo-controller moves the actuator in such a way that the specimen follows very accurately the target displacement, due to the lack of discontinuity in the motion. The forces are measured at every sampling period of the digital servo-controller and the equations of motion are integrated on the fly (without hold period) at the sampling rate. The next target displacement is determined and the motion proceeds without any interruption (see Figure 3). The execution of every integration step takes just one sampling period (δT) of the digital controller of the control system, e.g. 2ms in the ELSA implementation (Magonette et al., 1998). The ramp and stabilising periods are reduced to zero duration and the measuring plus the computation periods must be feasible within those few milliseconds of experimental time. The surprising fact is that, since the hydraulic control system is unable to respond significantly at frequencies in the range of the sampling frequency of its controller, e.g. 500 Hz, the missing periods of ramp and stabilising are not needed at all.

In practice, from each discrete value a_{g}^{n} of the ground acceleration, a sequence of M acceleration values $a_{\mathrm{g}}^{n+m/M}$ $(0 < m \leq M)$ is computed by interpolation between a_{g}^{n} and a_{g}^{n+1}. The PsD procedure is executed M times at the sampling rate δT of the controller performing M sub-steps $\delta t = \Delta t / M$ inside one classical accelerogram step as indicated in Figure 3. Thus the time scale expansion factor of the continuous PsD test is given by

$$\lambda = \frac{M\delta T}{\Delta t} \tag{2.4}$$

For a continuous PsD test performed on large scale multi-floor structures, one may have $M = 500$, $\delta T = 2\text{ms}$, $\Delta t = 5\text{ms}$ giving $\lambda = 200$ which means that 1 second of the real earthquake takes 200 seconds in the test. For a specified test, δT and Δt are constant, so the time scale expansion can be increased or decreased interactively simply by changing M during the experiment. As a rule, λ is decreased until the control error exceeds a prescribed tolerance limit depending on the nature and complexity of the test. In any case, by similitude with the classical PsD theory, these errors must always remain very small and it is worth noting that the smooth motion of the continuous PsD certainly contributes in this sense.

Practically λ can range from a value near 1 for simple substructuring tests with negligible inertial effect in the experimental set-up, to values as large as 250 or 500 for tests on full-scale multi-storey buildings. Moreover, in continuous PsD testing the restoring force measurements are updated at the sampling rate of the controller (typically 500 Hz). This procedure not only generates a smooth displacement of the structure and a continuous load application reducing relaxation, but also increases the signal/noise ratio with respect to a classical PsD test in which much less measurements enter in the integration algorithm.

Considering substructuring, it should be understood that, in principle, the continuous PsD test is thus never stopped and the effective laboratory duration δT of each algorithmic time step δt is constant and very short. Even if the displacement steps could be very small, somehow simplifying the convergence of a non-linear numerical process, it becomes very tedious to use the same substructuring strategy as before. In fact, it seems impossible, within the physical time allowed (2ms), to perform a reasonably complex computation, exchange data and proceed, keeping a distributed heterogeneous hardware. Then, a radically new substructuring strategy should be introduced.

2.4 Hardware and Software at ELSA

The implementation of the continuous PsD method has been achieved in order to synchronise and complete inside the control sampling time δT (2ms at ELSA), the main tasks of a PsD cycle: measurement, motion computation and displacement control. An advanced hardware configuration has been set up to ensure a strong coupling and a very high-speed data communication between the servo-controllers and the main computor solving the equations of the motion.

In practice, the hardware consists of three main parts as described in Figure 4 (Buchet, 2006a, and Casciati et al., 2006): the master card, the slave cards (they are usually more than one) and the ISA passive bus connection. The master card contains the kernel of the pseudo-dynamic algorithm. For this reason it is equipped with a fast processor (Pentium class) and enough memory in order to store the necessary data.

The slave card consists of a main board equipped with three main components plugged: a PC104 central processing unit card, a PC104 controller I/O card including specific interfaces such as the one needed for reading digital sensors or for generating servo-valve or electric motor commands, and an analog input/output card with AD and DA converters (see Figure 5). It is also equipped with a dual-port memory card enabling high-speed access and easy sharing of data between the local PC104 bus and the ISA bus connected to the master. The ISA passive bus connects the master and up

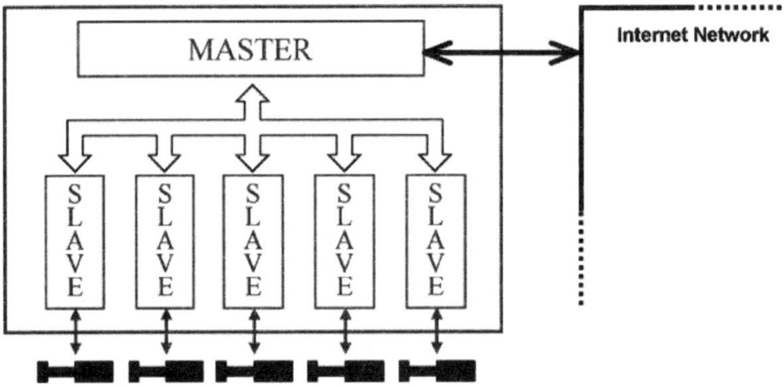

Figure 4. The various parts of the controller.

to 7 slaves cards in the current configuration, but in principle the limitation on the number of cards that can be connected with a passive bus is 16. The assembled controller must then be put into a rack and the peripherals (USB drive, LCD screens and so on) connected to it. It is possible to reset either the master CPU or the slaves CPUs separately.

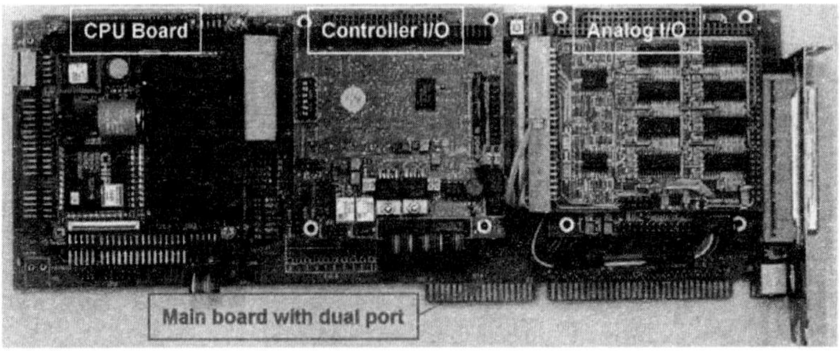

Figure 5. View of the controller slave board.

The control software (Buchet, 2006b, and Casciati et al., 2006) reflects the architecture of the hardware: there is one master program that communicates with several slaves programs. Both master and slaves programs handle *backgound* and *foreground* processes.

The *backgound* processes manage various services such as keyboard access, the uploading of control parameters, the user interface (on-line graphics, console), the hard disk management (data acquisition), the LAN connection, the remote services (under NT platform) and the data exchange between master and remote station. Since these services are not strictly necessary (the display refresh, for example, can be delayed a little bit, if it is needed), these processes have a lower priority than those in the foreground process.

The *foreground* processes are the core of the control software: they perform at a fixed sample rate the data acquisition and the computation of the control variables. For this reason they must have priority over the background processes because, obviously, delays cannot be accepted in the control algorithm. They work as follows (see Figure 6)

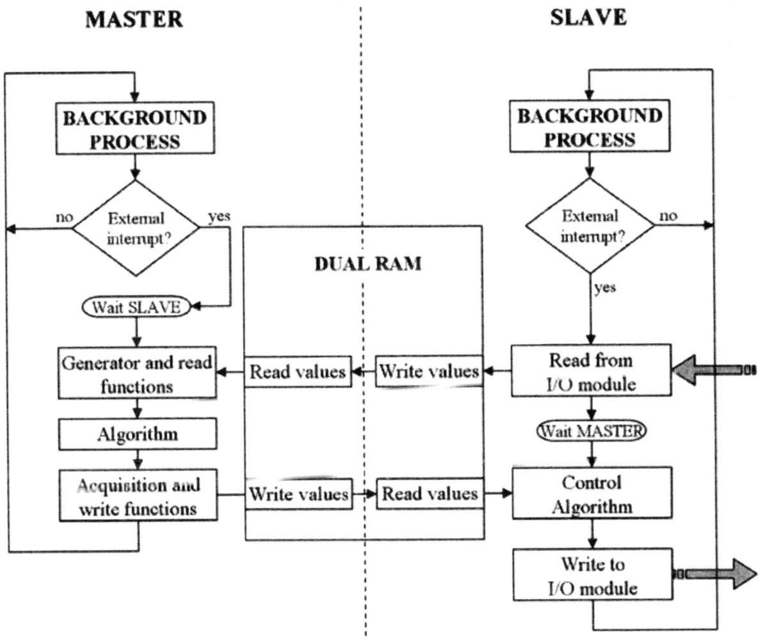

Figure 6. The master and the slave foreground processes.

1. When an external interrupt occurs (i.e. every 2ms, given by an external clock) simultaneously for the master and the slave, they start their foreground process (typically the hardware interrupt level is 10).
2. The master waits for the slave operation.
3. The slave reads the I/O module and writes the I/O value on the Dual-RAM, then waits for the master operation.
4. The master reads values on the Dual-RAM.

5. In the case of the pseudo-dynamic algorithm (the one of interest here), the master calculates the new target displacement according to its defined algorithm. This value is calculated taking into account the next base acceleration point, the masse of the structure and the measured restoring forces (read on the Dual-RAM).
6. The master writes the new target displacement on the Dual-RAM.
7. The slave reads values on the Dual-RAM.
8. The slave executes the control algorithm, PID for the pseudo-dynamic tests.
9. The slave writes the command signal to be passed to the actuator in the I/O modules and returns into the background process.

The data exchange is based on an internal database called AcqCtrlData. This database is composed by several memory blocks organized in several groups. These memory blocks contain the definition and value of several signals. Each signal value can be used via a variable. These signals are the key of the system, because the user can modify them in a starting file, in real time via a console windows or command windows and via the remote control program installed on the remote station. These memory blocks are exchanged to or from the master in mode asynchronous via a background process.

Only two memory blocks are exchanged at each interrupt (foreground process). They contain the data exchanged between the master and the slaves in foreground via the dual RAM.

In case of substructuring involving processes running on the master (experimental part) and on a remote computer (analytical part), the remote process will access the data using a (Microsoft) Distributed COM object. This access is asynchronous. For instance, when the master is updating one signal, this information will be sent to the internet port by a background process (the transmission is not guaranteed within 2ms). Then the progression over the internet depends on the connectivity available at that instant. Finally, even if the other process is in a waiting loop, the availability of the data at the level of the DCOM object is again depending on another background process.

The C++ software (Master.exe) running on the master is composed of several modules following the hierarchy of Figure 7. It loads at runtime the UserAlgorithm Dynamic-Link Library (DLL). This gives the final users the possibility to write their own algorithm without recompiling and touching the application Master.exe. The DLL can be used by Matlab to test the algorithm in off-line mode and by the application Master.exe in on-line mode (real-time). The openness of the system allows using it for any multi-actuator configurations such as PsD testing, cyclic & fatigue testing or structural control implementation.

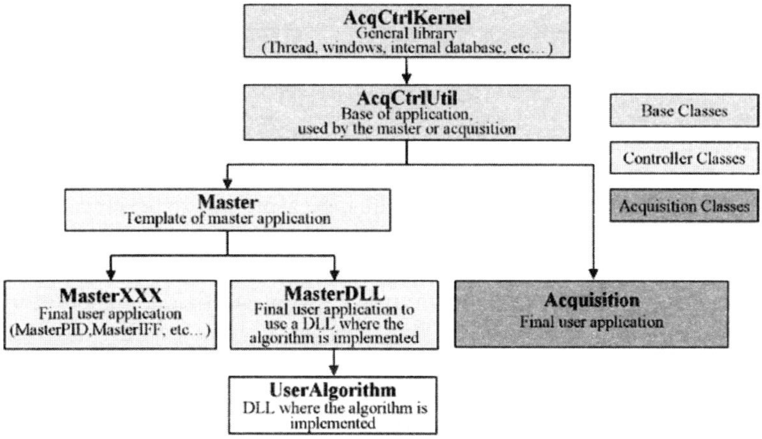

Figure 7. Real time software architecture

2.5 The Quality of a Continuous PsD Test

In terms of control, the main difference between the classical and the continuous PsD is that the former is performed in regulation mode, whereas the later is run in tracking mode. Regulation mode control (with undershoot depending on the sign of the force) is almost neutral in terms of damping whereas tracking mode control tends to introduce a negative damping (Combescure et al., 1995). Using a linear model, it is possible to show that the absolute error on the damping ratio is proportional to the delay and inversely proportional to the time magnification factor (Molina et al., 2002). This shows how the PsD method can be highly sensitive to time delays at the control system. It is clear that for low-damped high-frequency modes of the structure the response may easily show negative total damping whenever the test is not conducted slowly enough. It is thus important to limit the control error. The energy absorbed by the structure during the test may be computed from the measured restoring forces $\mathbf{r}$ and displacements $\mathbf{d}_\mathrm{m}$ as

$$E = \int \mathbf{r}^\mathrm{T} d\mathbf{d}_\mathrm{m} \tag{2.5}$$

whereas considering the target or reference displacement $\mathbf{d}$ given by the time integration scheme, it is possible to contruct a reference energy as

$$E_{\mathrm{r}} = \int \mathbf{r}^{\mathrm{T}} d\mathbf{d} \tag{2.6}$$

The energy error is the difference between them, which is also the work done by the restoring forces at the error displacements

$$\Delta E = E_{\mathrm{r}} - E = \int \mathbf{r}^{\mathrm{T}} d(\mathbf{d} - \mathbf{d}_{\mathrm{m}}) \tag{2.7}$$

It is then possible to show (Molina et al., 2002) that the lower is ΔE with respect to E, the lower is the damping error.

When applying this variable as an indicator, our experience is that in general, for an acceptable quality in the test results, the energy error should not exceed, say, a 5% of the total absorbed energy at any moment. It is thus advisable to perform, at the beginning of a test campaign, a short-duration very-low-intensity test, varying the number of substeps M (and thus λ), in order to check this criterion. If the criterion is satisfied for that small test in the linear range of the structure, the quality of the results should in general improve for larger amplitudes because, with the degradation of the structure, its damping ratios should grow at the same time that the damping error introduced by the control should decrease due to the stiffness loss.

Other, more elaborated, estimators of the quality of the results are based on the identification of linear models using a short-time portion of the response. The identified parameters are then transformed into frequency and damping characteristics. By gaining experience in the application of these analysing techniques, it is possible to have a better knowledge of how the feasibility and the accuracy of the experiment depend on the type of structure and the applied PsD testing set-up (Molina and Géradin, 2007).

3 Substructuring

3.1 Introduction

Substructuring is one of the main advantages of PsD testing. It is often called "hybrid testing" in the US since it combines even more obviously modelling and experiment, and since "Pseudo"-dynamic can be understood in a denigrative way.

Substructuring leads to full scale testing since it allows to concentrate the experimental model only on the critical part of a structure. In contrast with conventional cyclic testing, substructuring represents more realistically the demand on critical parts of a structure. Tests on substructures usually reduce the number and the interaction of the DoFs controlled at the laboratory. They are thus potentially able to be run fast, close to or even at real time.

The scope of this chapter is to present an algorithm which allows to keep the high quality of the continuous approach at the laboratory and which runs in parallel with another process handling the DoFs of the modelled part of the structure.

There are various types of substructuring, and in order to be more precise it is important to introduce some vocabulary.

1. The time integration scheme can be *explicit* or *implicit*: The earliest PsD time integration scheme has been the Newmark Central Difference (CD) scheme. This explicit scheme does not need any modelling of the stiffness of the tested structure but is conditionally stable. It is often felt that, for substructuring, due to the multiplication of the DoFs, an unconditionally implicit scheme should be used (Pan et al., 2006). These schemes, for instance the Newmark trapezoidal rule or the α-method, with or without iterations, as with the Operator Splitting (OS) method, always introduce a model of the stiffness of the tested structure. The OS approach in particular delivers better results if the tangent stiffness of the structure is known, which could be a challenge during a test. Then although the CD method often appears as the best scheme to use for PsD testing, it is often claimed that "only an implicit scheme can be used in case of substructuring". In this section we will present an alternative approach which combines the best of what is available. Use of the CD approach for the DoFs controlled at the laboratory and use of an implicit solver allowing an exact evaluation of the tangent stiffness for the numerical DoFs.
2. The time integration algorithm can be *parallel* or *monolithic*: In the monolithic case, all the degrees of freedom of the structures are handled by a unique process whereas in the parallel approach, at least two processes are running in parallel: one responsible for the analytical DoFs and one for the experimental DoFs. The monolithic approach has been widely used at the early stage of the substructuring, in combination with a concentrated hardware (see the next bullet). Some attempts to parallelise the process have been made in ELSA (Buchet and Pegon, 1994), leading to the first campaign of large bridge tests using the classical PsD method (Pegon and Pinto, 2000). Further on, due to the difficulty of maintaining a parallel approach together with the continuous PsD method, a monolithic approach has been adopted for the second campaign of large bridge tests (Pinto et al., 2004) distributed over the local internet. This approach paved the way for the first implementation of geographically distributed testing (MOST, 2003) and is the one in use in the currently proposed environment for distributed testing UI-SIMCOR (Kwon et al., 2005), OpenFresco (Takahashi and

Fenves, 2006) and ISEE (Yang et al., 2007).

3. The implementation can be *concentrated* or *distributed*: In the early stage of the PsD method, the implementation was concentrated on one hardware handling everything. Gradually, and in particular after the introduction of digital control systems, the implementation has become gradually distributed on various hardware resources distributed over a network.
4. The processes can be *synchronous* or *asynchronous*: In the case of a synchronous implementation, all the time steps of the accelerogram have the same laboratory duration: there is a uniform time scaling (λ is constant). This is the case of continuous testing, and this is the only way to reach real time capabilities. When the involved processes are asynchronous, the laboratory duration of one time step can vary from time to time and the time scale is variable. This is the case of the classical testing, and by extension of the current implementations of distributed testing.
5. The substructuring approach can be *fast* or *slow*: during a fast test the dilation parameter λ is lower than 10. During a slow test, a typical value of λ can be 100 or more.
6. The network used for implementing a distributed approach can be *deterministic* or *non-deterministic*: on a deterministic network the needed information are guaranteed to be transferred within a known time interval. This is the case when a bus and dual RAM or a Shared Common RAM Network (SCRAMNet®) are used to connect the resources performing the time integration and the control of the experiment. When a reasonable time interval cannot be guaranteed, such as on the internet, we are facing a non-deterministic network.

Most of the current implementations of the PsD testing and in particular the one involving distributed tests use an implicit time integration algorithm, a monolithic time integration, a distributed implementation, an asynchronous processes leadind to a variable time scale, running over a non-deterministic network resulting always in a slow test.

In this case, the uniform performance and the test speed are not an issue: it is thus possible to add as many software layers as needed and to rely on well assessed implicit algorithm (such as the α-OS scheme). It allows a high level of modularity and in particular the complete software separation between algorithms and control. In the monolithic implementation, the experimental parts can be considered as a collection of special elements: it is thus possible to talk of hybrid *simulation*.

The positive aspect of this approach, is that it allows standardization. Different laboratories running completely different control system can be connected through the simulation software layer, as long as a convenient interface has been implemented. Finally, security issues can be also addressed without adding a dramatic overhead. This is thus the natural framework for long distance geographically distributed testing.

When uniform performance is the issue, and in particular for fast testing, a synchronous approach (constant time scaling) is needed. A possibility is to introduce a monolithic approach using an iterative scheme and constructing on-line the trajectory of the experimental DoFs according to a pre-defined interpolation (Jung et al., 2007). What is proposed here is to maintain the use of continuous PsD method: the trajectory of the experimental DoFs will depend on the time integration. However, it is difficult to cope with such situations, where the number of DoFs describing the analytical structure is high, or when non-linear phenomena are modelled, because the laboratory time needed for computing one accelerogram time step may easily exceed the small value (1 or 2ms) of the continuous system. The solution to overcome this difficulty is to rely on a parallel algorithm using sub-stepping. The analytical process will work with an accelerogram time step larger than the experimental process. In this case where control and algorithms are still closely tight together, hybrid *testing* is performed.

3.2 The Basic Coupling Algorithm

Newmark Schemes Consider the following system of semi-discrete differential equations where a linear system has been chosen for convenience (see subsection 3.4 for the non-linear case)

$$\mathbf{M}\mathbf{a} + \mathbf{C}\mathbf{v} + \mathbf{K}\mathbf{d} = \mathbf{f} \tag{3.1}$$

which describes the motion of an elastic structure with m degrees-of freedom. $\mathbf{a}$, $\mathbf{v}$ and $\mathbf{d}$ are the acceleration, the velocity and the displacement vectors, $\mathbf{f}$ the external force vector, $\mathbf{M}$, $\mathbf{C}$ and $\mathbf{K}$ the mass, the damping and the stiffness matrices.

The system of equations given by Eq.(3.1) is integrated with respect to time using the classical β-Newmark time integration scheme. In such scheme, knowing the displacement vector $\mathbf{d}^n$, the velocity vector $\mathbf{v}^n$ and the acceleration vector $\mathbf{a}^n$ at time t^n, the values of these vectors are obtained at time $t^{n+1} = t^n + \Delta t$ according to

$$\begin{aligned} \mathbf{d}^{n+1} &= \tilde{\mathbf{d}}^{n+1} + \beta\Delta t^2\mathbf{a}^{n+1} & \tilde{\mathbf{d}}^{n+1} &= \mathbf{d}^n + \Delta t\mathbf{v}^n + (\frac{1}{2} - \beta)\Delta t^2\mathbf{a}^n \\ \mathbf{v}^{n+1} &= \tilde{\mathbf{v}}^{n+1} + \gamma\Delta t\mathbf{a}^{n+1} & \tilde{\mathbf{v}}^{n+1} &= \mathbf{v}^n + (1-\gamma)\Delta t\mathbf{a}^n \end{aligned} \tag{3.2}$$

and through verification of the time discrete system of equilibrium equations

$$\mathbf{M}\mathbf{a}^{n+1} + \mathbf{C}\mathbf{v}^{n+1} + \mathbf{K}\mathbf{d}^{n+1} = \mathbf{f}^{n+1} \tag{3.3}$$

$\mathbf{d}^{n+1}$ and $\mathbf{v}^{n+1}$ are thus derived from the predictor values of the displacement $\tilde{\mathbf{d}}^{n+1}$ and the velocity $\tilde{\mathbf{v}}^{n+1}$, and from a correction based on the acceleration $\mathbf{a}^{n+1}$, solution of the linear system

$$\hat{\mathbf{M}}\mathbf{a}^{n+1} = \hat{\mathbf{f}}^{n+1} \tag{3.4}$$

where the pseudo mass matrix $\hat{\mathbf{M}}$ and the pseudo force vector $\hat{\mathbf{f}}^{n+1}$ are given by

$$\begin{aligned} \hat{\mathbf{M}} &= \mathbf{M} + \gamma\Delta t\mathbf{C} + \beta\Delta t^2\mathbf{K} \\ \hat{\mathbf{f}}^{n+1} &= \mathbf{f}^{n+1} - \mathbf{K}\tilde{\mathbf{d}}^{n+1} - \mathbf{C}\tilde{\mathbf{v}}^{n+1} \end{aligned} \tag{3.5}$$

It is well known that the choice $\gamma = 1/2$ leads to second order accurate schemes. In addition, two values of β are of particular interest: $\beta = 0$, which corresponds to the explicit, conditionally stable, central difference scheme and $\beta = 1/4$, which corresponds to the implicit, unconditionally stable, so-called trapezoidal rule.

The Basic Coupling Algorithm Following (Gravouil and Combescure, 2001), the structure under consideration is split into two subdomains A and B. The continuity of the velocity is assumed between the two domains and ensured using a Lagrange multiplier technique. The semi-discrete problem (3.1) thus becomes

$$\begin{gathered} \mathbf{M}_{\mathrm{A}}\mathbf{a}_{\mathrm{A}} + \mathbf{C}_{\mathrm{A}}\mathbf{v}_{\mathrm{A}} + \mathbf{K}_{\mathrm{A}}\mathbf{d}_{\mathrm{A}} = \mathbf{f}_{\mathrm{A}} + \mathbf{L}_{\mathrm{A}}^{\mathrm{T}}\mathbf{\Lambda} \quad \mathbf{M}_{\mathrm{B}}\mathbf{a}_{\mathrm{B}} + \mathbf{K}_{\mathrm{B}}\mathbf{d}_{\mathrm{B}} = \mathbf{f}_{\mathrm{B}} + \mathbf{L}_{\mathrm{B}}^{\mathrm{T}}\mathbf{\Lambda} \\ \mathbf{L}_{\mathrm{A}}\mathbf{v}_{\mathrm{A}} + \mathbf{L}_{\mathrm{B}}\mathbf{v}_{\mathrm{B}} = 0 \end{gathered} \tag{3.6}$$

where $\mathbf{L}_{\mathrm{A}}$ and $\mathbf{L}_{\mathrm{B}}$ are connectivity matrices for the perfect connection between the subdomains A and B, and $\mathbf{\Lambda}$, the vector of Lagrange multipliers. Note that this formulation put into duality the connected kinematic quantities (velocities) and the resulting reaction forces (Lagrange multipliers) modifying the equilibrium of each subdomain.

Typically, in the context of the substructuring technique, A would be the analytical part of the structure, integrated in time using the trapezoidal rule, whereas B would be the experimental part, integrated using the central difference scheme. We can thus assume that A is associated with the coarse time scale Δt whereas B uses the fine time step δt with $\Delta t = M\delta t$ (see Figure 8).

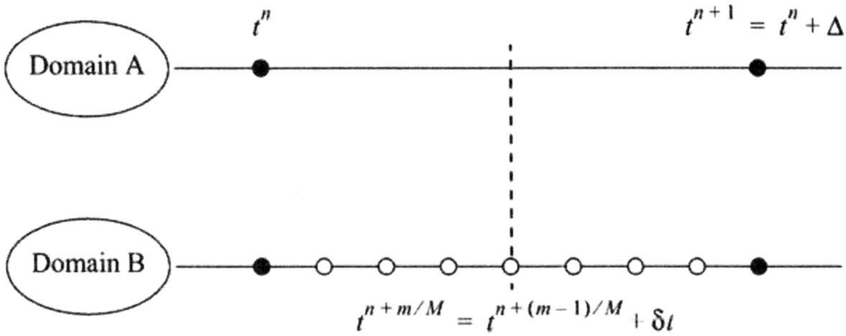

Figure 8. Time stepping in the two subdomains.

Note that a damping matrix $\mathbf{C}_{\mathrm{A}}$ has been introduced only for the analytical part. Since domain A corresponds to an analytical structure, it is important to introduce a viscous equivalent damping to represent micro-hysteresis at low amplitude.

The solution in both subdomains is split into an unconstrained (free) and a constrained (link) part, corresponding to the solution on each subdomain without and only considering the interface force, respectively. Then we have

$$\begin{array}{lll} \mathbf{d}_f^{n+1} = \tilde{\mathbf{d}}^{n+1} + \beta\Delta t^2\mathbf{a}_f^{n+1} & \mathbf{d}_l^{n+1} = \beta\Delta t^2\mathbf{a}_l^{n+1} & \mathbf{d}^{n+1} = \mathbf{d}_f^{n+1} + \mathbf{d}_l^{n+1} \\ \mathbf{v}_f^{n+1} = \tilde{\mathbf{v}}^{n+1} + \gamma\Delta t\mathbf{a}_f^{n+1} & \mathbf{v}_l^{n+1} = \gamma\Delta t\mathbf{a}_l^{n+1} & \mathbf{v}^{n+1} = \mathbf{v}_f^{n+1} + \mathbf{v}_l^{n+1} \end{array} \tag{3.7}$$

For example, on subdomain A, the free solution is immediately derived from

$$\hat{\mathbf{M}}\mathbf{a}_{\mathrm{A}f}^{n+1} = \hat{\mathbf{f}}_{\mathrm{A}}^{n+1} \tag{3.8}$$

whereas the link solution is obtained by solving

$$\hat{\mathbf{M}}\mathbf{a}_{\mathrm{A}l}^{n+1} = \mathbf{L}_{\mathrm{A}}^{\mathrm{T}}\mathbf{\Lambda}^{n+1} \tag{3.9}$$

and

$$\mathbf{L}_{\mathrm{A}}(\mathbf{v}_{\mathrm{A}l}^{n+1} + \mathbf{v}_{\mathrm{A}f}^{n+1}) + \mathbf{L}_{\mathrm{B}}(\mathbf{v}_{\mathrm{B}l}^{n+1} + \mathbf{v}_{\mathrm{B}f}^{n+1}) = 0 \tag{3.10}$$

which, recalling

$$\mathbf{v}_{\mathrm{A}l}^{n+1} = \gamma_{\mathrm{A}}\Delta t\mathbf{a}_{\mathrm{A}l}^{n+1} \tag{3.11}$$

can be cast in the following matrix form

$$\alpha_{\mathrm{A}}\left[\begin{array}{cc}\hat{\mathbf{M}}_{\mathrm{A}} & -\mathbf{L}_{\mathrm{A}}^{\mathrm{T}} \\ -\mathbf{L}_{\mathrm{A}} & 0\end{array}\right]\left[\begin{array}{c}\mathbf{a}_{\mathrm{A}l}^{n+1} \\ \mathbf{\Lambda}^{n+1}\end{array}\right]=\left[\begin{array}{c}0 \\ \mathbf{L}_{\mathrm{A}}\mathbf{v}_{\mathrm{A}f}^{n+1}+\mathbf{L}_{\mathrm{B}}(\mathbf{v}_{\mathrm{B}l}^{n+1}+\mathbf{v}_{\mathrm{B}f}^{n+1})\end{array}\right]$$
$$\alpha_{\mathrm{A}}=\gamma_{\mathrm{A}}\Delta t \tag{3.12}$$

The same operations can be performed on domain B and leads to

$$\hat{\mathbf{M}}_{\mathrm{B}}\mathbf{a}_{\mathrm{B}f}^{n+m/M}=\hat{\mathbf{f}}_{\mathrm{B}}^{n+m/M} \tag{3.13}$$

and

$$\alpha_{\mathrm{B}}\left[\begin{array}{cc}\hat{\mathbf{M}}_{\mathrm{B}} & -\mathbf{L}_{\mathrm{B}}^{\mathrm{T}} \\ -\mathbf{L}_{\mathrm{B}} & 0\end{array}\right]\left[\begin{array}{c}\mathbf{a}_{\mathrm{B}l}^{n+m/M} \\ \mathbf{\Lambda}^{n+m/M}\end{array}\right]=\left[\begin{array}{c}0 \\ \mathbf{L}_{\mathrm{B}}\mathbf{v}_{\mathrm{B}f}^{n+m/M}+\mathbf{L}_{\mathrm{A}}(\mathbf{v}_{\mathrm{A}l}^{n+m/M}+\mathbf{v}_{\mathrm{A}f}^{n+m/M})\end{array}\right]$$
$$\alpha_{\mathrm{B}}=\gamma_{\mathrm{B}}\delta t \tag{3.14}$$

The problem is now to relate $\mathbf{v}_{\mathrm{A}}^{n+m/M}$ to $\mathbf{v}_{\mathrm{A}}^{n}$ and $\mathbf{v}_{\mathrm{A}}^{n+1}$, and $\mathbf{\Lambda}^{n+m/M}$ to $\mathbf{\Lambda}^{n}$ and $\mathbf{\Lambda}^{n+1}$. Thus, the following time linearizations are introduced

$$\begin{aligned}\mathbf{L}_{\mathrm{A}}\mathbf{v}_{\mathrm{A}}^{n+m/M}&=(1-\frac{m}{M})\mathbf{L}_{\mathrm{A}}\mathbf{v}_{\mathrm{A}}^{n}+\frac{m}{M}\mathbf{L}_{\mathrm{A}}\mathbf{v}_{\mathrm{A}}^{n+1}\\ \mathbf{\Lambda}^{n+m/M}&=(1-\frac{m}{M})\mathbf{\Lambda}^{n}+\frac{m}{M}\mathbf{\Lambda}^{n+1}\end{aligned} \tag{3.15}$$

Using the linearization of the velocity, and being aware that the free solution on A (Eq.(3.8)) and B (Eq.(3.13)) can be readily obtained, the only obstacle for solving the constrained problem on B comes out from the term $\mathbf{L}_{\mathrm{A}}\mathbf{v}_{\mathrm{A}l}^{n+m/M}$ in Eq.(3.14).

Using the linearization (3.15)a on $\mathbf{v}_{\mathrm{A}l}^{n+m/M}$, then expressing $\mathbf{v}_{\mathrm{A}l}^{n}$ and $\mathbf{v}_{\mathrm{A}l}^{n+1}$ in terms of $\mathbf{\Lambda}^{n}$ and $\mathbf{\Lambda}^{n+1}$ using Eqs. (3.11) and (3.9), and taking profit of the linearization (3.15)b, the following expression for $\mathbf{L}_{\mathrm{A}}\mathbf{v}_{\mathrm{A}l}^{n+m/M}$ can be easily derived

$$\mathbf{L}_{\mathrm{A}}\mathbf{v}_{\mathrm{A}l}^{n+m/M}=\alpha_{\mathrm{A}}\mathbf{L}_{\mathrm{A}}\hat{\mathbf{M}}_{\mathrm{A}}^{-1}\mathbf{L}_{\mathrm{A}}^{\mathrm{T}}\mathbf{\Lambda}^{n+m/M} \tag{3.16}$$

Then, introducing this expression in system (3.14) and condensing it on the interface, leads to the following problem for $\mathbf{\Lambda}^{n+m/M}$

$$\mathbf{H}_{\mathrm{AB}}\mathbf{\Lambda}^{n+m/M}=-(\mathbf{L}_{\mathrm{A}}\mathbf{v}_{\mathrm{A}f}^{n+m/M}+\mathbf{L}_{\mathrm{B}}\mathbf{v}_{\mathrm{B}f}^{n+m/M}) \tag{3.17}$$

with

$$\mathbf{H}_{\mathrm{AB}}=\alpha_{\mathrm{A}}\mathbf{L}_{\mathrm{A}}\hat{\mathbf{M}}_{\mathrm{A}}^{-1}\mathbf{L}_{\mathrm{A}}^{\mathrm{T}}+\alpha_{\mathrm{B}}\mathbf{L}_{\mathrm{B}}\hat{\mathbf{M}}_{\mathrm{B}}^{-1}\mathbf{L}_{\mathrm{B}}^{\mathrm{T}} \tag{3.18}$$

Finally, the link solution on subdomain B is obtained by solving

$$\hat{\mathbf{M}}_{\mathrm{B}}\mathbf{a}_{\mathrm{B}l}^{n+m/M} = \mathbf{L}_{\mathrm{B}}^{\mathrm{T}}\mathbf{\Lambda}^{n+m/M} \tag{3.19}$$

The solution of system (3.17) is the key for the success of the coupling method. It ensures on each substep of the solution process on subdomain B the complete verification of the velocity relationship (3.6)c at the interface between the subdomains, according to the linearizations of Eqs. (3.15).

The solution of the coupled problem between A and B is given in the flowchart:

i.	Solve the unconstrained probem on subdomain A at time t^{n+1}	see(14) and (15)
ii.	Loop on the Mtime steps of subdomain B at time $t^{n+m/M}$	
iii.	Solve the unconstrained problem on subdomain B	see(14) and (20)
iv.	Compute $\mathbf{L}_{\mathrm{A}}\mathbf{v}_{\mathrm{A}f}^{n+m/M}$	see(22a)
v.	Calculate the Lagrange multipliers at the interface	see(24)
vi.	Solve the constrained problem on subdomain B	see(26)
vii.	Calculate the kinematic quantities on subdomain B	$(\ ) = (\ )_f + (\ )_l$
viii.	If $m = M$ then end loop	
ix.	Solve the constrained problem on subdomain A	see(16)
x.	Calculate the kinematic quantities on subdomain A	$(\ ) = (\ )_f + (\ )_l$

The kinematic continuity between the two subdomains can be expressed in terms of acceleration, velocity or displacement. It is shown in (Gravouil and Combescure, 2001) that the continuity in velocities does not introduce any restriction in the choice of the parameters β and γ, in contrast with the continuity in terms of acceleration or displacement. In addition, it is shown that continuity in displacement leads to almost always unstable schemes, if one domain is explicit and the other one implicit. This explains why the method proposed for instance in (Pegon and Magonette, 1999), based on a continuity of displacement, worked only with very few degrees of freedom.

Stability and Convergence of the Basic Coupling Algorithm In (Gravouil and Combescure, 2001) it is shown that the stability of the global problem depends on the stability conditions of the Newmark schemes considered in each subdomain. Then, if $\gamma_{\mathrm{A}} = 1/2$ and $\beta_{\mathrm{A}} = 1/4$ (trapezoidal rule) and $\gamma_{\mathrm{B}} = 1/2$ and $\beta_{\mathrm{B}} = 0$ (central difference), and providing that the number of substeps M is large enough to allow stability on subdomain B, the basic algorithm is stable.

The global convergence rate is defined as the minimum of the convergence rates for each subdomains. For the set of parameters chosen above, the scheme is globally of the second order.

Although the algorithm is stable, it can present numerical dissipation at the interface. It is shown in (Gravouil and Combescure, 2001) that this dissipation equals zero when the velocity on subdomain A is constant or when $M = 1$. Otherwise, some dissipation is observed at the interface: when the displacement of subdomain A is not linear with time, the smaller the local radius of curvature and the higher the numerical dissipation.

3.3 From Staggered to Inter-Field Procedure

Staggered and Inter-Field Procedures Partitioned procedures, i.e. methods aimed at splitting the general solution procedure of a complex problem into smaller, more dedicated procedures working in parallel have been already widely used. One field of particular interest is the one of fluid/structure interaction because the fluid solvers are often explicit as the experimental process in our case. In particular, algorithms with subcycling (for the fluid part) have already been introduced and it appears very interesting to also benefit from this experience. A short overview of such partitioned procedures was made in (Pegon and Magonette, 1999) and has shown that they can be classified in two groups: staggered and inter-field procedures.

1. *Staggered procedure*: The solution is first advanced in time on domain A. Then the interface informations are transferred to domain B which is advanced in turn. Finally, the interface informations are transferred back to domain A (see Figure 9).

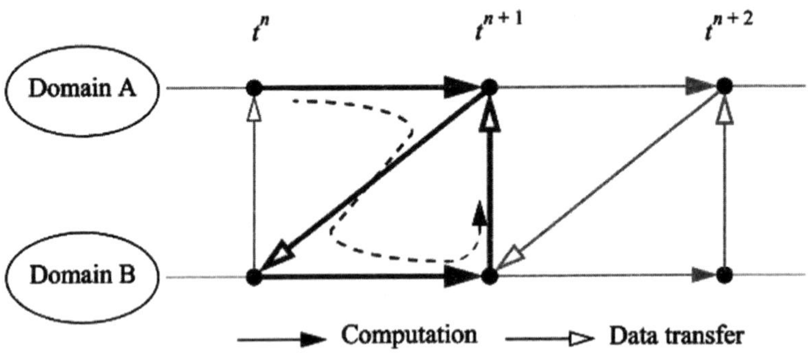

Figure 9. The classical staggered procedure.

This procedure is inherently sequential: one domain is advanced, and then the other. The staggered procedure allows to implement intra-field parallelism (parallel computations within each domain), but inter-field parallelism is inhibited. The processes in charge of the computations on A and B are intrinsically asynchronous since they always have to wait one for the other.

It is important to note that the basic algorithm presented in section 3.2.2 is staggered. With reference to the steps of Figure 9, it can be roughly said that steps ix, x and i correspond to the advancement of the solution on subdomain A. Then the free velocity at the interface is transmitted to subdomain B where the loop of step ii is realized. Then the values of the Lagrange multipliers at the end of the loop are transferred back to subdomain A.

As it is, the basic algorithm is not suitable for the continuous PsD method since the experimental process has to be stopped from time to time in order to wait for the results coming from the analytical process.

2. *Inter-field procedure*: The solution is advanced in time on both domains and the informations are exchanged at the end of the time step (see Figure 10).

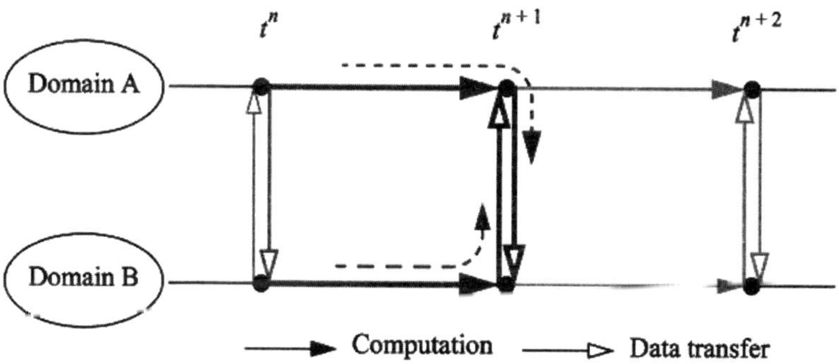

Figure 10. The inter-field procedure.

Using this procedure, it is clear that computations on each domain can run in parallel during the time interval $\left[t^n, t^{n+1}\right]$. Inter-field communication and I/O transfer are needed only at the beginning of each time interval. The processes in charge of the solution on the subdomaisn can be synchronous providing that the period of I/O transfer is shorter than the maximum computation time on each domain.

When compared with the classical staggered procedure, the parallelism is usually achieved at the expense of amplifying the errors in the subdomains and this result holds also when subcycling is introduced in one subdomain. An inter-field procedure would be ideal for continuous PsD testing. Knowing an upper bound of the CPU time of computation of a large step for the analytical process, the number of (synchronized) substeps for the experimental process would follow, together with stability considerations.

An Inter-Field Modification of the Basic Scheme Based on the work already performed in (Pegon and Magonette, 1999), the staggered basic scheme has been modified in order to be implemented in an inter-field manner. Two basic ideas are used:

1. The computation of the free velocity $\mathbf{v}_{\mathrm{A}f}$ on domain A is performed *in advance* with respect to what is made on domain B, by using a Newmark scheme with a *double* large time step $2\Delta t$. Typically $\mathbf{v}_{\mathrm{A}f}^{n+2}$ is computed in parallel with the determination of $\mathbf{v}_{\mathrm{B}}^{n+m/M}$, or in other words, $\mathbf{v}_{\mathrm{A}f}^{n+1}$ is known before starting any calculation on B after t^n.
2. The knowledge of $\mathbf{\Lambda}^{n+1}$ at the end of the loop on B is used during the next computation interval on A to recompute the solution on subdomain A using the large time step Δt.

The inter-field modification of the basic scheme (IF scheme) works as follows (see Figure 11):

1. Domain A: knowing $\mathbf{\Lambda}^n$, first advance the solution from t^{n-1} to t^n, and then compute $\mathbf{v}_{\mathrm{A}f}^{n+2}$ from t^n to t^{n+2}.
2. Domain B: knowing $\mathbf{v}_{\mathrm{A}f}^{n+1}$, perform the substeps and compute $\mathbf{\Lambda}^{n+1}$.

Although the two schemes are not working at the same time level (A is spanning from t^{n-1} to t^{n+2} while B works between t^n and t^{n+1}), they respect the inter-field format: parallel computation and synchronized exchange of information.

A simplified version of the scheme, presented in Figure 12 (SIF scheme), is derived by using $\mathbf{\Lambda}^n$ in order to compute $\mathbf{v}_{\mathrm{A}l}^n$ and thus complete the calculation performed previously with the double time step $2\Delta t$. In that case, no correction step between t^{n-1} to t^n is needed, and the velocity of the two domains at the connection points is guaranteed by construction, which is not the case for the IF scheme. The problem with the SIF scheme is that all the calculations on A are split into two almost independent interlaced schemes, one working on the odd time levels, and the other on the even ones, and leap-frogging over each other during the time steps. Then, as a first drawback, the accuracy of the scheme on A is related to $2\Delta t$ instead of Δt. A more severe drawback comes from the fact that 2 almost completely independent sets of internal variables would be introduced along the time steps when non-linear computations are to be performed. Only one set is used when the correction of the solution is allowed between t^{n-1} to t^n for the IF scheme, even if a prediction of them is used to compute $\mathbf{v}_{\mathrm{A}f}^{n+2}$.

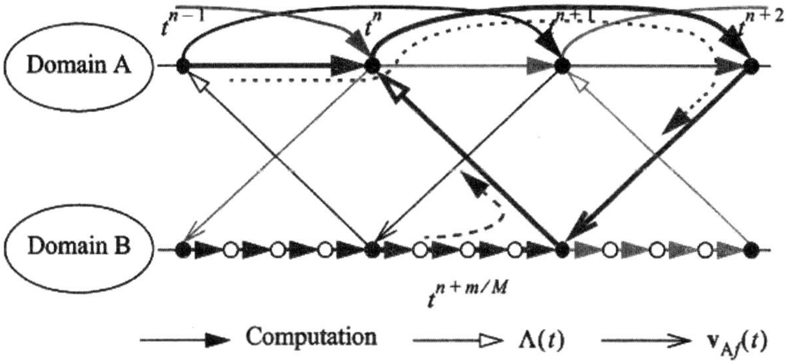

Figure 11. The inter-field modification of the basic scheme.

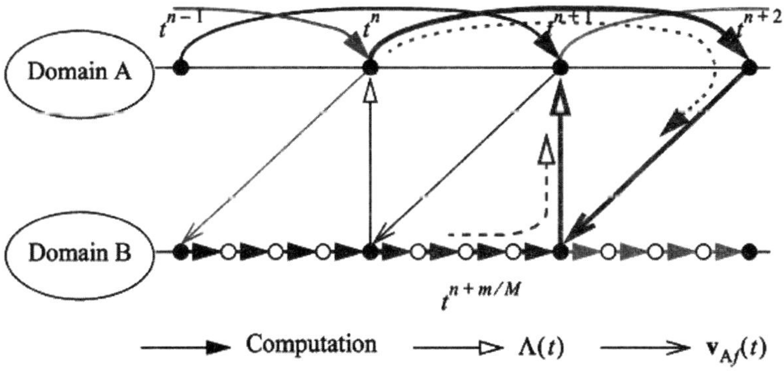

Figure 12. A simplified inter-field modification of the basic scheme.

An attempt was made to couple more strongly domain A and B for the IF scheme. Instead of imposing $\mathbf{\Lambda}^n$ for solving the constrained problem from t^{n-1} to t^n, it is possible in practice to impose directly to the connection points the velocity $\mathbf{v}_{\mathrm{B}}^n$ of domain B: unfortunately the resulting scheme is almost always unstable. Imposing the acceleration $\mathbf{a}_{\mathrm{B}}^n$ or the displacement $\mathbf{d}_{\mathrm{B}}^n$ results also in unstable schemes.

For performing the correction step on domain A between t^{n-1} and t^n, since $\mathbf{\Lambda}^n$ is already known, there is no need to distinguish between $\mathbf{a}_{\mathrm{A}f}^n$ and

$\mathbf{a}^n_{Al}$. From Eqs.(3.8) and (3.9), $\mathbf{a}^n_{A}$ can be derived in one step as the solution of

$$\hat{\mathbf{M}}_A \mathbf{a}^n_A = \hat{\mathbf{f}}^n_A + \mathbf{L}^T_A \mathbf{\Lambda}^n \tag{3.20}$$

The linearization of the velocity leading to $\mathbf{v}^{n+m/M}_{Af}$, and used in order to solve the constrain problem on B, is performed with respect to the estimation of $\mathbf{v}_{Af}$ at t^n and t^{n+1} at the end of two successive integrations with the double time step $2\Delta t$. In fact, the value of $\mathbf{v}^n_{Af}$ at the end of the correction step could not be sent to the process performing the time integration on domain B between t^n and t^{n+1}, since the integration on B already started before the result of the correction on $\mathbf{v}^n_{Af}$ is even known. In addition, using this corrected quantity in the linearization would introduce a jump at the beginning of the loop on B.

As will be shown in the numerical experiment of section 4, The IF scheme shares the same accuracy characteristics as the basic scheme but results to be less robust. On the other hand, the SIF scheme is less accurate than the basic scheme but enjoys the same robustness. In addition, since the SIF scheme is simpler than the IF scheme, it has been first implemented at the laboratory. This is why the next section dealing with the non-linear case put more emphasis on SIF than on IF.

3.4 The Non Linear Case

In case of non-linearities, the system (3.6) takes the form

$$\begin{array}{c} \mathbf{M}_A \mathbf{a}_A + \mathbf{C}_A \mathbf{v}_A + \mathbf{r}_A(\mathbf{d}_A) = \mathbf{f}_A + \mathbf{L}^T_A \mathbf{\Lambda} \quad \mathbf{M}_B \mathbf{a}_B + \mathbf{r}_B(\mathbf{d}_B) = \mathbf{f}_B + \mathbf{L}^T_B \mathbf{\Lambda} \\ \mathbf{L}_A \mathbf{v}_A + \mathbf{L}_B \mathbf{v}_B = 0 \end{array} \tag{3.21}$$

where $\mathbf{r}$ is the internal force vector.

The Experimental Domain B Since the experimental domain B is handled with the central difference scheme ($\beta_B = 0$), the determination of the free solution of Eq.(3.13) at any substep $m+1$ is readily obtained through

$$\mathbf{d}^{n+(m+1)/M}_{Bf} = \tilde{\mathbf{d}}^{n+(m+1)/M}_{B} = \mathbf{d}^{n+m/M}_{B} + \delta t \mathbf{v}^{n+m/M}_{B} + \frac{\delta t^2}{2} \mathbf{a}^{n+m/M}_{B} \tag{3.22}$$

allowing to compute (or, as in the case of the continuous PsD method, to measure) $\mathbf{r}_B(\mathbf{d}^{n+(m+1)/M}_{Bf})$ and thus derive $\mathbf{a}^{n+(m+1)/M}_{Bf}$, which, together with $\mathbf{a}_{\mathrm{B}l}^{n+(m+1)/M}$ solution of Eq.(3.19), will give the value of $\mathbf{a}_{\mathrm{B}}^{n+(m+1)/M}$ from which the displacement at the next substep will be derived.

The experimental part of the structure is thus handled by the following 3-step procedure

First step: solving the unconstrained problem

$$\begin{aligned}
\mathbf{d}_{\mathrm{B}f}^{n+(m+1)/M} &= \tilde{\mathbf{d}}_{\mathrm{B}f}^{n+(m+1)/M} \\
\tilde{\mathbf{d}}_{\mathrm{B}f}^{n+(m+1)/M} &= \mathbf{d}_{\mathrm{B}}^{n+m/M} + \delta t \mathbf{v}_{\mathrm{B}}^{n+m/M} + \frac{\delta t^2}{2}\mathbf{a}_{\mathrm{B}}^{n+m/M} \\
\mathbf{v}_{\mathrm{B}f}^{n+(m+1)/M} &= \tilde{\mathbf{v}}_{\mathrm{B}f}^{n+(m+1)/M} + \frac{\delta t}{2}\mathbf{a}_{\mathrm{B}f}^{n+(m+1)/M} \\
\tilde{\mathbf{v}}_{\mathrm{B}f}^{n+(m+1)/M} &= \mathbf{v}_{\mathrm{B}}^{n+m/M} + \frac{\delta t}{2}\mathbf{a}_{\mathrm{B}}^{n+m/M} \\
\mathbf{a}_{\mathrm{B}f}^{n+(m+1)/M} &= \mathbf{M}_{\mathrm{B}}^{-1}\left\{\mathbf{f}_{\mathrm{B}}(t^{n+(m+1)/M}) - \mathbf{r}_{\mathrm{B}}(\tilde{\mathbf{d}}_{\mathrm{B}f}^{n+(m+1)/M})\right\}
\end{aligned} \tag{3.23}$$

Second step: solving the interface problem

$$\mathbf{\Lambda}^{n+(m+1)/M} = -\mathbf{H}_{\mathrm{AB}}^{-1}(\mathbf{L}_{\mathrm{A}}\mathbf{v}_{\mathrm{A}f}^{n+(m+1)/M} + \mathbf{L}_{\mathrm{B}}\mathbf{v}_{\mathrm{B}f}^{n+(m+1)/M}) \tag{3.24}$$

with

$$\begin{aligned}
\mathbf{L}_{\mathrm{A}}\mathbf{v}_{\mathrm{A}f}^{n+(m+1)/M} &= \left(1 - \frac{(m+1)}{M}\right)\mathbf{L}\mathbf{v}_{\mathrm{A}f}^{n} + \frac{(m+1)}{M}\mathbf{L}\mathbf{v}_{\mathrm{A}f}^{n+1} \\
\mathbf{H}_{\mathrm{AB}} &= \mathbf{H}_{\mathrm{A}} + \frac{\delta t}{2}\mathbf{L}_{\mathrm{B}}\mathbf{M}_{\mathrm{B}}^{-1}\mathbf{L}_{\mathrm{B}}^{\mathrm{T}}
\end{aligned} \tag{3.25}$$

Third step: solving the constrained problem

$$\begin{aligned}
\mathbf{d}_{\mathrm{B}}^{n+(m+1)/M} &= \mathbf{d}_{\mathrm{B}f}^{n+(m+1)/M} \\
\mathbf{v}_{\mathrm{B}}^{n+(m+1)/M} &= \mathbf{v}_{\mathrm{B}f}^{n+(m+1)/M} + \frac{\delta t}{2}\mathbf{a}_{\mathrm{B}l}^{n+(m+1)/M} \\
\mathbf{a}_{\mathrm{B}}^{n+(m+1)/M} &= \mathbf{a}_{\mathrm{B}f}^{n+(m+1)/M} + \mathbf{a}_{\mathrm{B}l}^{n+(m+1)/M} \\
\mathbf{a}_{\mathrm{B}l}^{n+(m+1)/M} &= \mathbf{M}_{\mathrm{B}}^{-1}(\mathbf{L}_{\mathrm{B}}^{\mathrm{T}}\mathbf{\Lambda}^{n+(m+1)/M})
\end{aligned} \tag{3.26}$$

The following remarks can be performed:

1. The first step corresponds obviously to the standard central difference scheme for the experimental structure. Note also that, following Eq.(3.7) (detailed in Eq.(3.23) and Eq.(3.26), $\mathbf{d}_{\mathrm{B}} = \mathbf{d}_{\mathrm{B}f} = \tilde{\mathbf{d}}_{\mathrm{B}f}$. Thus, the displacement to be imposed to the structure is also the usual explicit displacement of the central difference scheme without substructuring.

2. The second and the third steps can be seen as a correction to the standard central difference scheme, accounting for the analytical structure. Since they only affect the derivation of $\mathbf{v}_{\mathrm{B}}$ and $\mathbf{a}_{\mathrm{B}}$, these steps are purely numerical.
3. The solution of the second step requires to know the unconstrained solution on both the experimental and the analytical domains. If $\mathbf{v}_{\mathrm{B}f}^{n+(m+1)/M}$ was just obtained during the first step, it should be underlined that not only $\mathbf{L}_{\mathrm{A}}\mathbf{v}_{\mathrm{A}f}^{n}$ but also $\mathbf{L}_{\mathrm{A}}\mathbf{v}_{\mathrm{A}f}^{n+1}$ should be known already at the beginning of the loop on m. This is why the unconstrained solution on the analytical structure should be known Δt in advance with respect to the experimental solution.
4. Finally, it should be understood that, according to this scheme, the trajectory of the DoFs controlled in the laboratory inside the large time step Δt has the same degree of smoothness as if no substructuring were introduced. Neither interpolation nor extrapolation is introduced. No velocity discontinuity occurs. It is also expected that the stability of the scheme, with or without subtructuring is the same. For an identical choice of M and $\mathbf{M}_{\mathrm{B}}$, there is no difference, from the stability point of view, whether the tested structure is part or not of a larger structure. However the accuracy of the time integration is quite different. In the case without substructuring, it is related to $\delta t = \Delta t/M$, in the case with substructuting, to Δt or $2\Delta t$.

The Analytical Domain A In the original basic coupling algorithm (Gravouil and Combescure, 2001), the non-linearities of the domain A are solved by means of iterations. However the substepping loop of domain B being within the iteration process of domain A, it was decided to use an alternative approach, already used for the classical PsD testing with substructuring, the Operator Splitting (Nakashima and Kato, 1987). According to this approach, the free solution in domain A is obtained by using the following first order approximation of the restoring force $\mathbf{r}_{\mathrm{A}f}^{n+1}$

$$\begin{aligned} \mathbf{r}_{\mathrm{A}f}^{n+1}(\mathbf{d}_{\mathrm{A}f}^{n+1}) &\approx \tilde{\mathbf{r}}_{\mathrm{A}f}^{n+1}(\tilde{\mathbf{d}}_{\mathrm{A}f}^{n+1}) + \mathbf{K'}_{\mathrm{A}}^{n+1}(\mathbf{d}_{\mathrm{A}f}^{n+1} - \tilde{\mathbf{d}}_{\mathrm{A}f}^{n+1}) \\ &\approx \mathbf{K'}_{\mathrm{A}}^{n+1}\mathbf{d}_{\mathrm{A}f}^{n+1} + (\tilde{\mathbf{r}}_{\mathrm{A}f}^{n+1}(\tilde{\mathbf{d}}_{\mathrm{A}f}^{n+1}) - \mathbf{K'}_{\mathrm{A}f}^{n+1}\tilde{\mathbf{d}}_{\mathrm{A}f}^{n+1}) \end{aligned} \tag{3.27}$$

where $\mathbf{K'}_{\mathrm{A}}^{n+1}$ is a stiffness matrix (in principle the tangent stiffness matrix) for the displacement $\tilde{\mathbf{d}}_{\mathrm{A}f}^{n+1}$.

Since domain A is only numerical, it is always possible to compute this tangent stiffness matrix, even if it is not necessarily useful in practice (as illustrated in section 4). This is in contrast with the use of the OS-scheme for monolithic substructuring. In this case, at least a part of $\mathbf{K'}^{n+1}$, the one representative of the experimental structure, should be measured. Since it is not always easy to obtained a reliable value of this matrix, the derivation of $\mathbf{r}^{n+1}$ is often based on an elastic approximation of it, measured at the very beginning of the test, with some theoretical degradations of the accuracy of the scheme as described in (Combescure and Pegon, 1997).

Introducing the approximation given by Eq.(3.27), it is possible to derive $\mathbf{a}_{\mathrm{A}f}^{n+1}$ as in Eq.(3.8), but with a different expression of $\hat{\mathbf{M}}_{\mathrm{A}}$ and $\hat{\mathbf{f}}_{\mathrm{A}}^{n+1}$ given by Eq.(3.5) for domain A. In fact we have now

$$\begin{aligned} \hat{\mathbf{M}}_{\mathrm{A}}^{n+1}\mathbf{a}_{\mathrm{A}f}^{n+1} &= \hat{\mathbf{f}}_{\mathrm{A}}^{n+1} \\ \hat{\mathbf{M}}_{\mathrm{A}}^{n+1} &= \mathbf{M}_{\mathrm{A}} + \frac{\Delta t}{2}\mathbf{C}_{\mathrm{A}} + \frac{\Delta t^2}{4}\mathbf{K'}_{\mathrm{A}}^{n+1} \\ \hat{\mathbf{f}}_{\mathrm{A}}^{n+1} &= \mathbf{f}_{\mathrm{A}}^{n+1} - \tilde{\mathbf{r}}_{\mathrm{A}f}^{n+1} - \mathbf{C}_{\mathrm{A}}\tilde{\mathbf{v}}_{\mathrm{A}f}^{n+1} \end{aligned} \tag{3.28}$$

In appearance, the central difference scheme and the trapezoidal-rule/OS scheme look like very similar: in particular they both do not involve iteration. However, they are in fact very different in nature. The central difference scheme should work with small time step in order to accommodate the conditional stability of the scheme but the time discrete equilibrium of the structure is always exactly verified. The trapezoidal-rule/OS scheme does not suffer from any time step restriction, but, since the approximation (3.27) is introduced in the derivation of the internal forces, the equilibrium equations are just approximately verified, apart if the structure remains linear elastic. If too many DoFs are involved in this approximate behaviour, the errors can propagate dramatically and the solution be completely corrupted. It may be important to introduce further appropriate condensations, as explained in (Pegon and Magonette, 2005) for the worst cases.

Introducing now the simplified version of the modification of the basic scheme (SIF) introduced in the previous subsection, the algorithm which is used by the analytical domain is the following: knowing the displacement vector $\mathbf{d}_{\mathrm{A}}^{n}$, the velocity vector $\mathbf{v}_{\mathrm{A}}^{n}$ and the acceleration vector $\mathbf{a}_{\mathrm{A}}^{n}$ at time t^n, the values of these vectors are obtained at time $t^{n+2} = t^n + 2\Delta t$ according to a the following 2-step procedure

First step: solving the unconstrained problem

$$\begin{aligned} \mathbf{d}_{\mathrm{A}f}^{n+2} &= \tilde{\mathbf{d}}_{\mathrm{A}f}^{n+2} + \Delta t^2\mathbf{a}_{\mathrm{A}f}^{n+2} \quad & \tilde{\mathbf{d}}_{\mathrm{A}f}^{n+2} &= \mathbf{d}_{\mathrm{A}}^{n} + 2\Delta t\mathbf{v}_{\mathrm{A}}^{n} + \Delta t^2\mathbf{a}_{\mathrm{A}}^{n} \\ \mathbf{v}_{\mathrm{A}f}^{n+2} &= \tilde{\mathbf{v}}_{\mathrm{A}f}^{n+2} + \Delta t\mathbf{a}_{\mathrm{A}f}^{n+2} & \tilde{\mathbf{v}}_{\mathrm{A}f}^{n+2} &= \mathbf{v}_{\mathrm{A}}^{n} + \Delta t\mathbf{a}_{\mathrm{A}}^{n} \\ \mathbf{a}_{\mathrm{A}f}^{n+2} &= \hat{\mathbf{M}}_{\mathrm{A}}^{-1}\left\{\mathbf{f}_{\mathrm{A}}(t^{n+2}) - \tilde{\mathbf{r}}(\tilde{\mathbf{d}}_{\mathrm{A}f}^{n+2}) - \mathbf{C}_{\mathrm{A}}\tilde{\mathbf{v}}_{\mathrm{A}f}^{n+2}\right\} \end{aligned} \tag{3.29}$$

with $\hat{\mathbf{M}}_{\mathrm{A}} = \mathbf{M}_{\mathrm{A}} + \Delta t\mathbf{C}_{\mathrm{A}} + \Delta t)^2\mathbf{K'}_{\mathrm{A}}$.

Second step: solving the constrained problem

$$\begin{aligned} \mathbf{d}_{\mathrm{A}}^{n+2} &= \mathbf{d}_{\mathrm{A}f}^{n+2} + \Delta t^2 \mathbf{a}_{\mathrm{A}l}^{n+2} \\ \mathbf{v}_{\mathrm{A}f}^{n+2} &= \mathbf{v}_{\mathrm{A}f}^{n+2} + \Delta t \mathbf{a}_{\mathrm{A}l}^{n+2} \\ \mathbf{a}_{\mathrm{A}}^{n+2} = \mathbf{a}_{\mathrm{A}f}^{n+2} + \mathbf{a}_{\mathrm{A}l}^{n+2} \quad \mathbf{a}_{\mathrm{A}l}^{n+2} &= \hat{\mathbf{M}}_{\mathrm{A}}^{-1} (\mathbf{L}_{\mathrm{A}}^{\mathrm{T}} \mathbf{\Lambda}^{n+2}) \end{aligned} \tag{3.30}$$

The following remarks can be performed:

1. The analytical process uses a time step which is the double of the large time step Δt. This allows to derive $\mathbf{v}_{\mathrm{A}f}^{n+2}$ (and thus $\mathbf{L}_{\mathrm{A}}\mathbf{v}_{\mathrm{A}f}^{n+2}$) during the time the experimental process is passing from $\mathbf{d}_{\mathrm{B}}^{n}$ to $\mathbf{d}_{\mathrm{B}}^{n+1}$ using synchronized substepping. If the number of substeps M is large enough, the information $\mathbf{L}_{\mathrm{A}}\mathbf{v}_{\mathrm{A}f}^{n+2}$ needed by the experimental process to go from $\mathbf{d}_{\mathrm{B}}^{n+1}$ to $\mathbf{d}_{\mathrm{B}}^{n+2}$ could be sent before $\mathbf{d}_{\mathrm{B}}^{n+1}$ is reached. Then the experimental process can proceed without stopping.
2. The correction process requires the value of $\mathbf{\Lambda}^{n+2} = \mathbf{\Lambda}^{n+1+M/M}$, which is known only when the experimental process executes the last substep between $\mathbf{d}_{\mathrm{B}}^{n+1}$ to $\mathbf{d}_{\mathrm{B}}^{n+2}$. This means that the analytical process should deliver the value of $\mathbf{v}_{\mathrm{A}f}^{n+2}$ before the experimental process reaches $\mathbf{d}_{\mathrm{B}}^{n+1}$ but then have to wait until the experimental process reaches $\mathbf{d}_{\mathrm{B}}^{n+2}$ in order to correct $\mathbf{d}_{\mathrm{A}f}^{n+2}$ to $\mathbf{d}_{\mathrm{A}}^{n+2}$ (second step).
3. Obviously, this waiting period will be used by a second analytical process, first using the value of $\mathbf{\Lambda}^{n+1} = \mathbf{\Lambda}^{n+M/M}$ in order to correct $\mathbf{d}_{\mathrm{A}f}^{n+1}$ to $\mathbf{d}_{\mathrm{A}}^{n+1}$ and then compute $\mathbf{v}_{\mathrm{A}f}^{n+3}$. The time integration of the analytical part is thus made by two interlaced schemes, computing in turn the odd and the even large time steps, and linked together by the continuously moving experimental structure.
4. The computation and the exchange of data between the two domains is represented in Figure 12. Note that in this graphic, the time axis represents the algorithmic time and not the laboratory time. The tasks that are executed in parallel have the same colour.
5. The two interlaced schemes are coupled through the experimental structure and this can be explained in Figure 12. The value of $\mathbf{v}_{\mathrm{A}f}^{n+2}$ computed at the analytical level from t^n is used by the experimental process for the time integration between t^{n+1} and t^{n+2}. Although the analytical schemes are both working with a double time step $2\Delta t$, they are contributing in rotation for the motion of the experiment for the substeps over a time interval of simulation duration Δt.
6. The value of $\mathbf{\Lambda}^{n+2}$, which is returned to the analytical process in order to complete its double time step by solving the constraint problem,

should not be considered just as a local value (the value of $\mathbf{\Lambda}$ at time $t^{n+1+M/M}$). In fact the system of equations (3.24) to solve at the interface derives from the projection operators for $\mathbf{\Lambda}$ (and $\mathbf{v}_{\mathrm{A}}$) given by Eq.(3.15). This is not to be considered as an interpolation because during the sub-stepping process for the domain B, neither $\mathbf{\Lambda}^{n+1}$ nor $\mathbf{v}_{\mathrm{A}}^{n+1}$ are known. This is not to be considered as an extrapolation because the quantities $\mathbf{\Lambda}^{n+1}$ or $\mathbf{v}_{\mathrm{A}}^{n+1}$ cannot be considered as being constant with m. The best is to consider it as a recurrence relationship, allowing to ensure the equilibrium of the global system (domains A and B) at the end of each large time step as well as the continuity of the velocities at the interface for the small time step. The constrain defined by Eq.(3.15) expresses also the decreasing influence of the quantities known at t^{n+1}. The final value of $\mathbf{\Lambda}^{n+2}$ which is returned to the process holding domain A results from an integration. It is then natural to infer that the impact of the experimental errors on the final value of $\mathbf{\Lambda}$ at the end of a large time step will decrease when the number of substeps will increase.

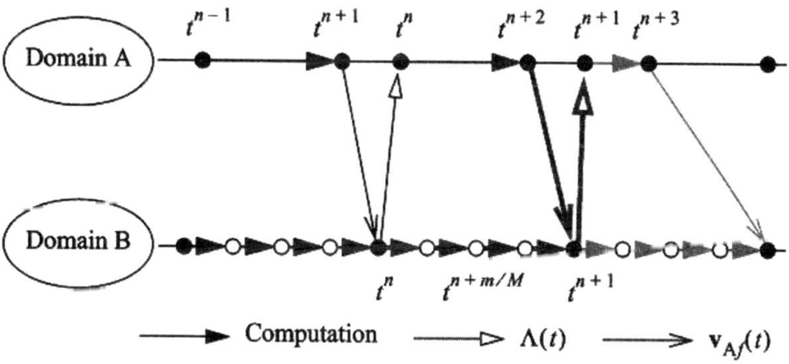

Figure 13. A simplified inter-field modification of the basic scheme, in the laboratory time.

7. If the same graphic as the one of Figure 12 is plotted using the laboratory time as axis, Figure 13 is obtained. It represents the ideal situation where the experimental process is never waiting the analytical process, thus being able to proceed smoothly in a synchronized way. In fact M should be tuned in such a way the analytical process is always (at least almost always) waiting for the experimental process as soon as the relevant $\mathbf{v}_{\mathrm{A}f}$ is known.
8. In Figure 13 the laboratory duration of the analytical process is not regular. This is to recall that there are non-linearities in the first step

of the analytical process. The determination of $\mathbf{r}_A(\tilde{\mathbf{d}}_{Af})$, which in fact represent the internal state of the structure may be complicated, in particular if a non-linear condensation procedure have been introduced to reduce the number of analytical DoF, thus leading to a better OS integration. Another heterogeneity in the computation time may be related to the determination of the tangent stiffness matrix $\boldsymbol{K'}_A$. The domain A being purely numerical, this matrix can be quite precisely evaluated. However, this evaluation can be time consuming, in particular when some local unloadings occur.

4 Numerical Experiments: a Simple 4-Storey Building

A shear resisting one-bay 4-storey structure, proposed as an example in (Clough and Penzien, 1993), is used here for performing extensive parametric studies and for delivering a detailed overview of the capacity of the inter-field procedure SIF/OS and IF/OS to cope with the linear and the non-linear range.

Other validation studies can be found elsewhere, in purely elastic cases (Pegon and Magonette, 2002) or for already tested non-linear structures such as the VAB bridge and an isolated structure using bracings (Pegon and Magonette, 2005).

4.1 Presentation of the Problem

Consider the 4-storey shear frame of Figure 14 idealized with rigid girders and supporting weightless columns. The mass of the structure is thus lumped and the columns in each storey act as simple lateral springs. In addition, the storey mass m and the inter-storey shear stiffness k are considered as being uniform along the height of the structure, and equal to

$$k = 1.75 \times 10^8 \text{N/m} \quad m = 3.5 \times 10^5 \text{Kg} \tag{4.1}$$

The structure is thus made of 4 lateral translational DoFs connected by shear springs. Its natural frequencies are

$$f_1 = 1.2360\text{Hz} \quad f_2 = 3.5588\text{Hz} \quad f_3 = 5.45240\text{Hz} \quad f_4 = 6.6884\text{Hz} \tag{4.2}$$

The purpose of the test is to study the performances of the schemes in the non-linear range by splitting the structure in two parts, as illustrated also in Figure 14. The lower part of the structure, corresponding to the two first storeys forms the "experimental structure" (domain B), whereas the two upper storeys will constitute the "analytical structure" (domain A). Since the connection between the two sub-structures is performed at the level of the 2nd DoF, a fraction of the mass lumped at the 2nd floor should be allocated to domain A. Apart in section 4.3, half of the mass is given to the two domains:

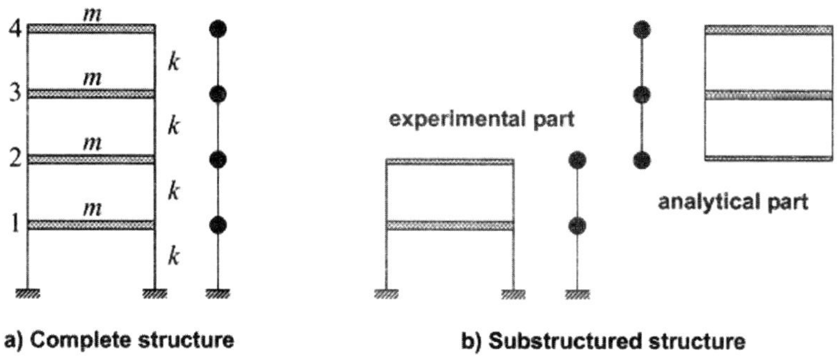

Figure 14. Presentation of the problem.

$$M_{A22} = \frac{m}{2} \quad M_{B22} = \frac{m}{2} \tag{4.3}$$

The natural vibration frequencies of domain B are given by

$$f_{B1} = 2.7238\text{Hz} \quad f_{B2} = 6.5758\text{Hz} \tag{4.4}$$

The stability time step for the explicit scheme running on B is thus given by

$$\Delta t_c = \frac{2}{2\pi f_{B2}} = 4.84061 \times 10^{-2}\text{s} \tag{4.5}$$

The natural frequencies of domain A are obtained by blocking the base (DoF no. 2)

$$f_{A1} = 2.1995\text{Hz} \quad f_{A2} = 5.7583\text{Hz} \tag{4.6}$$

By inspecting the frequencies given in (4.2), (4.4) and (4.6), and comparing them, it is evident that none of the substructures are able to substitute the whole structure, at least for the two first modes. Thus, the loading has been chosen to correspond to a base accelerogram exciting the first two modes of the global structure and the response of the structure is studied in a time interval corresponding to 4 periods of the first vibration mode.

$$a_b = a_\infty \left[\sin(2\pi f_1 t) + \sin(2\pi f_2 t)\right] \quad t \in [0, 4/f_1] \tag{4.7}$$

The inter-storey shear/drift behaviour is assumed elastoplastic with a kinematic hardening, as illustrated in Figure 15. The plastic part is characterized by the yielding shear force and the post yielding stiffness. In order to allow both domains A and B to enter in the non-linear range, the yielding force at each storey has been chosen in such a way that all the storeys plasticized at the same moment when imposing the so-called equivalent dynamic load (lateral load proportional to the mass and altitude of the point of application). Accordingly the chosen values of the constitutive parameters are

$$\begin{array}{cccc} & f_y = 1.75 \times 10^5\text{N} & k_y = k/1000 & \\ f_{y1} = f_y & f_{y2} = 0.9 \times f_y & f_{y3} = 0.7 \times f_y & f_{y4} = 0.4 \times f_y \end{array} \tag{4.8}$$

The result of such a loading is shown in Figure 16. Finally no viscous damping has been introduced: considering that the structure is shaken at two resonance frequencies, the only possibility to limit the motion will be to plastify.

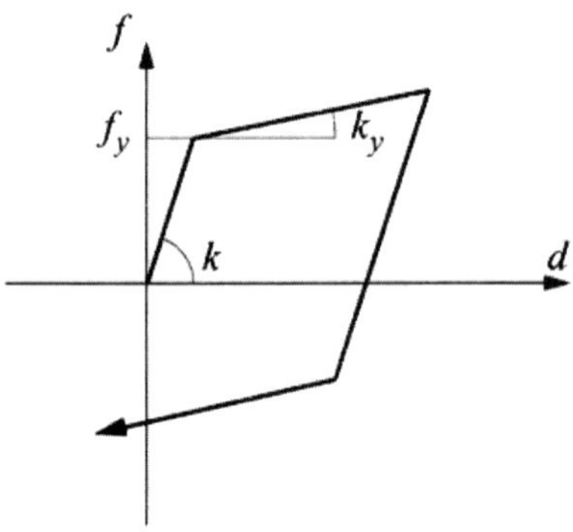

Figure 15. Elastoplastic behaviour.

Since the whole structure and the two substructures A and B do not need any further static condensation, all the computations performed with the OS schemes do not need any iterations. It is thus possible to perform intensive parametric studies and the following aspects have been investigated:

1. *Base acceleration*: four levels of base acceleration have been considered, ranging from almost elastic to deep plastification of all the storeys.

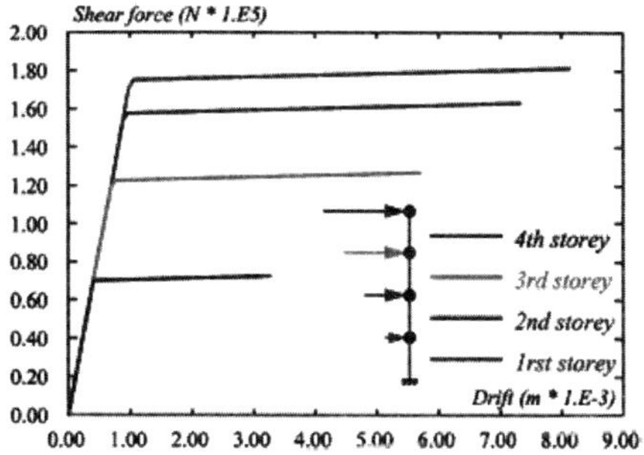

Figure 16. Effective storey behaviour.

2. *Large time step* Δt: the time step for domain A, Δt, is varied so that the convergence curves of the schemes can be obtained for all levels of base acceleration. These curves are presented in section 4.2.
3. *Mass distribution*: the mass distribution at the connection between the two domains is varied and the resulting errors evaluated. These curves are presented in section 4.3.
4. *Small time step* δt: finally, the effect of increasing the number of sub-steps for domain B, or decreasing δt, is investigated. The results are shown in section 4.4.

Three monolithic schemes are used: Reference corresponds to central difference with very small time steps, TR/iterative and TR/OS are associated respectively to an iterative and an operator splitting monolithic implementation of the trapezoidal rule. Three partitioned schemes are also studied, all based on the operator splitting for the domain A: Basic/OS, SIF/OS and IF/OS. All the OS implementations have been made using an elastic approximation of $\mathbf{K'}_{\mathrm{A}}$ in Eq.(3.27).

The four levels of intensity of the base acceleration in Eq.(4.7) has been chosen as

$$a_\infty = 0.01m/s^2 \quad a_\infty = 0.025m/s^2 \quad a_\infty = 0.05m/s^2 \quad a_\infty = 0.25m/s^2 \tag{4.9}$$

The first level induces an almost elastic motion whereas the other three induce an increasing amount of plasticity, as can be seen in Figure 17, where the shear/drift curves has been represented for the various base accelerations.

In the following subsections, various error curves will be presented. They are all derived from the relative error

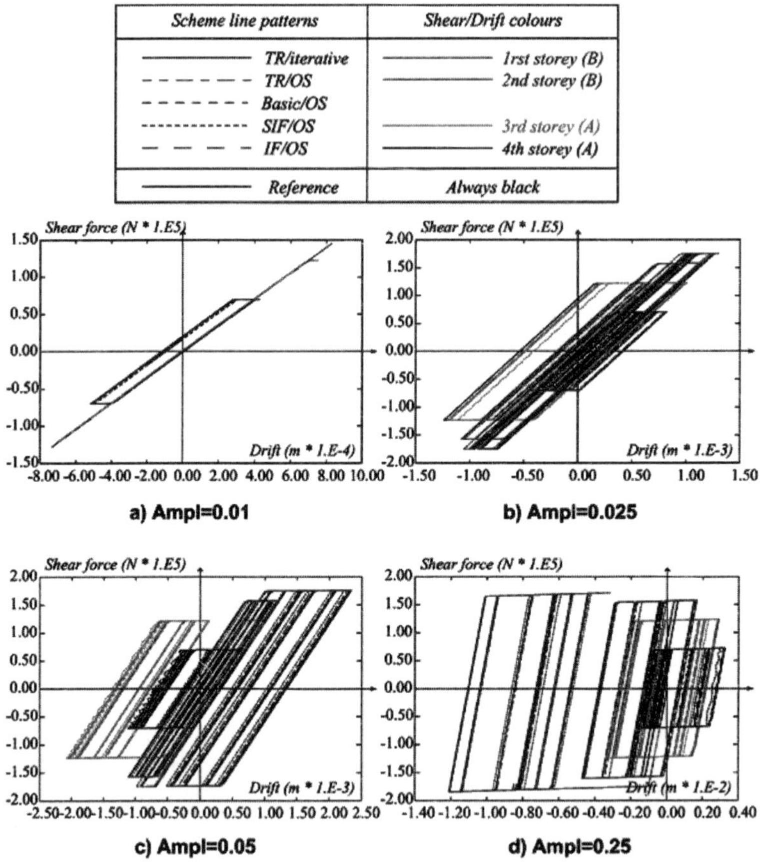

Figure 17. Shear/drift response for the four intensities of the loading.

$$e(f,g) = \frac{\int_0^T \mid f(t) - g(t) \mid dt}{\int_0^T \mid f(t) \mid dt} \tag{4.10}$$

where the interval of integration spans the whole computational interval. The effective integration is performed using the trapezoidal formula

$$\int_0^T h(t)dt = \sum_{i=1}^{n_T - 1} \frac{h(t^i) + h(t^{i+1})}{2} \Delta t \tag{4.11}$$

The considered errors are:

1. *Displacement*: logarithmic average of the displacement errors for the scheme Ψ , $e_d^L(\Psi)$.

$$e_d^L(\Psi) = \left(\prod_{i=1}^{4} e(d_i(\Psi), d_i(Ref)) \right)^{\frac{1}{4}} \tag{4.12}$$

   where d_i, $i = 1, ..., 4$ are the storey displacements (see Figure 14). When the scheme Ψ is partitioned, the two first displacements belongs to the experimental (B) part of the structure, $d_{\mathrm{B}1}$ and $d_{\mathrm{B}2}$ whereas the two last ones belongs to the analytical part of the structure, $d_{\mathrm{A}3}$ and $d_{\mathrm{A}4}$.
2. *Interface*: interface displacement error for the partitioned scheme Ψ , $e_d^I(\Psi)$

$$e_d^I(\Psi) = e(d_{\mathrm{B}2}(\Psi), d_{\mathrm{A}2}(\Psi)) \tag{4.13}$$

4.2 Convergence Curves With Decreasing Values of the Large Time Steps Δt

The same computation is repeated by increasing the number of time steps per period of forcing acceleration ($n_T = 10, 25, 50, 100, 200, 400$ and 1000), while keeping the number of sub-steps for the domain B equal to 5 and an equal distribution of mass at the interface

$$\Delta t = 1/(n_T \times f_1) \quad \delta t = \Delta t/5 \quad M_{\mathrm{A}22} = M_{\mathrm{B}22} = \frac{m}{2} \tag{4.14}$$

For the 4 levels of ground acceleration given in Eq.(4.9), the convergence curves in terms of displacement error e_d^L and interface error e_d^I are given in Figures 18 and 19, respectively.

Keeping in mind that the lowest level of base acceleration ($a_\infty = 0.01 m/s^2$) leads to almost elastic results, it can be said qualitatively that *the performances of the schemes are not affected by the amount of non-linearity.*

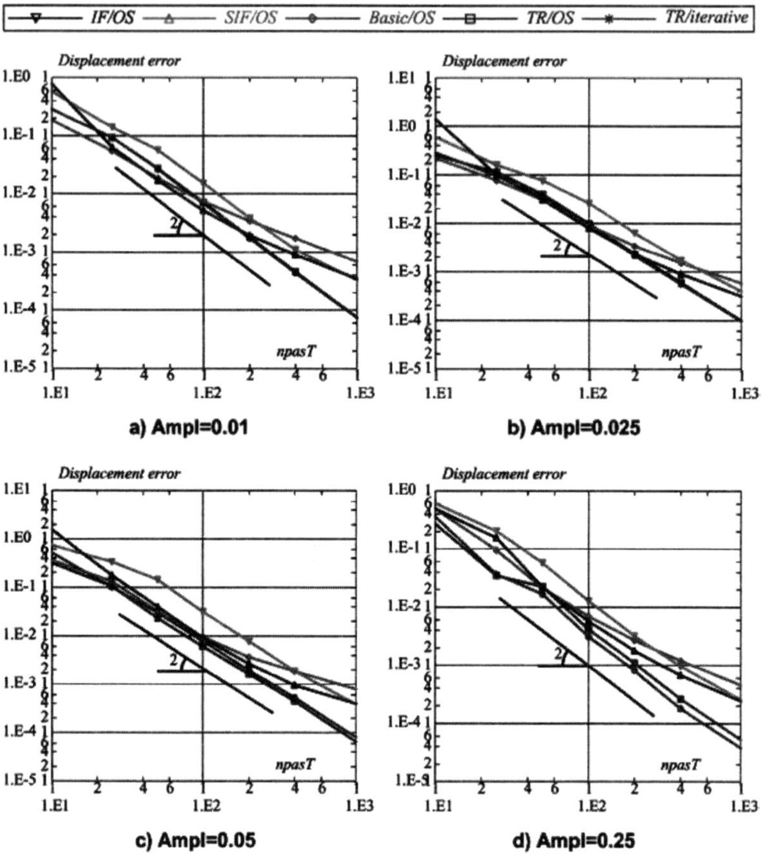

Figure 18. Logarithmic average of the displacement errors for various time steps and various acceleration intensities.

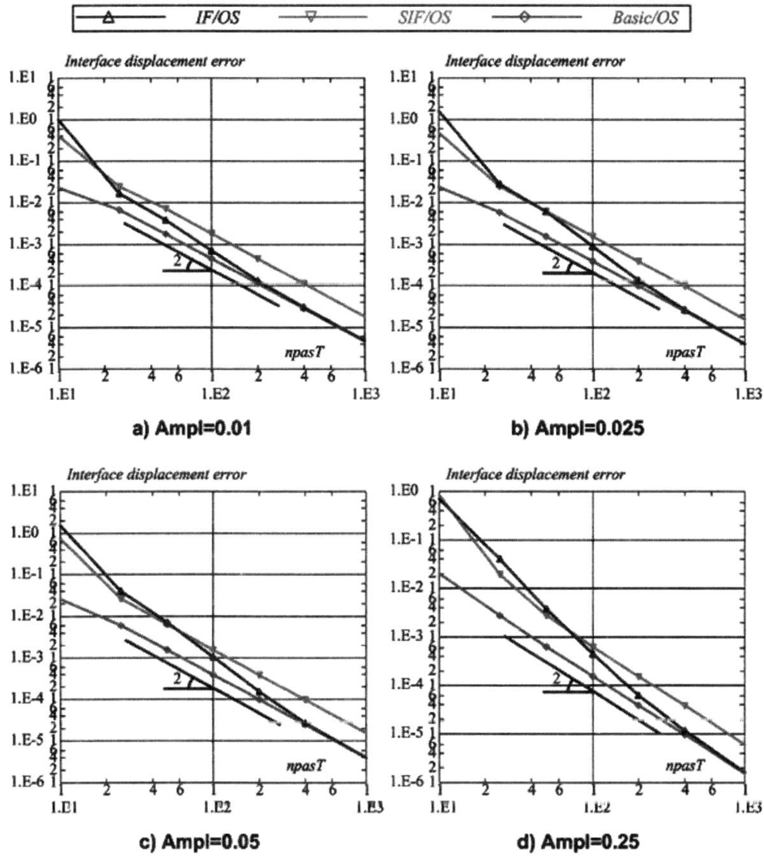

Figure 19. Interface displacement errors for various time steps and various acceleration intensities.

Qualitatively, all the schemes are second order accurate (convergence slope of the displacement errors close to 2). The curves associated with SIF/OS are translated with respect to the others (e_d(SIF/OS) $\approx 4e_d$(other schemes)) recalling that the error in SIF/OS are $O((2\Delta t)^2)$ instead of $O(\Delta t^2)$. Note also that the interface error is also always lower that the displacement error for the partitioned schemes.

Going a little bit into the details, it can be stated that:

1. There is a growing difference between TR/iterative and TR/OS when the level of non-linearity is increased. Although the convergence curves have comparable shapes and remain parallel, the iterative scheme tends to be *ideally* more accurate that the OS schemes. There is in fact an interesting difficulty with the TR/iterative scheme. The computations were first performed with a relative convergence criterion on the value of the residual equal to 1×10^{-4}. In this case the convergence curves reached a plateau for $n_T > 100$. The convergence curves presented in the figures have been obtained with a relative convergence criterion equal to 1×10^{-8}. For this level of accuracy, the CPU cost of the TR/iterative computation is about 15 times higher than the one of the TR/OS scheme, for similar results. Thus it would be very interesting to relate the value of the convergence criterion for the iterative scheme to the value of the time step Δt in order to maintain the accuracy of the scheme but at the same time avoid undue iterations. The difficulty is: how to do it a priori?
2. When increasing the level of non-linearity, the partitioned schemes tend to become less accurate than the monolithic ones. This is not surprising when considering that the coupling between the two domains (matrix $\mathbf{H}_{\mathrm{AB}}$ in Eq.(3.18)) uses the elastic stiffness of the analytical domain A, which is performing a large number of plastic steps for the highest level of base acceleration. What is surprising is that this effect is almost negligible. In fact, in terms of displacement (see Figure 18), IF/OS and Basic/OS are still able to compete fairly well with TR/OS for all the base accelerations. Only for $a_\infty = 0.25 m/s^2$ these schemes are unable to give better results than TR/iterative for any value of n_T.
3. The slopes of all the convergence curves are decreasing for large value of n_T, and this effect increases with the level of non-linearity. This is almost indistinguishable for the monolithic scheme in terms of displacement (see Figure 18), which are always exhibiting a second order accuracy (decrease of two decades of the error when n_T increases of one decade). This is particularly evident for Basic/OS and IF/OS. Despite the fact that SIF/OS is in general quantitatively less accu-

rate than the two other partitioned schemes (the basic time step to consider for this scheme is $2\Delta t$ rather than Δt), its slope decrease is less pronounced and it exhibits asymptotically a better convergence rate. This can be attributed to the fact that the partitioning does introduce a damping at the interface as soon as the number of substeps M is higher than 1 (see section 3.2.3). This effect is slight, because the total number of DoFs of the problem is already higher that the number of connection nodes, but is noticeable because they are of the same order of magnitude.

4. IF/OS and Basic/OS behave in a quite similar manner. As soon as n_T is large enough (typically when more than 100 time steps are introduced to represent one oscillation of the first mode of the system), IF/OS is in general even more accurate than Basic. Remember however that, in addition to the fact that Basic is a staggered scheme and IF/OS an inter-field scheme, IF/OS compute 2 times the internal state in domain A rather than one time for Basic.
5. The fact that the continuity of the velocity is ensured by SIF/OS (and Basic/OS) and not ensured by IF/OS is particularly evident when considering the behaviour of the interface (see Figure 19). Despite the fact that SIF/OS is structurally less accurate than IF/OS, they are showing comparable errors at the interface, at least up to $n_T < 100$. For higher values of n_T, IF/OS tends smoothly toward the performance of Basic. Note that this behaviour is only slightly modified by the increase of non-linearity.
6. The convergence of the displacement at the interface for the partitioned scheme is always faster (at least quadratic) than the displacement of the DoF's towards the accurate reference solution. The value of the interface displacement error for the same number of step n_T is always lower. In other words, the fact that the displacements are not equal at the interface does not impair the overall performance of the partitioned schemes.

4.3 The Distribution of Mass at the Interface

In this subsection the distribution of mass at the interface is varied, while keeping an average accuracy for the schemes by considering 100 time steps by period and still introducing 5 sub-steps for the domain B

$$M_{\mathrm{A}22} = (1-b) \times m \quad M_{\mathrm{B}22} = b \times m \quad \Delta t = 1/(100 \times f_1) \quad \delta t = \Delta t/5 \tag{4.15}$$

For the 4 levels of ground acceleration given in Eq.(4.9) and for various values of b, the displacement error e_d^L, and interface error e_d^I are given in

Figures 20 and 21, respectively. Since they have no internal interface, a constant value of the errors with respect to b has been assumed for the two monolithic schemes.

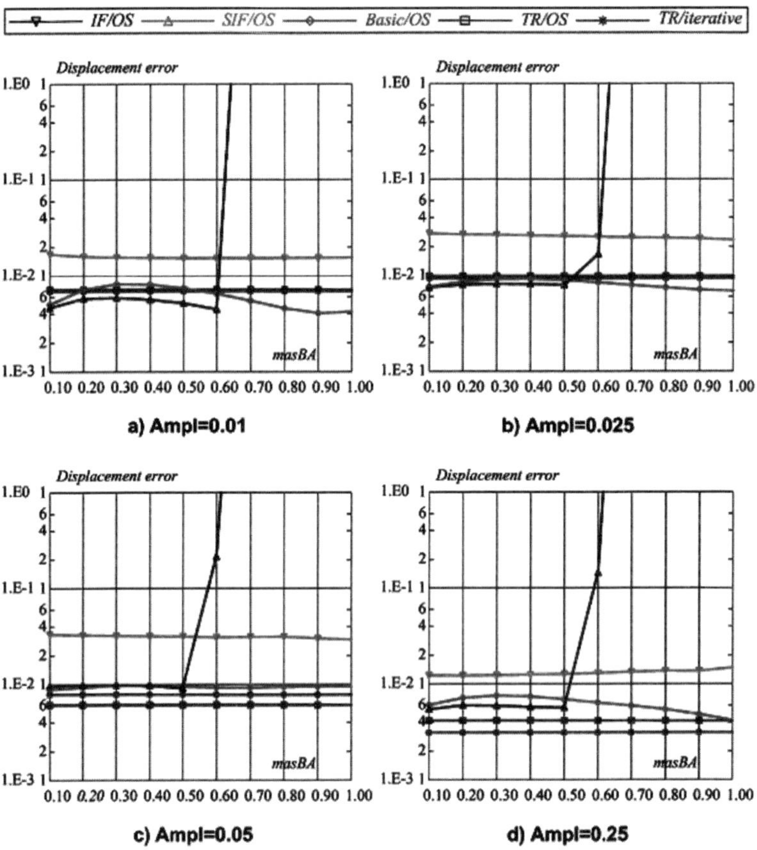

Figure 20. Logarithmic average of the displacement errors for various mass distribution and various acceleration intensities.

Again, it can be said that *the level of non-linearity almost does not affect the results.* However, some interesting differences can be put into evidence.

1. IF/OS becomes instable when the mass becomes too much important on the B domain. Increasing the level of non-linearity, the critical value of b decreases, passing roughly from 0.6 to 0.5. This is an interesting example where, in contrast with common thinking, the instabil-

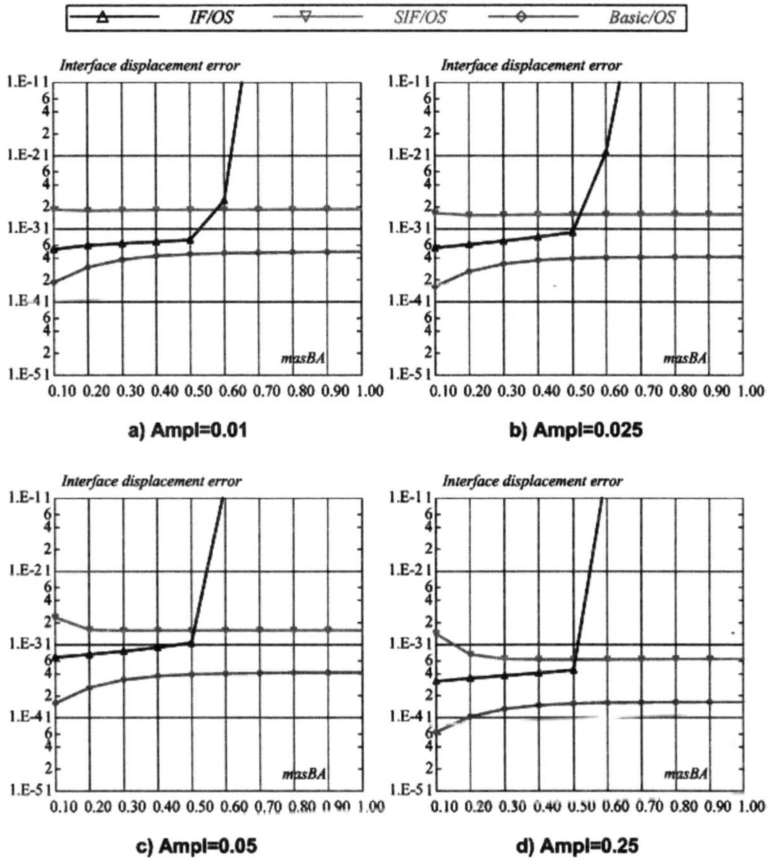

Figure 21. Interface displacement errors for various mass distribution and various acceleration intensities.

ity domain increases with the amount of non-linearity and dissipation, instead of decreasing.

2. The value of the displacement errors are more affected by b in the case of Basic/OS than for SIF/OS (see Figure 20). In fact, the curves for SIF/OS are almost flat. Note also that both schemes remain stable, even if no mass are put on the connection node of the implicit analytical domain A.
3. When decreasing b, that is dropping the mass at the interface of the explicit experimental domain B, the error at the interface tends to increase for SIF/OS, remain constant for IF/OS and decrease for Basic/OS. The increasing of the error is more pronounced for SIF/OS when the non-linearity becomes larger, even if, for all the partitioned schemes, the average value of the error tends to diminish with the increasing of the base acceleration.

4.4 Convergence Curves With Increasing Number of Substeps (Decreasing Values of the Small Time Steps δt)

In this subsection the number of small time steps performed by the experimental domain B is varied ($M = 1, 2, 3, 4, 5, 10, 100, 1000$ and 10000), still keeping a reasonable accuracy for the schemes by considering 100 time steps by period, and introducing again an equal distribution of mass at the interface between the domains

$$\delta t = \Delta t/M \quad \Delta t = 1/(100 \times f_1) \quad M_{\mathrm{A}22} = M_{\mathrm{B}22} = \frac{m}{2} \tag{4.16}$$

For the 4 levels of ground acceleration given in Eq.(4.9) and for various values of M, the displacement error e_d^L and interface error e_d^I are given in Figures 22 and 23, respectively. Since they have no internal interface, a constant value of the errors with respect to M has been assumed for the two monolithic schemes.

As already noticed in the purely elastic case (Pegon and Magonette, 2002), all the errors curves shows an asymptotic behaviour as soon as $M >$ 10 and, usually, the SIF scheme exhibits curves that are flatter than the ones obtained for the two other partitioned schemes, at least in terms of displacement (see Figure 22). In particular, the difference between $M = 1$ and $M = 2$ is very tiny. Concerning the interface error, all the curves show a constant increase of accuracy (see Figure 23). Still here, it can be said that *the level of non-linearity almost does not affect qualitatively the results*. Moreover, as soon as the number of sub-steps is sufficient, no further change of the solution is observed. As already outlined in the introduction, this property is very important: The fact that M can be large or even very large opens the possibility for the analytical process to perform the time step Δt and to receive and send data in the needed laboratory time ΔT. Tt allows to consider complex configurations for the analytical domain A without needing necessarily a powerful CPU server, and to implement PsD testing with non-linear structuring, without taking too much care on the speed of the communication between processes.

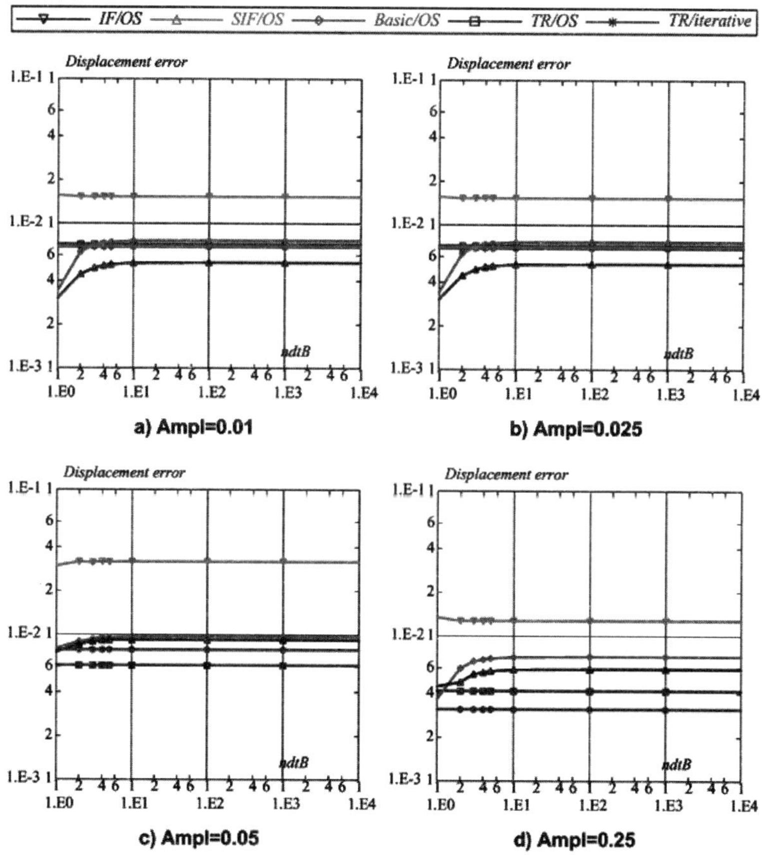

Figure 22. Logarithmic average of the displacement errors for various sub-steping time steps and various acceleration intensities.

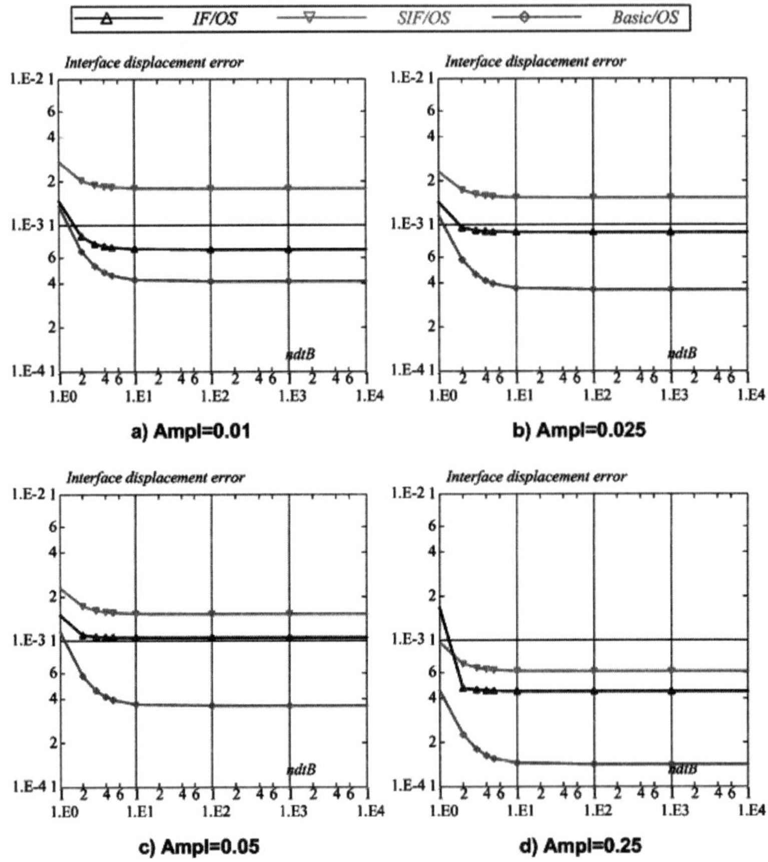

Figure 23. Interface displacement errors for various substeping time steps and various acceleration intensities.

4.5 Conclusions

As repeatedly done when presenting the results, it can be stated that *the level of non-linearity almost does not affect the performance of the schemes.* This analysis, together with others performed in the linear and the non-linear case, shows that

1. It is possible to combine an explicit and an implicit scheme without loss in accuracy. It is thus possible to use an implicit scheme *only* for the analytical part and the experimental process can still rely on the central difference scheme, at the expense of solving the problem in three steps for the experimental process.
2. No further loss of accuracy is observed when performing substeps in the experimental process, even if the number M of substeps becomes very large. Since the stability of the overall scheme depends only of the stability of the experimental process *alone*, it is always possible to chose M large enough in order to constrain the explicit time step to be lower than the critical stability time step. Furthermore, M can be tuned in such a way it allows the implicit process to complete all its tasks.
3. There is no staggered exchange of informations, forcing the processes to have a dialogue instead of just sending (and receiving) informations at the beginning of each large time step. The explicit experimental process can thus proceed continuously, without any hold period.

These characteristics are optimal for running the continuous PsD testing together with substructuring.

In addition, the following points have been illustrated:

1. The additional damping at the interface is marginal for the partitioned schemes.
2. The IF/OS scheme is similar in accuracy to the staggered Basic/OS scheme, but is less robust.
3. The SIF/OS scheme, is similar in robustness to the Basic/OS scheme, but is less accurate ($O((2\Delta t)^2)$ instead of $O(\Delta t^2)$).

Since the SIF/OS schemes is simpler than the IF/OS scheme, and is more robust, it was the first implemented at the laboratory. A setup using a friction device aimed to realize the seismic isolation of bridge was used as a first step of the experimental validation phase, as shown in the next section.

5 First Experimental Validation

5.1 Presentation of the Experiment

Substructuring is particularly attractive when performing subcomponent testing. Using the same testing device (usually a simple 1 DoF system), it is possible to characterize for instance an isolation system by imposing to it a predefined loading history, and then to test it in realistic seismic conditions combining it with a numerical substructure.

This situation is illustrated considering the setup of the left part of Figure 24. A special hysteretic dissipation device, UHYDE-fbr (Dorka, 2004) is put between two fixed steel plates. The lower part of the device is free to slide on a Teflon plate whereas the upper part is rigidly connected to the upper steel plate. Four actuators put at the top of the upper plate can impose a constant pressure to an internal membrane, thus varying the friction force. The UHYDE-fbr system had been studied within the Testing of Algorithms for Semi-Active Control of Bridges (TASCB) project under the European Consortium of Laboratories for Earthquake and Dynamic Experimental Research (ECOLEADER) programme. At that time, the substructured tests were performed using a monolithic algorithm, see (Bossi, 2003).

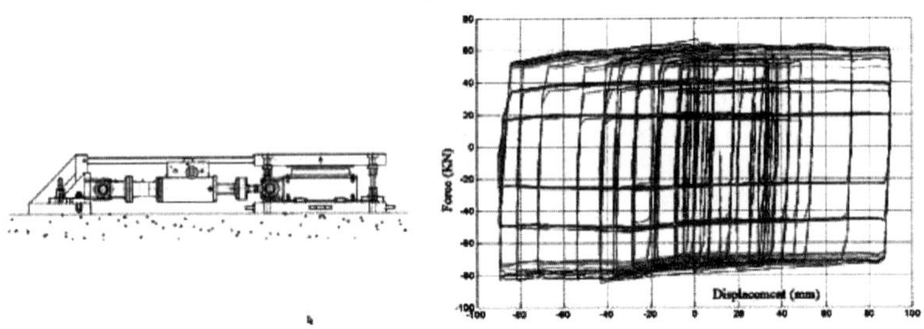

Figure 24. UHYDE testing setup (left) and characterization of the friction device (right), from (Bossi, 2003).

The results of the characterization of the devices are shown in the right part of Figure 24. A cyclic loading has been imposed with an internal pressure of 2 (blue), 4 (red), 6 (green) and 7 (black) bars. This elastoplastic behaviour is almost not modified by the velocity of the loading.

Now this device is used to isolate a structure, in this case the bridge of Figure 25. The structure (and the use of the isolators) being symmetrical, only half of the structure has been considered. Anticipating the use of isolators working mainly in shear at the top of the piers, elastomeric supports have been considered in order to support the weight of the deck in static condition and to re-align the deck after a seismic event. The structure (and the elastomeric support) has been modelled using finite elements (beams and plates), further statically condensed on 20 carefully selected DoFs. Thus the analytical substructure is characterized by its stiffness $\mathbf{K}_{\mathrm{A}}$, damping $\mathbf{C}_{\mathrm{A}}$ and mass $\mathbf{M}_{\mathrm{A}}$ matrices. Only the first configuration, where the UHYDE-fbr has been mounted at the top of the external piers is considered here and the test has been performed for an internal pressure of 6 bars.

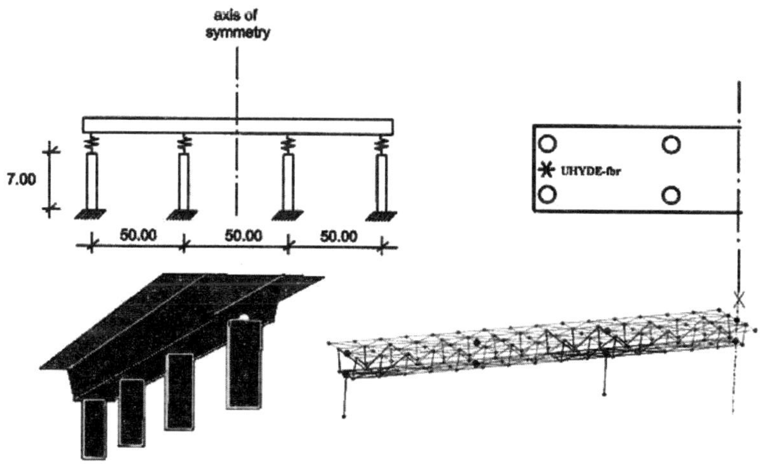

Figure 25. The bridge configuration.

As for the numerical validation test of the previous section, we considered a forced vibration problem

$$a_b = a_\infty \sin(2\pi f_1 t) \quad f_1 = 2\text{Hz} \quad t \in [0, 10s] \tag{5.1}$$

At the implementation level a new version of the UserAlgorithm DLL has been developed, in order to implement the algorithm of Figure 13 and to manage the communication with the analytical process. Remembering that the algorithm is called within an interrupt, these communication should be performed using a robust send/receive acknowledgement cycle that has been implemented using 2 semaphores. Extensive studies demonstrated that for the hardware used during the validation studies, a minimum of 75 substeps (150ms of laboratory time) were needed to complete this cycle. For $M < 75$, it is likely that the experimental process will not find the next value of $\mathbf{L}_{\mathrm{A}}\mathbf{v}_{\mathrm{A}f}$ available for proceeding continuously.

5.2 Increasing the Number of Substeps

The first study was performed by varying the number of substeps performed by the experimental domain B ($M = 75, 100, 125, 200, 250$ and 500), and considering 50 time steps by period of excitation

$$\delta t = \Delta t / M \quad \Delta t = 0.01\text{s} \tag{5.2}$$

The results in terms of relative displacement of the friction device, shear and shear/drift curves are shown in Figure 26, 27 and 28, respectively. In these figures, the whole curves are shown on the left with a zoom on the right. Note that the algorithmic time sampling of all the presented data is Δt. It can be observed that the response curves are qualitatively the same, independently from the number of substeps confirming the uniform properties at the algorithmic level using a setup almost unaffected by the velocity of the test ($\lambda = 15$ for $M = 75$ to $\lambda = 100$ for $M = 500$).

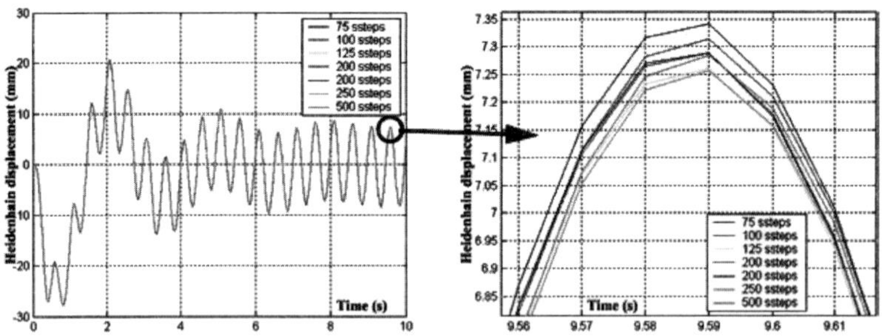

Figure 26. Displacement when increasing the number of substeps.

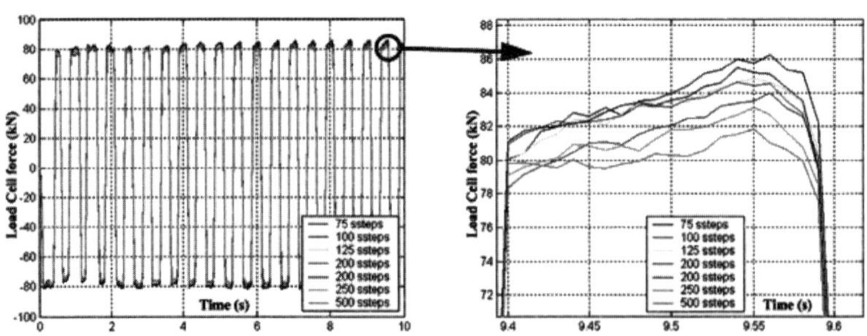

Figure 27. Shear force when increasing the number of substeps.

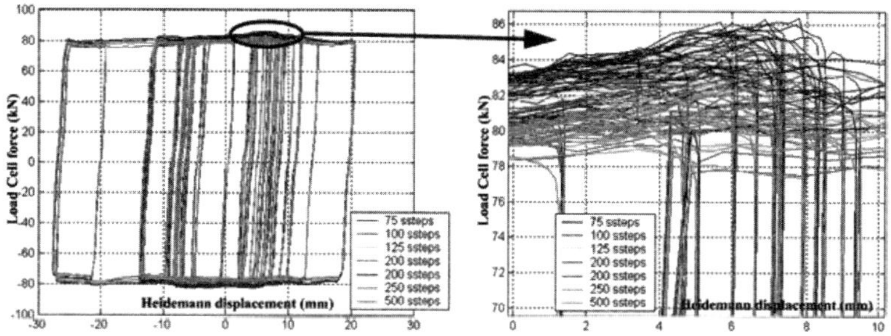

Figure 28. Shear/displacement when increasing the number of substeps.

It is however interesting to explain the differences between the curves, magnified in the right side of the figures.

1. There is a slight dependence of the response of the setup with the velocity of the test which manifests as an increase of the friction plateau (5% for the interval of variation of λ, see Figure 27). However, it can be seen that the test performed with $M = 100$ ($\lambda = 20$) shows a lower plateau than for $M = 200$ ($\lambda = 40$), raising some concerns on the reproducibility of the test.
2. The maximum of the almost stabilized forced response in terms of measured displacement is raising slightly (1.5% for the interval of variation of λ) with the velocity of the test, and this is in apparent contradiction with the raising of the force (and thus damping) in the device. However, the friction in the device is only one source of dissipation of the overall system, the other is provided by $\mathbf{C}_{\mathrm{A}}$.
3. In addition, the displacement in Figure 26 are the measured displacement and not the imposed (or reference) displacement. What is in fact visible is the growing impact of the displacement error with the velocity of the test, which manifests as a slight diminution of the apparent damping (see section 2.5).

5.3 Repetitiveness of the Test

As seen in the previous subsection, it is difficult to distinguish between all the sources of error. This task should also take into account the degree of reproducibility of the test. This is why the test with $M = 200$ ($\lambda = 40$) has been repeated and the two closer results has been kept, to give a lower bound. The differences in measured displacements and measured shear forces are given in Figure 29. The following points may be noted:

1. The difference of displacement is maximum when the value of the displacement is low (see in the left size of Figure 26, where the two curves obtained for $M = 200$ are almost indistinguishable). This difference is quite low when compared with the resolution (0.002 mm) of the Heidenhain measurement, but can explain some apparent discrepancies in the order of the displacement curves of Figure 26, in particular when considering that this difference can be easily 3 times higher.
2. The difference of forces is varying in an interval of 2 kN which can partly explain also the discrepancies observed in the shear curves of Figure 27.

5.4 Increasing the Number of Time Steps

In this study the number of substeps is kept constant but the number of large time steps Δt per period of forced vibration is varied

$$\Delta t = 0.01\text{s} \times m_T \quad \delta t = \Delta t/100 \tag{5.3}$$

with $m_T = 2, 1, 0.5, 0.25, 0.125, 0.0625$ corresponding to values of λ ranging from 10 to 320. The results concerning the displacement and the shear force are given in Figure 30 and Figure 31. Since the sampling for each curve is still Δt, there are 32 times more point for the slowest test than for the fastest test, which is particularly evident for the shear curves in Figure 31.

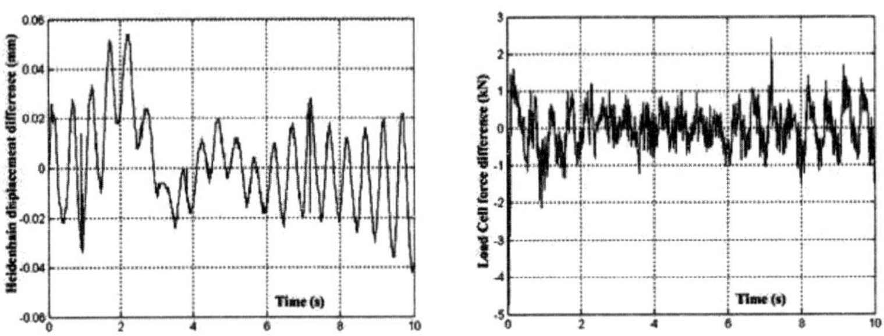

Figure 29. Displacement and force differences when repeating the same test.

What is visible at first sight is that the change in the time steps has, as expected, much more influence on the results that the change of the number of substeps. Together with the improvement of the control with the raising of λ, there is a much more noticeable improvement of the accuracy of the algorithm while diminishing Δt. The diminishing of the maximum displacements at the last peak is about 26% (to be compare with 1.5%)

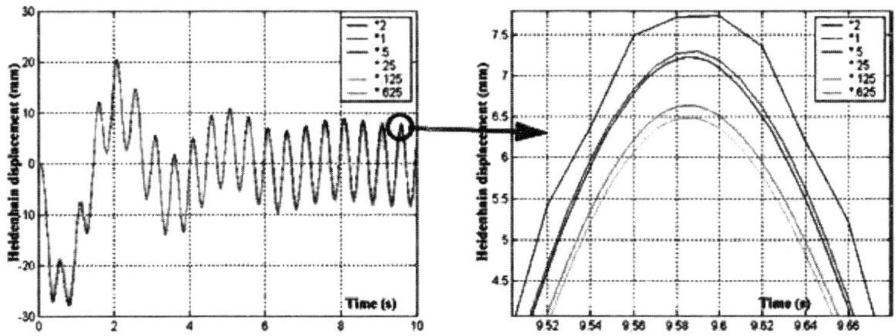

Figure 30. Displacement when increasing the number steps.

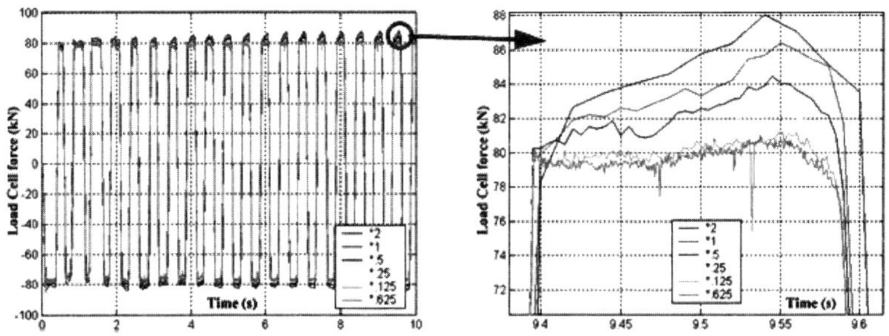

Figure 31. Force when increasing the number of steps.

for a maximum force variation at the "plateau" of the device of 9% (to be compared with 5%).

The following comments could be added

1. The increasing of λ tends to flatten the "plateau". This tendency was already observable in Figure 27 but is much more evident in Figure 31.

2. The zoom in Figure 31 gives also a good idea of the amplitude noise in the load cell (about 0.5 kN) to be compared with the typical discrepancy of 2kN observed when reproducing carefully a test.

3. It is almost impossible to assess a quadratic convergence in the displacement curve of Figure 31 (for instance the curve obtained for $m_T = 1$ and $m_T = 0.5$ are very close, the curves for $m_T = 0.0625$ are higher than the one for $m_T = 0.125$).

5.5 Trajectory Continuity

The aim of this last section is to illustrate various properties of the SIF scheme in an unfavourable case corresponding to the fastest test of the preceding series ($\Delta t = 0.02s$, $\delta t = \Delta t/100$, $\lambda = 10$). Not only the test is fast but it is supposed to represent the force response with only 25 points per cycle.

Figure 32 shows the measured displacement trajectory of the device for small time step δt. The points associated with the sampling Δt are materialized by circle. Four zoom windows are provided giving details on the trajectory for various arrangements of the displacement points at large time steps. It is underlined that these smooth trajectories has been obtained maintaining the continuous PsD technique, thus by means of a time integration scheme, without introducing any assumed interpolation or extrapolation of the displacement.

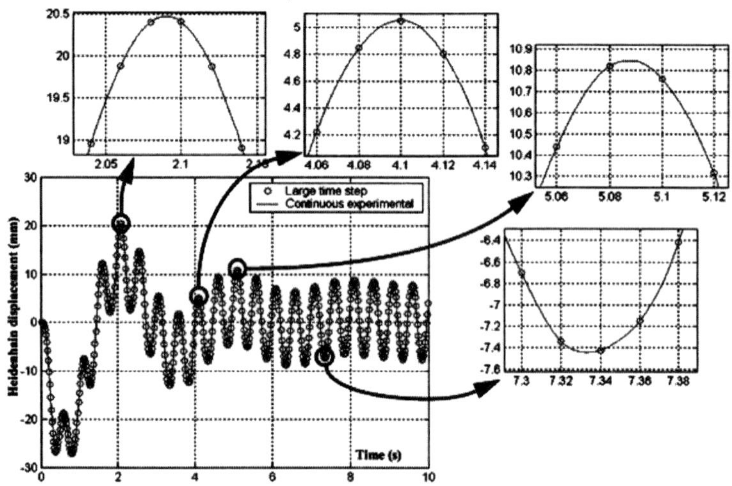

Figure 32. Displacement trajectories within the large time steps.

Figure 33 shows the displacement of the device together with the difference of displacements of the connection points for the numerical structure. The fact that the displacements are not equal is magnified by the large value of the time step Δt. Reducing the time step reduces this discrepancy as illustrated numerically in Figure 19 and in practice at the laboratory in Figure 34 for a time step 4 times smaller. What is interesting in Figure 33 are the apparent oscillations of the relative displacement of the numerical domain. In fact the time integration is made using *two* interlaced schemes working with $2\Delta t$. It would thus be more appropriate to connect the even and the odd points. It is expected that the distance between the curves, together with the amplitude of the oscillation could be used to derive on-line a quality estimator of the accuracy of the time integration. The other interesting point is that the large amplitude of these "oscillations" does not seem to affect the smoothness of the displacement of the device in the laboratory.

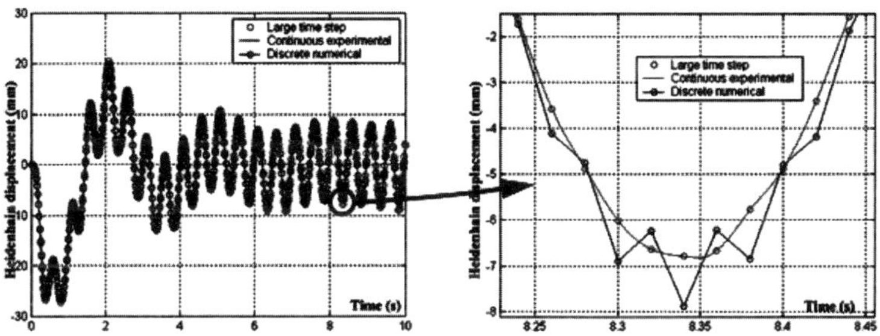

Figure 33. Displacements on domain A and B($\Delta t = 0.02$s).

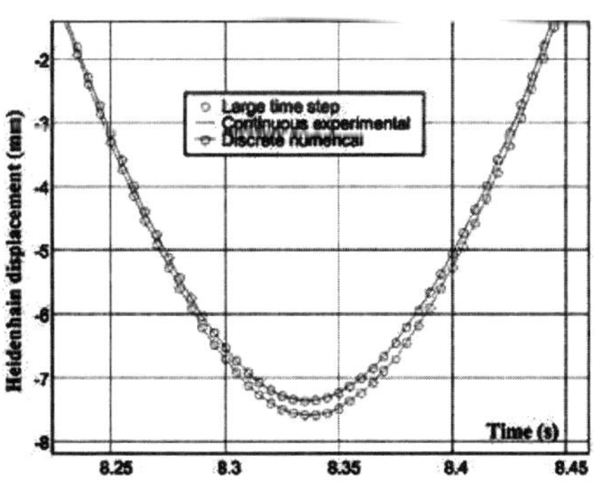

Figure 34. Displacements on domain A and B($\Delta t = 0.005$s).

5.6 Conclusions

This first campaign of tests demonstrates that the proposed method is applicable at the laboratory. In particular the presence of experimental errors (load cell, control system), together with sudden changes in the behaviour of the experimental structure, does not seem to affect the characteristics and the robustness of the SIF scheme.

Further developments are currently performed during the current upgrading of both hardware and software at ELSA. They aim at improving the starting conditions (and as consequence to improve test reproducibility), and increasing the speed of the transfer cycle between the analytical and the experimental structure.

6 Concluding Remarks

This chapter gave a detailed overview of the issues raised by combining continuous PsD testing and subtructuring and presented a solution to this problem, illustrated not only by means of numerical parametric studies but also using a practical implementation at the laboratory.

From a conceptual point of view, the proposed approach seems to be optimal.

1. The experimental structure is moved *continuously* according to an *explicit* time integration scheme. No (possibly hazardous) operator approximation based on the experimental results is needed and the global scheme (analytical + experimental structure) is stable as soon as the explicit scheme is stable. In other terms, testing the experimental structure alone or included on a wider one, possibly involving thousands of DoFs, does not change anything at the stability level. In contrast with what could be felt, it is possible to introduce substructuring keeping all the benefit of the central difference scheme for the experimental structure.
2. The analytical structure could be carefully modelled introducing as many DoFs as needed. As usual for Earthquake Engineering applications, an implicit time integration scheme is introduced. Non-linear condensations could be used to improve Operator Splitting approximations, or precise evaluation of tangent operators could be exploited. Being separated from the experimental process, any finite element software can be used. Finally, the overall accuracy of the scheme (analytical + experimental structure) is only related to the choice of the "large" time step of the analytical process.
3. The inter-field coupling of the two processes is able to preserve the synchronous character of the experimental process. In order to start again a series of small time steps, the experimental process does not need to send an information to the analytical process and then pause, waiting for an immediate answer. Receiving the information at the beginning of a series of small time steps, the experimental process is

able to finish this series on its own, without further interaction with the analytical process.

4. The same approach can be used for performing slow or fast tests. The limitations are not on the algorithmic side: they are related to the properties of the network and the computing hardware concerning data transfer and to the loading/control system characteristics.

Experimental validations are proceeding, considering more complex analytical structures (number of DoFs, non-linearities) and multiple DoFs experimental models.

In parallel, various theoretical studies are ongoing. The first one focuses on the assessment of the stability of the SIF scheme in the same condition as the Basic scheme (Bonelli et al., 2008). The second explores the possibility of adding user-controlled damping in the SIF scheme (He, 2008)(Bursi, 2008). The third investigates the opportunity of using the leap-frog nature of SIF or IF (introducing small oscillations in the analytical response), and/or the difference of displacements at the interface to derive a robust estimate of the time-integration error.

Bibliography

A. Bonelli, O. S. Bursi, L. He, G. Magonette, and P. Pegon. Convergence analysis of a parallel interfield method for heterogeneous simulations with dynamic substructuring. Accepted for publication in *Int. J. Numer. Meth. Engng.* 2008.

A. Bossi. Isolamento sismico: prove di dispositivi dissipativi con il metodo pseudo-dinamico continuo con sottostrutturazione. Tesi di laurea, Universita degli studi di, Perugia,Italy, 2003.

O. S. Bursi. Computational Techniques for Simulation of Monolithic and Heterogeneous Structural Dynamic Systems. In O. S. Bursi and D. J. Wagg, editors, *Modern Testing Techniques for Structural Systems, Dynamics and Control.* CISM, 2008.

Ph. Buchet. Controller users manual. *EUR-Report No. 22329 EN*, 2006a.

Ph. Buchet. Acquisition and control software architecture. *EUR-Report no. 22723 EN*, 2006b.

Ph. Buchet and P. Pegon. Psd testing with substructuring: Implementation and use. *JRC-Special Publication No. I.94.25*, 1994.

F. Casciati, G. Magonette, and F. Marazzi. *Technology of Semiactive Devices and Applications in Vibration Mitigation.* Chichester, UK, 2006.

D. Combescure and P. Pegon. α-operator splitting time integration technique for pseudodynamic testing: error propagation analysis. *Soil Dynamics & Earthquake Engineering*, 16:427–443, 1997.

D. Combescure, P. Pegon, and G. Magonette. Numerical investigation of the impact of experimental errors on various pseudo-dynamic integration algorithms. pp 2479–2484, Balkema, Rotterdam, The Netherlands, 1995. Proceedings of the 10th European Conference on Earthquake Engineering.

S. N. Dermitzakis and S. A. Mahin. Development of substructuring techniques for on-line computer controlled seismic performance testing. Technical report, Earth. Eng. Research Center, Univ. of California, Berkeley, CA, USA, 1985. Report No. UCB/EERC-85/04.

J. Donea, G. Magonette, P. Negro, P. Pegon, A. V. Pinto, and G. Verzeletti. Pseudodynamic capabilities of the ELSA laboratory for earthquake testing of large structures. *Earthquake Spectra*, 12:163–180, 1996.

U. Dorka. Erdbebensicherung durch Structural Control. *Stahlbau*, 73:661–667, 2004.

A. Gravouil and A. Combescure. Multi-time-step explicit-implicit method for non-linear structural dynamics. *Int. J. Numer. Meth. Engng*, 50:199–225, 2001.

L. He. Development of Partitioned Time Integration Schemes for Parallel Simulation of Heterogeneous Systems. PhD thesis, University of Trento, Italy, 2008.

R. -Y. Jung, P. B. Shing, E. Stauffer, and B. Thoen. Performance of a real-time pseudodynamic test system considering nonlinear structural response. *Earthquake Engineering & Structural Dynamics*, pages 36:1785–1809, 2007.

O. -S. Kwon, N. Nakata, A. Elnashai, and B. Spencer. A framework for multi-site distributed simulation and application to complex structural systems. *J. of Earthquake Engineering*, page 9:741–753, 2005. Technical Note.

G. Magonette. Development and application of large-scale continuous pseudo-dynamic testing technique. *Phil. Trans. R. Soc. Lond. A*, 359:1771–1799, 2001.

G. Magonette, P. Pegon, F. J. Molina, and Ph. Buchet. Development of fast continuous pseudodynamic substructuring tests. Kyoto, Japan, 1998. Proc. of the Second World Conf. on Struct. Control, 28 June - 1 July 1998.

A. Millard. *CASTEM 2000, Guide dutilisation.* Saclay, France, 1993. Rapport CEA 93/007.

F. J. Molina, G. Magonette and P. Pegon. Assessment of systematic experimental errors in pseudo-dynamic tests. proc. *12th European Conference on Earthquake Engineering*, 2002. Paper Reference 525, 9-13 September.

F. J. Molina and M. Géradin. Earthquake Engineering experimental Research at JRC-ELSA. *NATO workshop. Extreme Man-Made and Natural Hazards in Dynamics of Structures (NATO Security through Science Series C: Environmental Security)*. 2007. Springer, 311-351.

F. J. Molina, G. Verzeletti, G. Magonette, Ph. Buchet, and M. Géradin. Bidirectional pseudodynamic test of a full-size three-storey building. *Earthquake Engineering & Structural Dynamics*, pages 28:1541–1566, 1999.

G. Mosqueda. *Continuous Hybrid Simulation with Geographically Distributed Substructures*. PhD thesis, University of California, Berkeley, CA, 2003.

The MOST Experiment, July 30 2003. Neesgrid technical report. Technical report, 2003. http://it.nees.org/documentation/pdf/MOST_document_v1.0.pdf.

M. Nakashima and H. Kato. *Experimental Error Growth Behavior and Error Growth Control in On-line Computer Test Control Method.* Japan, 1987. BRI-Report.

M. Nakashima and N. Masaoka. Real-time on-line test for mdof systems. *Earthquake Engineering & Structural Dynamic*, 28:393–420, 1999.

M. Nakashima, T. Akazawa, and O. Sakaguchi. Integration method capable of controlling experimental error growth in substructure pseudo dynamic test. *AIJ J. of Struct. Constr. Engng.*, 454:61–71, 1993.

P. Pan, M. Tada, and M. Nakashima. Online hybrid test by internet linkage of distributed test-analysis domains. *Earthquake Engineering & Structural Dynamics*, page 35:15811583, 2006.

P. Pegon. Psd testing with substructuring: the case of asynchronous motion. *JRC-Special Publication No.I.96.42*, 1996.

P. Pegon and G. Magonette. Continuous psd testing with non-linear substructuring: presentation of a stable parallel inter-field procedure. *JRC-Special publication No. SPI.02.167*, 2002.

P. Pegon and G. Magonette. Continuous psd testing with non-linear substructuring: using the operator splitting technique to avoid iterative procedures. *JRC-Special publication No. SPI.05.30*, 2005.

P. Pegon and G. Magonette. Continuous psd testing with non-linear substructuring: recent developments for the vab project. *JRC Special Publication No.99.142*, 1999.

P. Pegon and A. V. Pinto. Pseudo-dynamic testing with substructuring at the elsa laboratory. *Earthquake Engineering & Structural Dynamics*, pages 29:905–925, 2000.

A. V. Pinto, P. Pegon, G. Magonette, and G. Tsionis. Pseudo-dynamic testing of bridges using non-linear substructuring. *Earthquake Engineering & Structural Dynamics*, pages 33:1125–1149, 2004.

R. W. Clough and J. Penzien. *Dynamics of structures.* McGraw-Hill, New-York, 1993. 2nd edition.

Y. Takahashi and G.L. Fenves. Software framework for distributed experimentalcomputational simulation of structural systems. *Earthquake Engineering & Structural Dynamics*, pages 35:267–291, 2006.

Y. -S. Yang, S. -H. Hsieh, K. -C. Tsai, S. -J. Wang, K. -J. Wang, W. -C. Cheng, and C. W. Hsu. Isee: Internet-based simulation for earthquake engineering–part I: Database approach. *Earthquake Engineering & Structural Dynamics*, 2007. In press.

Notations

T	Laboratory or experimental time
δT	Sampling period of the continuous control
t	Algorithmic, prototype or accelerogram time
Δt	Algorithmic time step
n	Generic index of the algorithmic steps
δt	Algorithmic substep
m	Generic index of the substeps
M	Number of substeps
λ	Time scale factor
$\mathbf{M}$	Mass matrix
$\hat{\mathbf{M}}$	Pseudo mass matrix resulting from the time integration
$\mathbf{C}$	Damping matrix
$\mathbf{K}$	Stiffness matrix
$\mathbf{K'}$	Tangent stiffness matrix, or an approximation of it
$\mathbf{a}$	Acceleration vector
$\mathbf{v}$	Velocity vector
$\mathbf{d}$	Displacement vector
$\mathbf{f}$	External force vector
$\hat{\mathbf{f}}$	Pseudo external force vector resulting from the time integration
$\mathbf{r}(\mathbf{d})$	Restoring force vector
$\mathbf{I}_g$	Incidence vector for the ground acceleration
$\mathbf{a}_g$	Ground acceleration
β, γ	parameters of the Newmark time integration scheme familly
A	Analytical domain
B	Experimental domain
$\mathbf{\Lambda}$	Lagrange multiplier vector resulting from the connection between A and B
$\mathbf{L}_A, \mathbf{L}_B$	Linear connectivity matrices
l,f	Subscripts used for link and free solutions in the partitioned schemes
$\mathbf{H}_{AB}$	Interface operator

Real-Time Hybrid Testing Techniques

P. Benson Shing[†][*]

[†] Department of Structural Engineering, University of California at San Diego, La Jolla, CA 92093-0085, USA

Abstract Real-time hybrid testing combines the efficiency of numerical simulation with the realism of experimental analysis. This chapter presents an overview of real-time hybrid test methods from a general perspective, and discusses schemes of structural partitioning and issues related to numerical computation, system dynamics, and system performance. Different approaches of hybrid testing are presented under a unified framework. A specific implementation, including the hardware configuration, numerical integration scheme, and delay compensation methods, is discussed in detail. Results of experimental validation and numerical simulation study are presented to illustrate the effectiveness of several delay compensation schemes and the subtleties of real-time testing. The possibility of conducting reliable real-time hybrid tests using an unconditionally stable implicit integration scheme is demonstrated.

1 Introduction

In recent years, a number of innovative techniques that combine numerical simulation with physical testing have been developed for evaluating the dynamic performance of structural systems and components, including response mitigation devices and secondary systems, under severe load conditions, such as those induced by earthquakes (Nakashima & Masaoka 1999, Horiuchi *et. al.* 2000, Blakeborough *et. al.* 2001, Magonette 2001, Shing *et. al.* 2004, Sivaselvan *et. al.* 2004, Wallace *et. al.* 2005, Bayer

[*]Some of the material presented in this chapter is based on the previous work of the author supported by the US National Science Foundation under the Cooperative Agreement No. 0086592 as part of the Network for Earthquake Engineering Simulation program. The author would also like to acknowledge the important contribution of his former students, Drs. Rae-young Jung and Zhong Wei, to some of the findings presented in this chapter, and the work of Dr. Eric Stauffer in the development of the real-time test system presented here. However, opinions expressed in this chapter are those of the author and do not necessarily represent those of the sponsor.

et. al. 2005, Wu *et. al.* 2006, Bursi *et. al.* 2008, Bonelli *et. al.* 2008). These methods have been referred to as real-time hybrid testing or simulation, real-time pseudodynamic testing, or real-time dynamic substructuring (Wagg *et. al.* 2008). All these techniques are based on a concept that is similar to the pseudodynamic test method (Takanashi *et al.* 1975) while some of them (such as dynamic substructuring) more resemble the hardware-in-the loop techniques that have been used for testing control systems and mechanical components as discussed by Wagg *et al.* (2008). They will be simply referred to as real-time hybrid testing in this article. The term hybrid reflects the fact that part of a structure is analytically modeled while the remainder is physically tested. It combines the efficiency of numerical simulation with the realism of physical testing. This technique is similar to partitioned structural analysis (substructuring) methods. However, unlike a conventional substructuring approach, the partition of a structure into analytical and physical substructures in a real-time hybrid test can be based on the stiffness, damping, and inertial properties of the structure. In fact, the inertial and damping properties of a physical substructure can be partially simulated analytically and partially represented in the physical specimen. For the extreme case that the inertia and damping properties of the physical substructure are entirely represented by an analytical model, one recovers the pseudodynamic test method. Similar to the pseudodynamic test method, real-time hybrid testing relies on a numerical time integration scheme to solve of the equations of motion formulated for the structural system.

This chapter provides an overview of real-time hybrid test methods in a general context, and discusses schemes of structural partitioning and issues related to numerical computation, system dynamics, and system performance. A specific implementation, including the hardware configuration, numerical integration scheme, and delay compensation methods, is discussed in detail. Results of experimental validation and numerical simulation study are also presented.

2 Structural Partitioning

The basic concept of hybrid test methods is best conveyed from a general *partitioned structural analysis* perspective with the premise that the dynamic behavior of a structural system can be accurately represented by a set of spatially discretized equations of motion, which can be expressed as follows.

$$\mathbf{Ma} + \mathbf{Cv} + \mathbf{r}_s = \mathbf{f} \tag{2.1}$$

in which $\mathbf{a}$ is the acceleration,$\mathbf{v}$ is the velocity, $\mathbf{r}_s$ is the vector of static restoring forces developed by the structure, $\mathbf{M}$and$\mathbf{C}$ are the structural mass and viscous damping matrices, and $\mathbf{f}$ is the excitation force. Static restoring forces $\mathbf{r}_s$ are results of structural displacements $\mathbf{d}$. For the case of hybrid testing, structural partitioning has a major advantage in that the efficiency of computational simulation can be combined with the realism of experimental analysis. For this purpose, one can partition a structural system into two main portions with the physical properties of one portion simulated by an analytical model (or models) and the other represented by one or more physical models that will be subjected to actual loads. The latter normally represents the portion of a structure that is difficult to model computationally either because of its complicated structural nonlinearity or dynamics. For example, this can be a structural subassemblage that is most vulnerable to damage and that can develop an inelastic load-deformation behavior which is difficult to simulate analytically. To convey this concept of structural partitioning, one can rewrite Equation 2.1 in the following form.

$$(\mathbf{M}^A + \mathbf{M}^E)\mathbf{a} + (\mathbf{C}^A + \mathbf{C}^E)\mathbf{v} + \mathbf{r}_s^A + \mathbf{r}_s^E = \mathbf{f} \tag{2.2}$$

in which the terms with superscript A represent the properties and response variables of the analytical model and the terms with superscript E are those of the physical component that is tested experimentally. Since both the inertia and damping forces developed by the physical component are measured quantities in a test, Equation 2.2 can be more appropriately expressed as

$$\mathbf{M}^A\mathbf{a} + \mathbf{C}^A\mathbf{v} + \mathbf{r}_s^A + \mathbf{r}^E = \mathbf{f} \tag{2.3}$$

with

$$\mathbf{r}^E = \mathbf{r}_I^E + \mathbf{r}_D^E + \mathbf{r}_S^E = \mathbf{M}^E\mathbf{a} + \mathbf{C}^E\mathbf{v} + \mathbf{r}_S^E \tag{2.4}$$

In the above equation, the damping forces $\mathbf{r}_D^E$ developed by the physical component do not necessarily have to be viscous in nature; and for most structures, it can be contributed by a number of possible energy dissipation mechanisms such as Coulomb and hysteretic damping. Other forms of nonlinear damping mechanisms are also possible, such as those due to special damping devices like magneto-rheological dampers. However, non-viscous type damping forces can be grouped into the term $\mathbf{r}_S^E$as well. Here, we adopt the notation that $\mathbf{r}_D^E$ represents viscous damping only. In a hybrid test, it may be desirable that part of the inertial and damping forces associated with the degrees of freedom of the physical component be accounted for analytically. One such situation may arise when only the self-weight of a structure is included in the physical model and the additional mass carried by the structure is represented in the analytical model for the sake of convenience. To this end, it is appropriate to partition the terms in Equations 2.3 and 2.4 in the following way.

$$\mathbf{a} = \left\{ \begin{array}{c} \mathbf{a}_A \\ \mathbf{a}_B \\ \mathbf{a}_E \end{array} \right\} \mathbf{v} = \left\{ \begin{array}{c} \mathbf{v}_A \\ \mathbf{v}_B \\ \mathbf{v}_E \end{array} \right\} \mathbf{r}_S^A = \left\{ \begin{array}{c} \mathbf{r}_{S,A} \\ \mathbf{r}_{S,B}^A \\ 0 \end{array} \right\} \mathbf{r}_S^E = \left\{ \begin{array}{c} 0 \\ \mathbf{r}_{S,B}^E \\ \mathbf{r}_{S,E} \end{array} \right\}$$
$$\mathbf{f} = \left\{ \begin{array}{c} \mathbf{f}_A \\ \mathbf{f}_B^A + \mathbf{f}_B^E \\ \mathbf{f}_E \end{array} \right\} \tag{2.5}$$

$$\mathbf{M}^A = \begin{bmatrix} \mathbf{M}_{AA}^A & \mathbf{M}_{AB}^A & 0 \\ \mathbf{M}_{BA}^A & \mathbf{M}_{BB}^A & \mathbf{M}_{BE}^A \\ 0 & \mathbf{M}_{EB}^A & \mathbf{M}_{EE}^A \end{bmatrix} \quad \mathbf{M}^E = \begin{bmatrix} 0 & 0 & 0 \\ 0 & \mathbf{M}_{BB}^E & \mathbf{M}_{BE}^E \\ 0 & \mathbf{M}_{EB}^E & \mathbf{M}_{EE}^E \end{bmatrix} \tag{2.6}$$

in which subscripts A and E denote the degrees of freedom internal to the analytical and physical substructures, respectively, and B denotes those at the boundary of the two, while the superscripts reflect the distribution of the different quantities between the analytical model and the physical structure. For the reason mentioned above, the mass associated with the degrees of freedom of the physical substructure can be divided into an analytical part, with superscript A, and a physical part, with superscript E. The latter may just represent the self-weight of the structure. This also applies to the viscous damping matrix.

In fact, a variety of test methods can be represented by Equations 2.3 through 2.6. When the physical substructure is loaded at a rate that is much lower than the actual dynamic response of the structure, the $\mathbf{r}_I^E$ and $\mathbf{r}_D^E$ developed by it will be negligibly small. As a result, to account for the dynamic effects, the inertia and velocity-dependent damping properties have to be modeled analytically during the test. Under this situation, we recover the conventional pseudodynamic test method (Takanashi *et. al.* 1975). On the other hand, when all the terms with superscript A disappear, i.e., when the analytical model vanishes, we have the condition of the effective force test method (Dimig *et. al.* 1999). Indeed, it has been demonstrated that under this extreme, a real-time hybrid test can be conducted in the same way as an effective force test when an equivalent force control strategy is used with an implicit integration scheme to solve the governing equations of motion (Wu *et. al.* 2007). Hence, one can see the generality of the hybrid test concept. Different hybrid testing approaches in terms of the structural partitioning, strategies for solving the resulting equations of motion, and hardware configurations will be discussed in the following section.

The notion of structural partitioning adopted for hybrid testing is more general than that for conventional substructure analysis in the sense that partitioning does not necessarily have to be performed along geometric boundaries but can also be performed on the mass and damping quantities associated with the internal degrees of freedom of the physical substructure so that each has a physical part and a numerical part.

3 Real-Time Hybrid Test Methods

The various real-time hybrid test methods that have been proposed in the literature differ in the way structural partitioning is performed and the partitioned equation, i.e., Equation 2.3, is solved in the simulation process as well as how the dynamic properties of the physical component are treated. Three main approaches can be identified here.

3.1 Total approach

The total approach can be used when the dynamics of the entire structural system being evaluated can be accurately described by Equation 2.3, and only the static resistance of a portion of the structure is difficult to model. This portion is isolated as the physical substructure. In this case, the most straightforward approach is to formulate the equations of motion for the entire structure as one integrated set and solve these equations using a single time integration scheme. This is illustrated with the four-story shear frame shown in Figure 1, for which Equation 2.3 can be expressed as follows.

$$\begin{bmatrix} M_1^A & 0 & 0 & 0 \\ 0 & M_2^A & 0 & 0 \\ 0 & 0 & M_3^A & 0 \\ 0 & 0 & 0 & M_4^A \end{bmatrix} \begin{Bmatrix} a_1 \\ a_2 \\ a_3 \\ a_4 \end{Bmatrix}$$

$$+ \begin{bmatrix} C_1^A + C_2^A & -C_2^A & 0 & 0 \\ -C_2^A & C_2^A + C_3^A & -C_3^A & 0 \\ 0 & -C_3^A & C_3^A + C_4^A & -C_4^A \\ 0 & 0 & -C_4^A & C_4^A \end{bmatrix} \begin{Bmatrix} v_1 \\ v_2 \\ v_3 \\ v_4 \end{Bmatrix} \tag{3.1}$$

$$+ \begin{Bmatrix} 0 \\ r_{S,2}^A \\ r_{S,3} \\ r_{S,4} \end{Bmatrix} + \begin{Bmatrix} r_1 \\ r_2^E \\ 0 \\ 0 \end{Bmatrix} = \begin{Bmatrix} f_1 \\ f_2 \\ f_3 \\ f_4 \end{Bmatrix}$$

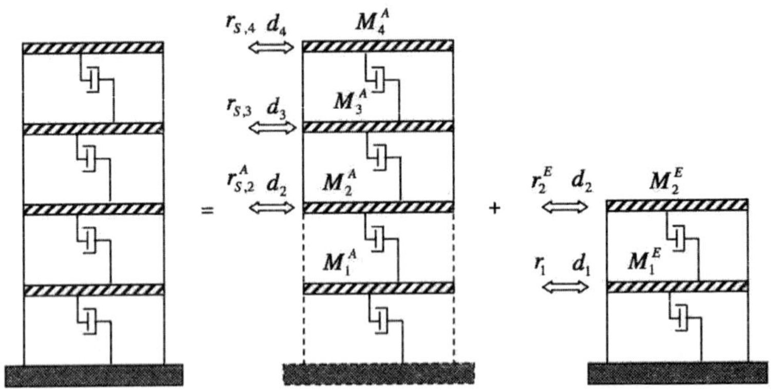

Figure 1. Partitioning of a structure into analytical and physical substructures in the total approach.

in which a damping coefficient C_i^A represents a dash-pot in story i to account for the damping that is not represented in the physical substructure. This approach is identical to a standard analysis procedure with the physical substructure treated as a *super-element* whose resistance is assembled into the global force vector of the entire structure. Equation 3.1 can be solved with a step-by-step numerical integration scheme. In each step, the displacements computed are imposed on the analytical and physical substructures, respectively. The static restoring forces, $r_{S,2}^A$, $r_{S,3}$, and $r_{S,4}$, developed by the analytical substructure are evaluated from appropriate structural element models and constitutive laws, while the static and dynamic resisting forces, r_1 and r_2^E , of the super-element are measured from the physical substructure after the computed displacements, d_1 and d_2, have been imposed in the time step. This approach has been used by Nakashima el at. (1999), Blakeborough *et. al.* (2001), Shing *et. al.* (2004), and Wu *et. al.* (2006) for real-time hybrid testing, and it is similar in concept to non-real-time pseudodynamic substructure testing (Dermitzakis and Mahin 1985, Nakashima *et. al.* 1990, Shing *et. al.* 1994). The only difference is that in a real-time test, the inertia and damping forces measured from the physical structure can be significant and should be accounted for in the analytical model.

3.2 Coupled subdomain approach

This approach is based on the domain-decomposition method that is frequently used in parallel computing. With this approach, one treats the analytical model and physical substructure as two separate domains that are coupled by the interaction forces f_B^I at the boundary of the two. In this case, Equation 2.3 is divided into two sets of equations as follows, with one representing the analytical model and the other the physical substructure.

$$\begin{bmatrix} \mathbf{M}_{AA}^A & \mathbf{M}_{AB}^A \\ \mathbf{M}_{BA}^A & \mathbf{M}_{BB}^{A1} \end{bmatrix} \begin{Bmatrix} \mathbf{a}_A \\ \mathbf{a}_B \end{Bmatrix} + \begin{bmatrix} \mathbf{C}_{AA}^A & \mathbf{C}_{AB}^A \\ \mathbf{C}_{BA}^A & \mathbf{C}_{BB}^{A1} \end{bmatrix} \begin{Bmatrix} \mathbf{v}_A \\ \mathbf{v}_B \end{Bmatrix} + \begin{Bmatrix} \mathbf{r}_{S,A} \\ \mathbf{r}_{S,B}^A \end{Bmatrix} = \begin{Bmatrix} \mathbf{f}_A \\ \mathbf{f}_B^A \end{Bmatrix} + \begin{Bmatrix} 0 \\ \mathbf{f}_B^I \end{Bmatrix} \tag{3.2}$$

$$\begin{bmatrix} \mathbf{M}_{BB}^{A2} & \mathbf{M}_{BE}^A \\ \mathbf{M}_{EB}^A & \mathbf{M}_{EE}^A \end{bmatrix} \begin{Bmatrix} \mathbf{a}_B \\ \mathbf{a}_E \end{Bmatrix} + \begin{bmatrix} \mathbf{C}_{BB}^{A2} & \mathbf{C}_{BE}^A \\ \mathbf{C}_{EB}^A & \mathbf{C}_{EE}^A \end{bmatrix} \begin{Bmatrix} \mathbf{v}_B \\ \mathbf{v}_E \end{Bmatrix} + \begin{Bmatrix} \mathbf{r}_B^E \\ \mathbf{r}_E \end{Bmatrix} = \begin{Bmatrix} \mathbf{f}_B^E \\ \mathbf{f}_E \end{Bmatrix} + \begin{Bmatrix} -\mathbf{f}_B^I \\ 0 \end{Bmatrix} \tag{3.3}$$

with $\mathbf{M}_{BB}^{A1} + \mathbf{M}_{BB}^{A2} = \mathbf{M}_{BB}^A$ and the same for the damping terms. As reflected in Equation 3.3, part of the inertia and damping forces of the physical structure is analytically accounted for and part of them is measured from the physical structure during the test. For the four-story frame example shown in Figure 1, the decomposition is illustrated in Figure 2 and Equations 3.2 and 3.3 become

$$\begin{bmatrix} M_2^{A1} & 0 & 0 \\ 0 & M_3^A & 0 \\ 0 & 0 & M_4^A \end{bmatrix} \begin{Bmatrix} a_2 \\ a_3 \\ a_4 \end{Bmatrix} + \begin{bmatrix} C_3^A & -C_3^A & 0 \\ -C_3^A & C_3^A + C_4^A & -C_4^A \\ 0 & -C_4^A & C_4^A \end{bmatrix} \begin{Bmatrix} v_2 \\ v_3 \\ v_4 \end{Bmatrix} + \begin{Bmatrix} r_{S,2}^A \\ r_{S,3} \\ r_{S,4} \end{Bmatrix} = \begin{Bmatrix} f_2^A \\ f_3 \\ f_4 \end{Bmatrix} + \begin{Bmatrix} f_2^I \\ 0 \\ 0 \end{Bmatrix} \tag{3.4}$$

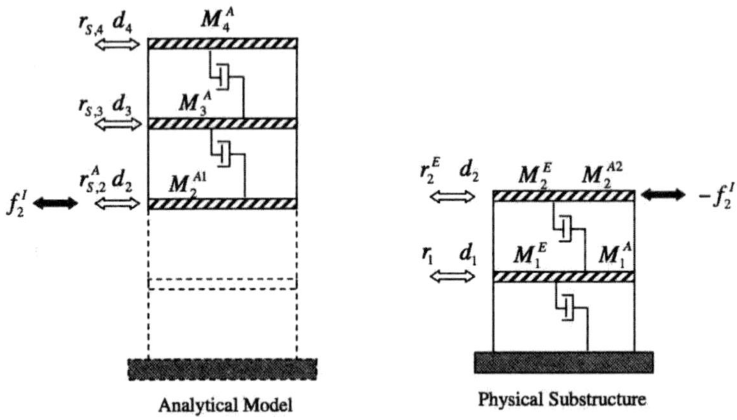

Figure 2. Coupled subdomain approach.

$$\begin{bmatrix} M_1^A & 0 \\ 0 & M_2^{A2} \end{bmatrix} \begin{Bmatrix} a_1 \\ a_2 \end{Bmatrix} + \begin{bmatrix} C_1^A + C_2^A & -C_2^A \\ -C_2^A & C_2^A \end{bmatrix} \begin{Bmatrix} v_1 \\ v_2 \end{Bmatrix} + \begin{Bmatrix} r_1 \\ r_2^E \end{Bmatrix} = \begin{Bmatrix} f_1 \\ f_2^E \end{Bmatrix} + \begin{Bmatrix} 0 \\ -f_2^I \end{Bmatrix} \tag{3.5}$$

One advantage of this approach is that Equations 3.4 and 3.5 can be solved separately using different integration schemes that are most suited for the problem at hand. In the solution process, both the compatibility and equilibrium conditions between the two subdomains must be satisfied. Magonette *et. al.* (1998) have proposed a staggered integration procedure to solve Equation 3.3 with an explicit scheme and Equation 3.2 with an implicit scheme. This approach has been further extended by Bonelli *et. al.* (2008) and is discussed in Pegon (2008).

3.3 Dynamic substructuring approach

The dynamic substructuring approach can be considered as a special case of the coupled subdomain method by having the dynamic response of the physical substructure obtained purely by actual testing. In this case, the mass and damping terms with superscript A in Equation 3.3 will all

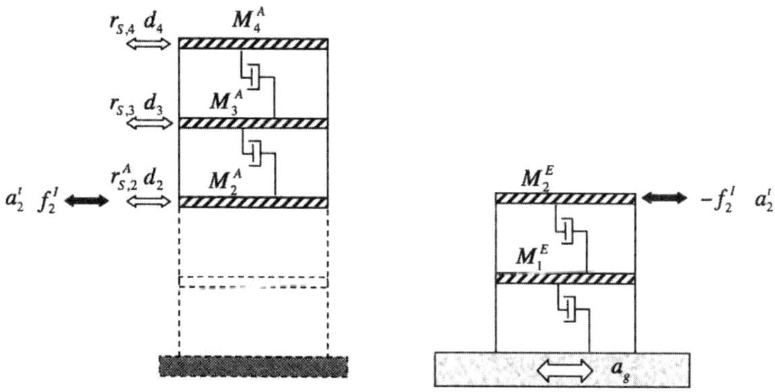

Figure 3. Dynamic substructuring approach – scenario 1.

vanish. As a result, only Equation 3.2 needs to be solved numerically. Two possible scenarios can be conceived with this approach. They are illustrated here with the four-story frame example shown in Figure 1.

One scenario is illustrated in Figure 3 with the frame subjected to a base acceleration. In this case, most of the structural nonlinearity may be expected at the lower stories. Furthermore, there could be a compelling situation that the dynamic response of the lower two stories can only be realistically obtained with dynamic testing, while the upper-story behavior can be simulated numerically. Hence, one may subject the lower stories of the frame to a base acceleration a_g by means of a shake table, and, at the same time, subject the second floor to the interface force f_2^I by means of a dynamic actuator. In this case, the equations of motion for the upper stories, which are modeled analytically, are best expressed in the following form

$$\begin{bmatrix} M_2^A & 0 & 0 \\ 0 & M_3^A & 0 \\ 0 & 0 & M_4^A \end{bmatrix} \begin{Bmatrix} a_2^t \\ \hat{a}_3 \\ \hat{a}_4 \end{Bmatrix} + \begin{bmatrix} 0 & -C_3^A & 0 \\ 0 & C_3^A + C_4^A & -C_4^A \\ 0 & -C_4^A & C_4^A \end{bmatrix} \begin{Bmatrix} v_2^t \\ \hat{v}_3 \\ \hat{v}_4 \end{Bmatrix}$$
$$+ \begin{Bmatrix} r_{S,2}^A \\ r_{S,3} \\ r_{S,4} \end{Bmatrix} = \begin{Bmatrix} 0 \\ -M_3^A a_2^t \\ -M_4^A a_2^t \end{Bmatrix} + \begin{Bmatrix} f_2^I \\ 0 \\ 0 \end{Bmatrix} \tag{3.6}$$

in which the response variables with superscript t denote the total motion and those with a hat represent the relative motions with respect to those at the second floor. Furthermore, it should be noted that $r_{S,2}^A = r_{S,3} + r_{S,4}$. Hence, the last two equations in Equation 3.6 can be solved numerically with a_2^t measured from the physical substructure, while the first equation can be used to compute f_2^I, which will be exerted on the physical substructure. This way, both the compatibility and equilibrium conditions between the two substructures are satisfied. This approach has been proposed by Sivaselvan *et. al.* (2004) for applications such as soil-pile-structure interaction problems, in which only the soil-pile subsystem needs to be tested on a shake table. This is an area that needs further research because of the difficulty of force control.

A second scenario of dynamic substructuring is illustrated in Figure 4. In this case, Equation 3.2 becomes

$$\begin{bmatrix} M_1^A & 0 \\ 0 & M_2^A \end{bmatrix} \begin{Bmatrix} a_1 \\ a_2 \end{Bmatrix} + \begin{bmatrix} C_1^A + C_2^A & -C_2^A \\ -C_2^A & C_2^A \end{bmatrix} \begin{Bmatrix} v_1 \\ v_2 \end{Bmatrix} + \begin{Bmatrix} r_{S,1} \\ r_{S,2}^A \end{Bmatrix} = \begin{Bmatrix} -M_1^A a_g \\ -M_2^A a_g \end{Bmatrix} + \begin{Bmatrix} 0 \\ f_2^I \end{Bmatrix} \tag{3.7}$$

which can be solved by measuring f_2^I from the physical substructure, while the physical substructure will be subjected to the total base motion d_2^t calculated with Equation 3.7. It should be noted that d_2^t is the total displacement at the second floor, which can be imposed on the physical substructure using a shake table. This approach is best suited for testing secondary systems mounted in structures (Horiuchi *et. al.* 2000, Bayer *et. al.* 2005). Such tests can also be conducted with dynamic actuators (Bursi *et. al.* 2008, Wagg *et. al.* 2008). It is evident that the distinction between this and the total approach is small in the sense that only one set of equations is solved in both cases. However, the main difference of this approach is that the dynamic response of the physical substructure is obtained purely from the test without any modeling assumption.

Different explicit and implicit integration schemes can be used to solve the equations of motion formulated above. The selection of a particular integration scheme or schemes depends on the properties of the analytical and physical substructures and how the two are coupled in the analysis.

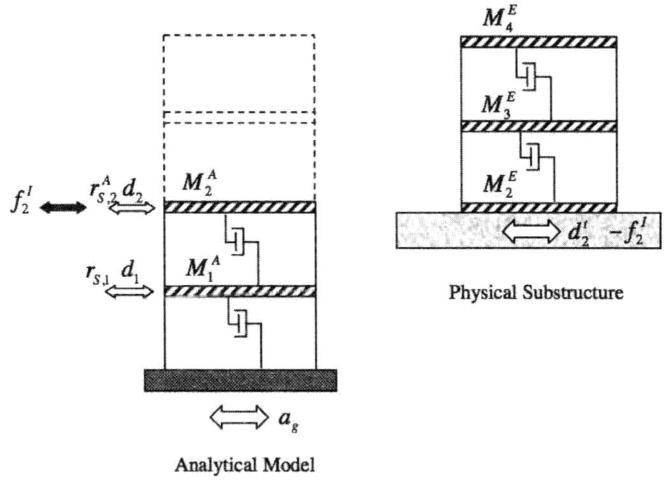

Figure 4. Dynamic substructuring approach – scenario 2.

4 Implicit Time Integration for Real-Time Testing

In any of the real-time hybrid testing approaches presented above, a step-by-step time integration scheme is essential for solving the resulting equations of motion to generate displacement or force commands to load the physical substructure. Such a scheme has to be robust and stable in the presence of numerical and experimental feedback errors (Shing and Mahin 1987, 1990). Conditionally stable explicit schemes are in general not suitable for structures with many degrees of freedom (which can be the case for an analytical substructure) and high-frequency modes. With a coupled subdomain approach, Magonette *et. al.* (1998) and Bonelli *et. al.* (2008) have proposed a staggered method to solve Equation 3.2 (for the analytical substructure) with an implicit scheme and Equation 3.3 (for the experimental substructure) with an explicit scheme based on the assumption that the latter does not have high-frequency modes present. Details of this method can be found in Pegon (2008).

Many implicit integration schemes are unconditionally stable. This is certainly an attractive feature when one has to deal with many degrees of freedom and high-frequency modes in the analytical substructure. Different methods of applying implicit integration schemes have been proposed for real-time hybrid tests (Shing *et. al.* 2004, Bayer *et. al.* 2005, Wu *et. al.* 2006 and 2007, Bursi *et. al.* 2008). The method by Shing *et. al.* (2004) is presented here. It is illustrated with the total approach of hybrid testing. However, as shown in the following discussion, it can also be readily adopted for other testing scenarios.

Implicit schemes normally require an iterative solution procedure for nonlinear structures. This poses a major challenge for real-time hybrid testing as the actuators controlling structural displacements have to move continuously with a smooth velocity, and the displacement commands to the actuators have to be sent at the sampling frequency of the digital controller. However, with a standard Newton-type iterative solution scheme, the displacement corrections will be progressively reduced as the trial displacements approach a converged solution. Hence, updating the displacement commands directly with the trial displacements will lead to undesired velocity fluctuations from one time step to the next. In addition, the number of iterative corrections required in each time step will vary depending on the degree of nonlinearity developed in the structure. This uncertainty is not acceptable for a real-time test where a converged solution has to be attained within a designated time interval.

To circumvent the aforementioned problem, a special iterative method that adopts a fixed number of iterations can be used. This method is presented here using the implicit Newmark integration method as an example. With Equation 2.3, the Newmark method can be expressed as follows.

$$\mathbf{M}^A\mathbf{a}_{i+1} + \mathbf{C}^A\mathbf{v}_{i+1} + \mathbf{r}^A_{s,i+1} + \mathbf{r}^E_{i+1} = \mathbf{f}_{i+1} \tag{4.1}$$

$$\mathbf{d}_{i+1} = \tilde{\mathbf{d}}_{i+1} + \Delta t^2\beta\mathbf{a}_{i+1} \tag{4.2}$$

$$\mathbf{v}_{i+1} = \tilde{\mathbf{v}}_{i+1} + \Delta t\gamma\mathbf{a}_{i+1} \tag{4.3}$$

where

$$\tilde{\mathbf{d}}_{i+1} = \mathbf{d}_i + \Delta t\mathbf{v}_i + \Delta t^2(\frac{1}{2} - \beta)\mathbf{a}_i \tag{4.4}$$

$$\tilde{\mathbf{v}}_{i+1} = \mathbf{v}_i + \Delta t(1-\gamma)\mathbf{a}_i \tag{4.5}$$

with β and γ being numerical parameters that determine the properties of the integration scheme. With $\beta = 1/4$ and $\gamma = 1/2$, we have the constant-average-acceleration (trapezoidal) method. In an iterative solution, if $d^{(k)}_{i+1}$ is the trial solution in iteration step k and $r^{A(k)}_{S,i+1}$ and $r^{E(k)}_{i+1}$ are the associated forces obtained from the analytical and physical substructures, respectively, the equilibrium error can be calculated with the following equation obtained by re-arranging Equations 4.1 through 4.3.

$$\mathbf{R}^{(k)}_{i+1} = \bar{\mathbf{M}}(\tilde{\mathbf{d}}_{i+1} - \mathbf{d}^{(k)}_{i+1}) - \mathbf{C}^A\tilde{\mathbf{v}}_{i+1} - \mathbf{r}^{A(k)}_{s,i+1} - \mathbf{r}^{E(k)}_{i+1} + \mathbf{f}_{i+1} \tag{4.6}$$

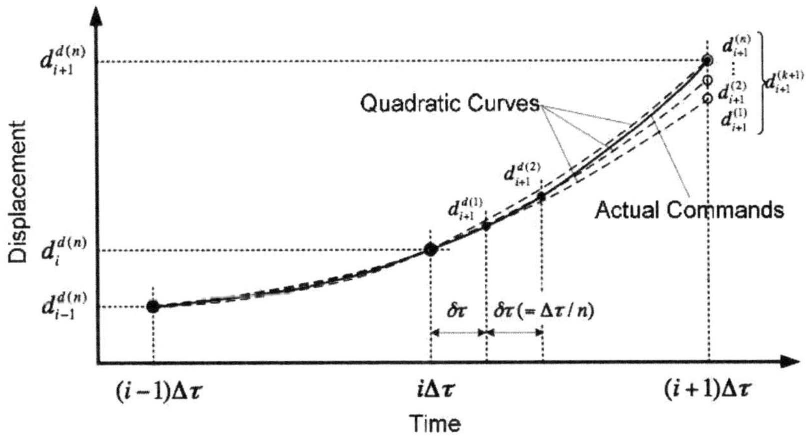

Figure 5. Iteration scheme.

where

$$\bar{\mathbf{M}} = \frac{\mathbf{M}^A + \Delta t \gamma \mathbf{C}^A}{\Delta t^2 \beta} \tag{4.7}$$

To eliminate the equilibrium error evaluated with Equation 4.6, we can adopt a modified Newton iteration method based on the initial stiffness of the structure. Assuming that damping in the physical substructure is viscous, we have the modified Newton method expressed as follows.

$$\mathbf{K}^* \Delta \mathbf{d}_{i+1}^{(k)} = \mathbf{R}_{i+1}^{(k)} \tag{4.8}$$

where the effective stiffness matrix is defined as

$$\mathbf{K}^* = \frac{(\mathbf{M}^A + \mathbf{M}^E) + \Delta t \gamma (\mathbf{C}^A + \mathbf{C}^E)}{\Delta t^2 \beta} + (\mathbf{K}_0^A + \mathbf{K}_0^E) \tag{4.9}$$

After solving Equation 4.8 for $\Delta \mathbf{d}_{i+1}^{(k)}$, one can compute the updated displacement.

$$\mathbf{d}_{i+1}^{(k+1)} = \mathbf{d}_{i+1}^{(k)} + \Delta \mathbf{d}_{i+1}^{(k)} \tag{4.10}$$

Since $\mathbf{M}^E$, $\mathbf{C}^E$, and $\mathbf{K}_0^E$ are only used to estimate the updated displacements in the iteration process, their exact values are not needed other than the fact that this may affect the convergence rate. In fact, it is desirable that their values be a little (few percent) higher than the exact ones to avoid overshoot and to preserve the stability of the numerical solution (Shing *et. al.* 1991a, 1991b).

As mentioned previously, for a real-time hybrid test, one must avoid undesired velocity fluctuations from one iteration step to the next and attain convergence within a fixed time interval. To achieve this, a special correction procedure that uses a fixed number of iterations in each time step can be adopted (Jung *et. al.* 2007). As shown in Figure 5, this method relies on an interpolation technique to assure a smooth motion of an actuator during iteration. Each iteration consists of two steps: (1) the trial displacement $d_{i+1}^{(k+1)}$ for each degree of freedom is evaluated as in a conventional Newton-type iterative process; and (2) a fraction of the trial displacement, which is termed the desired displacement $d_{i+1}^{\mathrm{d}(k+1)}$, is then computed with a quadratic interpolation and imposed on the structural specimen. Except for the first time step, three points are used for the construction of the interpolation function: the trial displacement $d_{i+1}^{(k+1)}$ in the current iteration, and the displacements $d_{i-1}^{\mathrm{d}(n)}$ and $d_{i}^{\mathrm{d}(n)}$, which are computed in the very last iteration of the two previous time steps, with n denoting the total number of iterations specified in each time step and k varying from 0 to $(n-1)$. In the first time step, the interpolation will be based on the initial velocity and displacement.

Once iteration is completed in a time step, the response needs to be updated so that the computation for the next time step can begin. Since the number of iterations in each time step is fixed, convergence errors are expected to be larger than those with a conventional Newton-type iteration. In addition, convergence errors can be exacerbated by the time delays in actuator response and in the data exchange between the controller/data-acquisition processor and the computer that handles the numerical computation. To minimize the effect of these errors and enforce equilibrium at the end of each time step, the following approximate correction can be introduced for updating the displacements and forces in the last iteration.

$$\mathbf{d}_{i+1} = \mathbf{d}_{i+1}^{d(n)} \tag{4.11}$$

$$\mathbf{r}_{i+1}^{E} = \mathbf{r}_{i+1}^{m(n-1)} + \mathbf{K}^{E*}\left[\mathbf{d}_{i+1}^{d(n)} - \mathbf{d}_{i+1}^{m(n-1)}\right] \tag{4.12}$$

in which $\mathbf{d}_{i+1}$ and $\mathbf{r}_{i+1}^{E}$ are the updated displacements and forces that are treated as the converged solutions that will be carried over to the next time step, $\mathbf{d}_{i+1}^{d(n)}$ is the desired displacement vector computed in the n-th iteration, which is the last iteration, $d_{i+1}^{\mathrm{m}(n-1)}$ and $\mathbf{r}_{i+1}^{\mathrm{m}(n-1)}$ are the displacements and restoring forces measured from the test structure at the beginning of the last iteration, and $\mathbf{K}^{E*}$ is the portion of the effective stiffness, as presented in Equation 4.9, that contains only the stiffness, damping, and mass

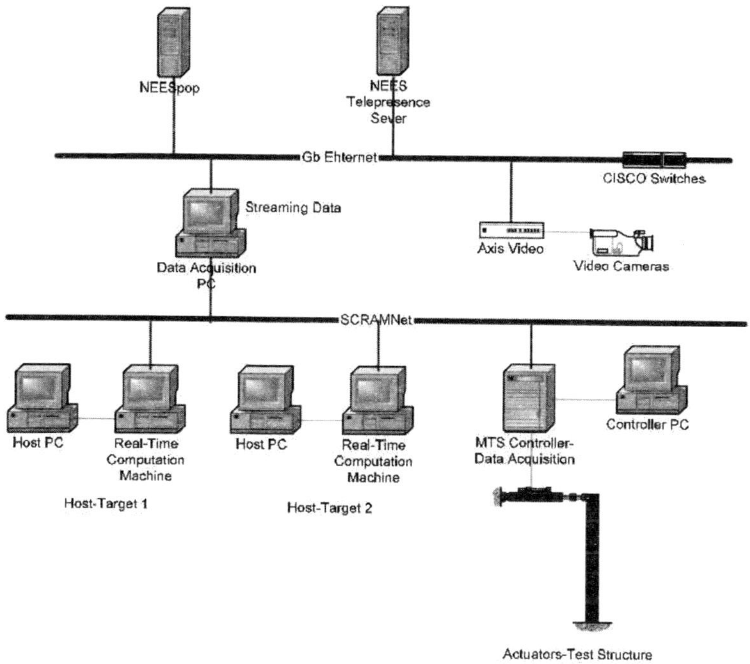

Figure 6. Real-time hybrid test system.

properties of the physical substructure. With $\mathbf{d}_{i+1}$ computed, the velocities and accelerations can then be updated with Equations 4.2 and 4.3 and the computation is advanced to the next time step.

Another issue to address is that during each iteration, the displacements imposed on the physical substructure can be different from the desired displacements computed by the interpolation method. To avoid displacement incompatibility between the physical and analytical substructures at the boundary degrees of freedom, it is necessary that the displacements and forces measured from the physical substructure and used to compute the residual errors (with Equation 4.6) be corrected in each iteration as follows (Wei 2005).

$$\mathbf{d}_{i+1}^{(k)} = \mathbf{d}_{i+1}^{d(k)} \tag{4.13}$$

$$\mathbf{r}_{i+1}^{E(k)} = \mathbf{r}_{i+1}^{m(k)} + \mathbf{K}^{E*}\left[\mathbf{d}_{i+1}^{d(k)} - \mathbf{d}_{i+1}^{m(k)}\right] \tag{4.14}$$

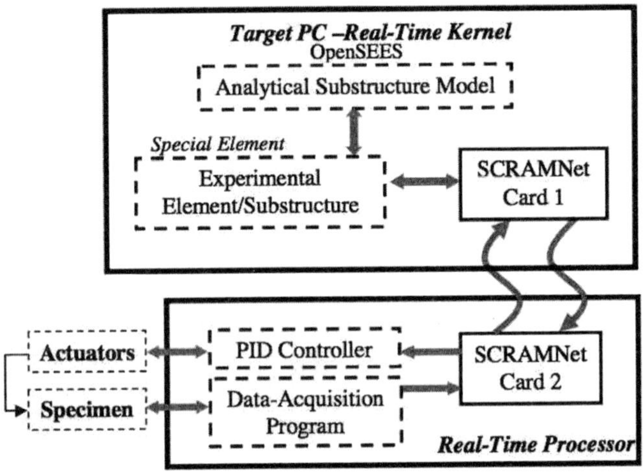

Figure 7. Real-time computation in Target 1 and interface with experimental hardware.

The iteration scheme presented here can be used with other implicit integration methods. Shing *et. al.* (2004) has applied this scheme with the α-method (Hughes 1983). However, the α-method presents a consistency issue when the inertia forces developed by the physical substructure become significant and cannot be ignored.

Several unconditionally stable schemes that do not require iteration have been proposed for non-real-time hybrid testing. They are referred to as either implicit or explicit schemes. However, these schemes are based on implicit formulations, and iteration is avoided by performing either a prediction or correction that is based on the initial stiffness properties of the structure. Hence, strictly speaking, these schemes should be considered as either predictor-corrector or implicit methods. One such scheme is the operator-splitting (OS) method that has been proposed by Nakashima *et. al.* (1990) for non-real-time pseudodynamic tests. It is based on an explicit prediction-implicit correction approach. It has been used by Wu *et. al.* (2006) for real-time tests.

The unconditionally stable "explicit" scheme of Chang (2002) adopts an explicit prediction that is based on the initial elastic stiffness of a structure. For a linearly elastic structure, this method is essentially identical to the constant-average-acceleration method. However, its accuracy for highly nonlinear structural behavior has not been thoroughly studied. This also applies to the Rosenbrock algorithms proposed by Bursi *et. al.* (2008). The Rosenbrock algorithms do not require iteration but the multi-stage algorithms require the computation of structural responses at intermediate steps.

5 System Configuration

A real-time hybrid test system requires four basic components: a robust computation scheme, which has already been discussed, a computer program to construct and evaluate the response of the analytical substructure, a high-performance loading system to exert displacements or forces on the physical substructure, and a high-speed network that links the computation machine to the load application hardware. Figure 6 shows a real-time hybrid test system that has been developed to serve such a purpose (Shing *et. al.* 2004). The system has also been designed for teleparticipitation and teleobservation by remote users through the NEES Point of Presence (NEESpop) and NEES Telepresence Server. The basic system for real-time testing is shown as the bottom layer in the figure. Servo-hydraulic actuators are used for load application. A digital PID (Proportional-Integral-Differential) servo-controller with a sampling frequency of 1,024 Hz is used. Because of inevitable mechanical delays in the servo-hydraulic system, a higher sampling rate is deemed unnecessary. Target Computer 1 is for real-time numerical computation to evaluate the response of the analytical substructure and solve the governing equations of motion. Target Computer 2 can be used to generate the displacement commands to be sent to the digital controller by means of interpolation using the trial displacements computed in Target 1 as described in the previous section. It can also be used in pretest simulations to model the dynamics of the test system including the servo-hydraulic actuators and test structure. For this purpose, the computer can run Simulink models of the servo-hydraulic system and test structure using Mathworks' xPC-Target as the real-time kernel to simulate a real-time test environment (Jung *et. al.* 2007). The target computers communicate with one another and with the digital controller that controls the hydraulic actuators through SCRAMNet (Shared Common RAM Network), which can share data among different processors in less than a micro-second's time through a fiber-optic network.

An open-source, object-oriented structural analysis framework, called OpenSEES, developed by McKenna and Fenves (2000), has been adapted to serve as the computational platform for the real-time system. The existing modeling and analysis capabilities of OpenSEES, including its library of nonlinear beam-column elements, are available for the construction of an an-

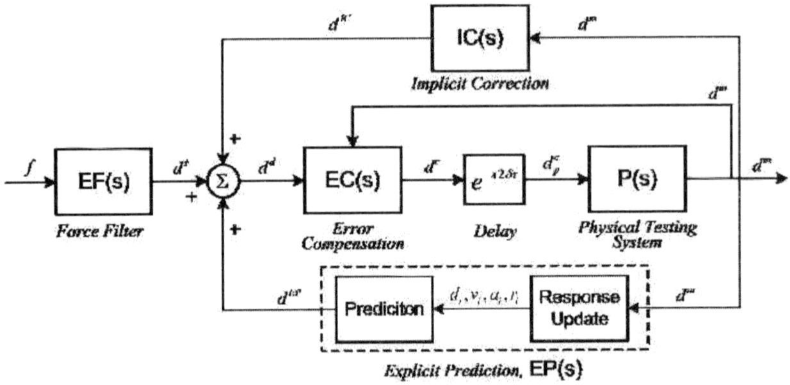

Figure 8. Block diagram for the real-time hybrid test system.

alytical substructure. The α-method (Hughes 1983) has been implemented in the program with the iterative solution scheme presented above for numerical integration. The α-method adopts the same approximations for the velocity and acceleration as the Newmark method (as shown in Equations 4.2 and 4.3) but has the equations of motion expressed in the following form.

$$\mathbf{M}^A\mathbf{a}_{i+1} + (1+\alpha)\mathbf{C}^A\mathbf{v}_{i+1} - \alpha\mathbf{C}^A\mathbf{v}_i + (1+\alpha)(\mathbf{r}^A_{s,i+1} + \mathbf{r}^E_{i+1}) - \alpha(\mathbf{r}^A_{s,i} + \mathbf{r}^E_i) = (1+\alpha)\mathbf{f}_{i+1} - \alpha\mathbf{f}_i \tag{5.1}$$

in which α is an additional numerical parameter that has to assume a value between -1/3 and 0 to arrive at a desired numerical damping property. It becomes the Newmark method when α is zero. In the above equation, it has to be assumed that the inertia forces developed by the physical substructure are negligible, i.e., $\mathbf{M}^A = \mathbf{M}$. Otherwise, there will be an inconsistency in the treatment of $\mathbf{M}^A$ and $\mathbf{M}^E$ in that the latter, through $\mathbf{r}^E_i$, will be modified by α just like the damping and stiffness terms. The impact of introducing the α parameter into the inertia term remains to be explored. With the iterative scheme described previously, one interpolation and one correction are performed within each sampling time interval. An interface has been developed in the software to interact with the experimental substructure through the digital control and data-acquisition system. To have a full and deterministic control over all of the control and computational resources during a high-speed test, OpenSEES is executed in Target Computer 1 using

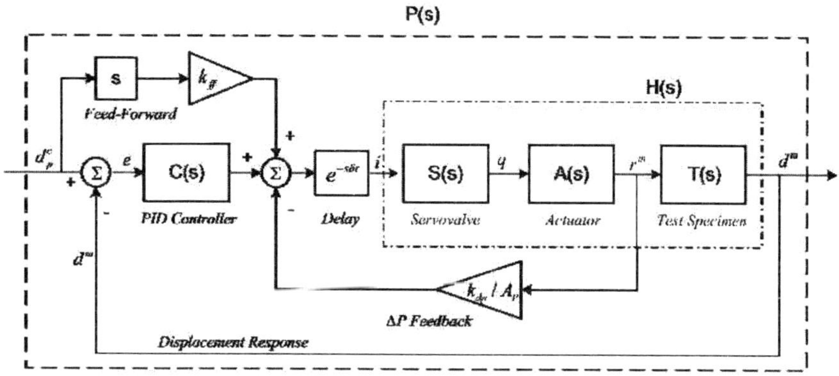

Figure 9. Block diagram for the physical test system.

an embedded real-time operating system. Host Computer 1 is for program development and the construction of the structural model for testing. Once constructed, a compiled program is downloaded from the host to the target for execution in the real-time operating system. The synchronization of the computations in the target computers with the control/data-acquisition functions is achieved through "interrupts" generated by the controller at a frequency of 1,024 Hz.

Figure 7 shows the functions of Target Computer 1 and its interface with the digital control/data-acquisition system. A special *Experimental Element* has been developed in OpenSEES to serve as the interface between the analytical model and the experimental substructure. The *Experimental Element* serves as a communication channel that reads the structural displacements and forces recorded by the data-acquisition system, and sends displacement commands to the controller, all through SCRAMNet. The *Experimental Element* also transforms the displacements and forces between the global coordinates of the structural model and the actuator coordinates, and applies the force correction (Equations 4.12 and 4.14) (Wei 2005).

6 System Dynamics and Delay Compensation

The performance of a real-time hybrid test system is influenced by the interaction of the numerical integration scheme, PID controller, servo-hydraulic actuators, and test structure. To evaluate the performance of an entire system, the dynamics of each physical component can be modeled in detail (Williams 2001, Jung *et. al.* 2007) including system nonlinearities. How-

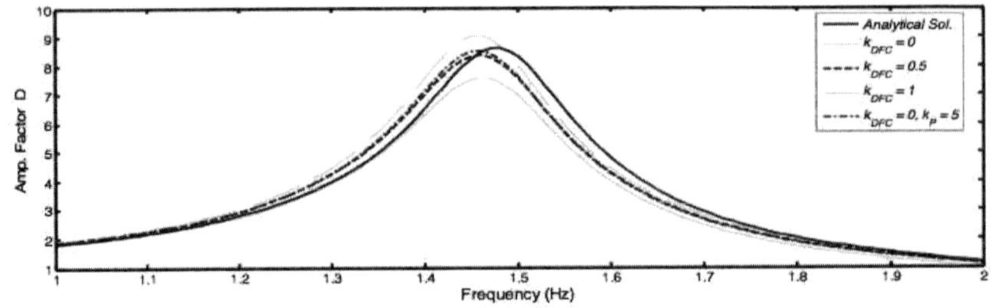

Figure 10. Effect of the DFC scheme on system performance.

ever, the system characteristics can be most conveniently identified with a transfer function. A transfer function for the system described in the previous section, including the integration scheme, has been derived by Jung and Shing (2006). This work is briefly summarized here.

A linear single-degree-of-freedom test structure is considered for this purpose. As a result, the iterative solution procedure presented in a previous section can be reduced to a one-step correction. Hence, the $d_{i+1}^{(k+1)}$ computed with Equation 4.10, with k equal to zero, becomes the desired displacement d_{i+1}^{d} that the test structure has to attain. With this simplification, the real-time hybrid test system can be expressed in the form of a block diagram as shown in Figure 8. The delay block in the figure is to simulate the time delay in the communication between the target computer (assuming only one target computer is used) and controller, which is $2\delta\tau$ for the system considered here, where $\delta\tau$ is the sampling time interval of the controller. The blocks $P(\mathrm{s})$ and $EC(\mathrm{s})$ represent the transfer functions of a physical test system and error compensator, respectively. The latter is to compensate for time delays in actuator response.

Based on Figure 8, the transfer function for the real-time hybrid test system can be expressed as

$$RTPD(s) = \frac{d^m(s)}{f(s)} = \frac{EF(s) \cdot EC(s) \cdot P(s)}{e^{2s\delta\tau} - IC(s) \cdot EC(s) \cdot P(s) - EP(s) \cdot EC(s) \cdot P(s)} \tag{6.1}$$

in which $EF(s)$, $IC(s)$, and $EP(s)$ represent the various parts of the numerical integration scheme. They can be derived from Equations 5.1, 4.2, and 4.3.

$$EF(s) = d^F(s)/f(s) = \left[(1+\alpha) - \alpha\, e^{-s\,\Delta t}\right]/K^* \tag{6.2}$$

$$IC(s) = d^{IC}(s)/d^m(s) = 1 - \bar{M}/(K^*\Delta t^2\beta) - (1+\alpha)/\left[K^*T(s)\right] \tag{6.3}$$

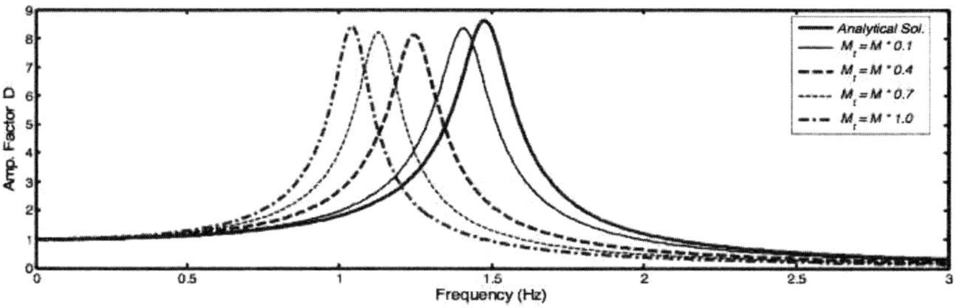

Figure 11. Effect of structural mass M_t on system transfer function.

$$\begin{aligned} EP(s) &= \frac{d^{EP}(s)}{d^m(s)} = \frac{\bar{M}}{K^*\Delta t^2\beta}\frac{e^{-s\Delta t}\,e^{2s\,\delta\tau}}{EC(s)\cdot P(s)} + \frac{1}{K^*}\left(\frac{\bar{M}}{\Delta t\beta} - C\right)VA(s)\times \\ &\quad AD(s)\frac{e^{-s\Delta t}e^{2s\,\delta\tau}}{EC(s)\cdot P(s)} + \frac{1}{K^*}\left[\frac{\bar{M}}{\beta}\left(\frac{1}{2}-\beta\right)-(1+\alpha)C\Delta t(1-\gamma)\right]\times \\ &\quad AD(s)\frac{e^{-s\Delta t}\,e^{2s\,\delta\tau}}{EC(s)\cdot P(s)} + \frac{\alpha}{K^*}\left(\frac{e^{-s\Delta t}}{T(s)} + \frac{e^{-s\Delta t}e^{2s\delta\tau}}{EC(s)\cdot P(s)} - Ke^{-s\Delta t}\right) \end{aligned} \tag{6.4}$$

with

$$VA(s) = v(s)/a(s) - \left[\Delta t\gamma + \Delta t(1-\gamma)e^{-s\Delta t}\right]/\left(1 - e^{-s\Delta t}\right) \tag{6.5}$$

$$AD(s) = \frac{a(s)}{d(s)} = \frac{(1-e^{-s\Delta t})}{[\Delta t^2\beta + \Delta t\,e^{-s\Delta t}\cdot VA(s) + \Delta t^2(1/2-\beta)e^{-s\Delta t}]} \tag{6.6}$$

in which K is the initial stiffness of the structural model considered and K^* is the effective stiffness in Equation 4.8. The transfer function, $T(s)= d^m(s)/r^m(s)$, of the test structure together with a flexible reaction frame that supports the actuator can be expressed as follows.

$$T(s) = 1/\left[s^2 M_t + s\,C_t + K_t\right] + 1/\left[s^2 M_r + s\,C_r + K_r\right] \tag{6.7}$$

in which the quantities with a subscript t represent the actual properties of the physical test structure and those with a subscript r are associated with the reaction frame. If the reaction frame is rigid, the last term on the right-hand side of the above equation becomes zero.

Many compensation schemes have been proposed to counteract delays in actuator response (Horiuchi *et. al.* 2000, Zhao *et. al.* 2003, Wagg *et. al.*

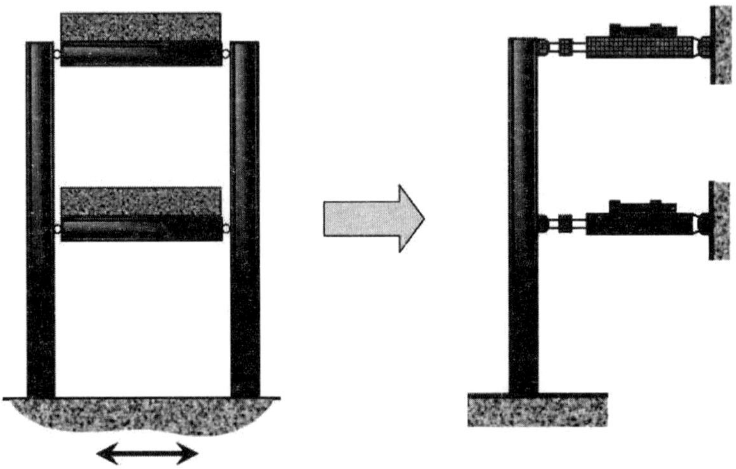

Figure 12. Structural model for two-degree-of-freedom tests.

2008). Three compensation schemes have been implemented in the system considered here. One is a simple feedforward compensation. The second is a discrete feed-forward compensation (DFC) scheme (Jung *et. al.* 2007), which is based on the assumption that the displacement control errors within a time step are more or less the same as those in the previous step. Thus, the desired displacement $d_{i+1}^{d(k+1)}$ is modified as follows to compensate for the anticipated error.

$$d_{i+1}^{c(k+1)} = d_{i+1}^{d(k+1)} + k_{DFC} \cdot \left[d_i^{c(n)} - d_i^{m(n)} \right] \tag{6.8}$$

in which k_{DFC} is a gain factor, and $d_i^{c(n)}$ and $d_i^{m(n)}$ are the corrected and measured displacements in the last iteration of the previous time step. For this case, the transfer function $EC(s)$ can be obtained from Equation 6.8 by ignoring the iteration and replacing $d^m(s)$ by $P(s)\, e^{-2s\,\delta\tau} \cdot d^c(s)$.

$$EC(s) = d^c(s)/d^d(s) = 1/\left(1 - k_{DFC} \cdot e^{-s\Delta t}\left[1 - e^{-2s\,\delta\tau}P(s)\right]\right) \tag{6.9}$$

The third compensation scheme is a first-order Phase-Lead Compensator (PLC), which is used quite extensively in mechanical control to improve system response. A phase-lead compensator is usually constructed in the form of a first-order transfer function as follows (Dorf and Bishop 2000).

$$PLC(s) = \frac{T_{\mathrm{d}}\, s \,+\, 1}{\alpha_{\mathrm{d}}\, T_{\mathrm{d}}\, s \,+\, 1} \tag{6.10}$$

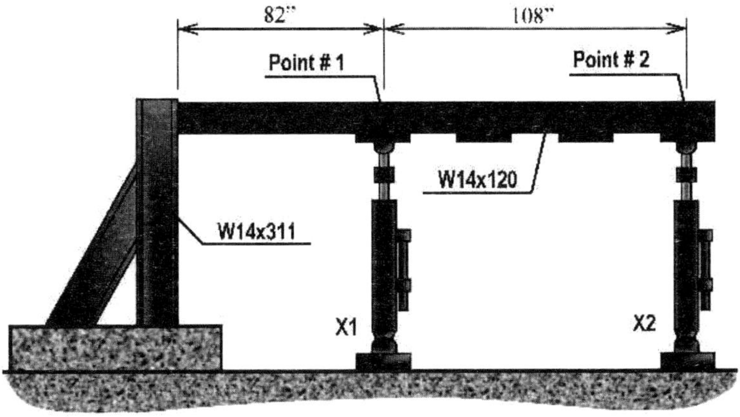

Figure 13. Two-DOF real-time hybrid test (1 in. = 24.5 mm).

in which α_{d} and T_{d} are parameters for the phase-lead network. The maximum phase angle introduced by the above compensator is decided by the constant α_{d}. In general, a single lead compensator can contribute a maximum of 60° to the phase (Goodwin *et. al.* 2001). This method has been proposed by Zhao *et. al.* (2003) for real-time structural testing. They have suggested that T_{d} be calculated as the inverse of the roll-off frequency obtained from a sine-sweep test and α_{d} be equal to 0.1.

6.1 Transfer Function for the Physical Test System

The block diagram for the physical test system that consists of a PID controller with the simple feed-forward compensation, servo-hydraulic actuator, actuator reaction frame, and test structure is shown in Figure 9. The transfer function for the physical test system is given by the following expression (Jung and Shing 2006).

$$P(s) = \frac{d^m(s)}{d_p^c(s)} = \frac{\left[sk_p + k_i + s^2(k_d + k_{ff})\right] \cdot H(s)}{s \cdot e^{s\,\delta\tau} + H(s) \cdot \left[s\,k_p + k_i + s^2 k_d + s\frac{k_{\Delta p}}{A_p}\frac{1}{T(s)}\right]} \tag{6.11}$$

where

$$H(s) = \frac{d^m(s)}{i(s)} = S(s) \Big/ \left[s2A_p + s\frac{V}{2\beta_E\,A_p\,T(s)}\right] \tag{6.12}$$

$$S(s) = \frac{q_t(s)}{i(s)} = \frac{\omega_s^2}{s^2 + 2\xi_s\omega_s s + \omega_s^2}\,k_o \tag{6.13}$$

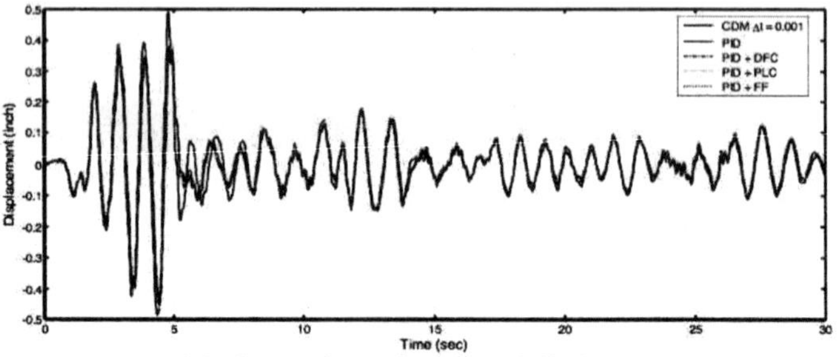

(a) Comparison of measured displacements.

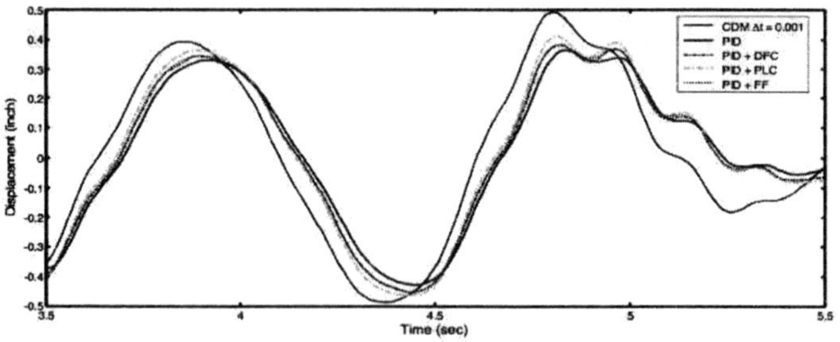

(b) Zoomed-in measured displacements.

Figure 14. Displacement responses at DOF 1 of 2-DOF linearly elastic structure (1 in. = 24.5 mm).

in which k_p, k_i, k_d, and $k_{\Delta p}$ are the proportional, integral, derivative, and delta pressure gains, k_{ff} is the feed-forward gain, d_p^c is the displacement command, $r^m$and $A_p$are the measured piston force and the piston area, respectively, k_o is a servo-valve gain factor that relates the oil flow to the spool displacement, and V is the total volume of the actuator oil chambers.

The transfer function for a linearized real-time hybrid test system is obtained by substituting Equations 6.2, 6.3, 6.4, 6.9, 6.11 into Equation 6.1. In the following section, the performance characteristics of the system are examined using this transfer function.

6.2 System Performance

The actuator considered in this study is a 50-ton (110-kip) actuator that has a stroke of ±127-mm (5-in.) and a 682-lpm (180-gpm) three-stage servo-valve. The parameters for the actuator model are, to a large extent, based on the technical specifications provided by the manufacturer. Their values are shown in Table 1 together with the properties of the structural model and test structure. The structural model is a single-degree-of-freedom system. As shown in the table, the mass of the test structure is assumed to be very small as compared to that of the model. This can often be the case in hybrid testing as the inertia effect can be simulated analytically rather than introducing a large physical mass. Therefore, in Equation 5.1, it is assumed that $M^A = M$. The impact of this for different values of M_t will be explored below.

Parameter	**Value**
Servo-valve:	
Damping ratio, ξ_s	0.7
Natural frequency, ω_s, rad/sec	817
Servo-valve gain, k_o, m^2/sec (in^2/sec)	0.537(832)
Actuator:	
Piston area, A_p, mm^2 (in^2)	2.483x10^4 (38.48)
Total volume, V, mm^3 (in^3)	6.936x10^6 (423.3)
Effective bulk modulus of oil, β_E, MPa (psi)	689.5 (100,000)
Test Structure:	
Mass, M_t, kg (lb-sec^2/in)	679.8 (3.882)
Damping ratio, ξ_t	0.05
Stiffness, K_t, kN/m (kips/in)	2,277 (13)
Structural Model:	
Mass, M, kg (lb-sec^2/in)	2.627x10^4 (150)
Damping ratio, ξ	0.05
Stiffness, K, kN/m, (kip/in)	2,277 (13)
Time Integration Scheme:	
Integration parameters α, β, γ	-0.25, 0.391, 0.75
Integration Time interval, Δt, sec	0.01

Table 1. Values of system parameters used in simulations.

To investigate the effectiveness of the DFC and simple feed-forward compensation, analyses are preformed with the system transfer function using various values of k_{DFC} and k_{ff}, and the results are evaluated against the transfer function of the structural model alone. For these analyses, k_p is set to 1 V/in., and the other gains, k_i, k_p, and $k_{\Delta p}$, are set to 0. It has been observed that the two compensation schemes have comparable performance and are effective in improving the performance of the system. For the DFC scheme, the influence of k_{DFC} is illustrated in Figure 10. It can be seen that $k_{DFC} = 0.5$ provides a good compensation resulting in a performance that is comparable to that with a proportional gain of 5 V/in. With a low proportional gain and no delay compensation, the response appears to be excessively damped as compared to the analytical result that represents the exact response of the structural model. This dissipation effect is consistent with the error-propagation properties of the integration scheme used here (Shing *et. al.* 1991a, 1991b). Furthermore, there is always a noticeable discrepancy in the vicinity of the resonance frequency of the structural model regardless of the proportional gain and delay compensation being used. This is due to the inertia and damping effects introduced by M_t and C_t that are ignored in the numerical solution, i.e., the measured force r^Eused in the numerical computation is assumed to be purely the static restoring force of the structure. However, such effects diminish as the excitation frequency is away from the resonance frequency of the model.

Sometimes, the mass and damping of a physical structure cannot be precisely estimated. Inertia and damping effects can also be introduced by the loading apparatus. As a result, one may not be able to compute $M^A$and C^A in an accurate manner to account for these influences. To examine the significance of the inertia effect further, analyses are conducted by varying the mass of the test structure M_t from $0.1M$ to M, where M is the mass of the structural model, and keeping the stiffness and damping coefficient the same as above. For these analyses, a set of optimum gain values of $k_p = 5$ V/in., $k_i = 1$ V/in.-sec, $k_d = 0$, and $k_{\Delta p} = 0$ are selected and no error-compensation scheme is used. Figure 11 compares the dynamic amplification factors obtained from the system transfer function with the exact analytical solution for the structural response. It can be observed that increasing M_t will lower the resonance frequency of the system. For seismic response, the impact of this frequency distortion depends on the damping in the structure and the shape of the response spectrum of the earthquake ground motion (Shing and Mahin 1985).

7 Two-Degree-of-Freedom Example Tests

Real-time hybrid tests were performed on single- and two-degree-of-freedom (SDOF and 2-DOF) structures using the system described in Section 5. For the 2-DOF tests, the structural model is shown in Figure 12.

The test specimen was a W14 × 120 cantilever beam, which was controlled by two actuators, X1 and X2, as shown in Figure 13. The beam was supported by a steel frame. Actuator X1 had a load capacity of 100 ton (220 kips) with a three-stage 947-lpm (250-gpm) servo-valve while X2 had a capacity of 50 ton (110 kips) with a three-stage 682-lpm (180-gpm) servo-valve. The mass of the structural modal was assumed to be concentrated at points where the actuators were attached to.

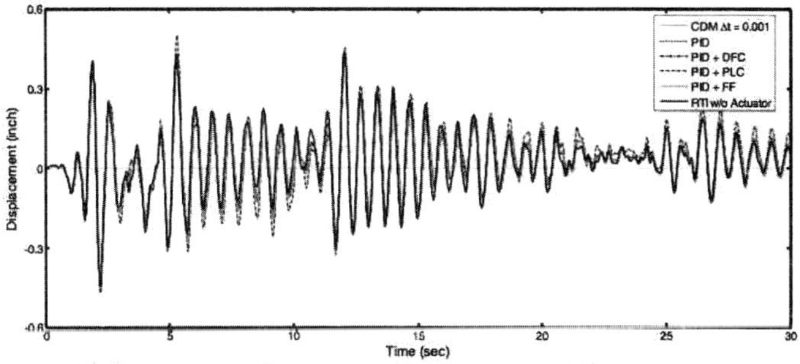

(a) Comparison of measured displacements at the first story.

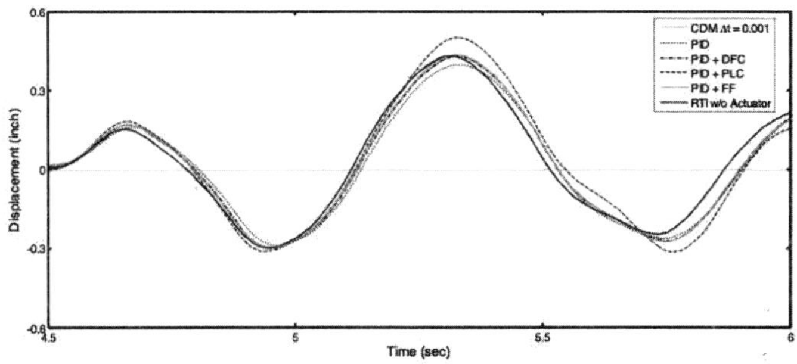

(b) Zoomed-in measured displacements.

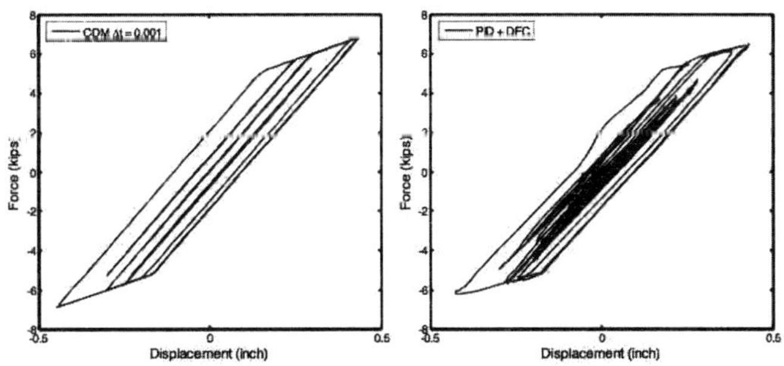

(c) First-story load - displacement hystereses.

Figure 15. 2-DOF structure with strain hardening (1 in. = 24.5 mm; 1 kip = 4.45 kN).

The stiffness of the 2-DOF test structure was measured by loading the beam with actuators X1 and X2 independently. The stiffness coefficients measured were K_{11} = 5.64 x 10^4 kN/m (322 kips/in), $K_{12} = K_{21}$ = -1.87

x 10^4 kN/m (-107 kips/in), and $K_{22} = 8.58$ x 10^4 kN/m (49 kips/in), with DOF 1 associated with actuator X1. The mass at each DOF was assumed to be 5.59 x 104 kg (326 $lb \cdot s^2/in$) for the structural model so that the structure had natural frequencies of 1.0 and 5.3 Hz. The actual mass of the specimen was about 0.5% of that of the structural model at each DOF and can, therefore, be ignored. Damping was assumed to be 5% of the critical for each mode. Tests were conducted to evaluate the effectiveness of the three error-compensation schemes presented previously. For these tests, the proportional, integral, derivative, and delta pressure gains were set to 340 V/m (8.7 V/in), 5 V/m-s (0.13 V/in-s), 0, and 0, for actuator X1; and 390 V/m (10 V/in), 4 V/m-s (0.1 V/in-s), 0, and 0 for actuator X2. With the discrete feed-forward compensation (DFC) scheme, k_{DFC} was set to 0.8. With the phase-lead compensator (PLC), the time constant T_d was set to 15 ms, which was obtained from a sine-sweep test, and $\alpha_d = 0.1$ for both degrees of freedom. The feed-forward gains were set to 0.17 and 0.14 for actuators X1 and X2, respectively. These and the PI gains were obtained by tuning the controller with harmonic displacement commands.

The α-method was used with the integration time step $\Delta t = 0.01$ s, and α, β, and γ set to -0.25, 0.391, and 0.75, respectively. Ten iterations were used for each integration time step. The initial stiffness of the structure used in the modified Newton iteration and response correction was set to be 5% higher than the actual stiffness to assure the stability of the numerical solution (Shing *et. al.* 1991a, 1991b). The NS component of the 1940 El Centro earthquake was scaled to 0.1 g. The structure remained linearly elastic. The responses obtained at DOF 1 from the different tests are shown and compared to the "exact" result, obtained with the central different method (CDM) using a very small time step, in Figure 14. Damping was assumed to be 6% for each mode in the exact result to account for the additional damping introduced by the specimen. It can be seen that the test results match the "exact" response reasonably well except for a slight mismatch of the response frequency probably due to the very small inertia forces developed by the specimen. All of the compensation schemes showed a positive influence as compared to the PI control alone.

8 Nonlinear Structures

To evaluate the performance of the real-time system in testing severely nonlinear structures, real-time simulations are conducted with a Simulink model of an entire hybrid test system including the dynamics of the servo-hydraulic system and the test structure (Jung *et. al.* 2007). Here, a 2-DOF shear frame structure that has bilinear strain-hardening and strain-softening

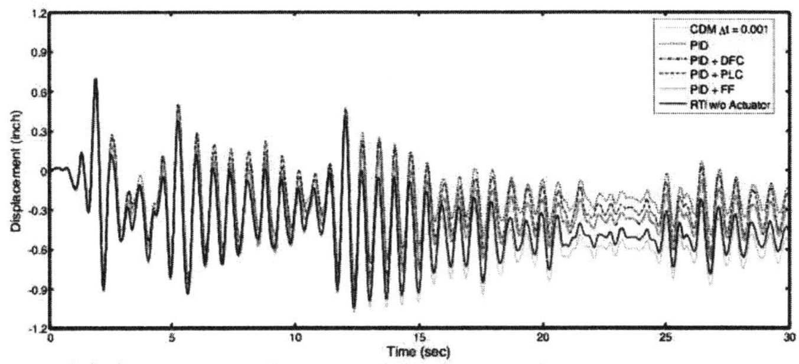

(a) Comparison of measured displacements at the top story.

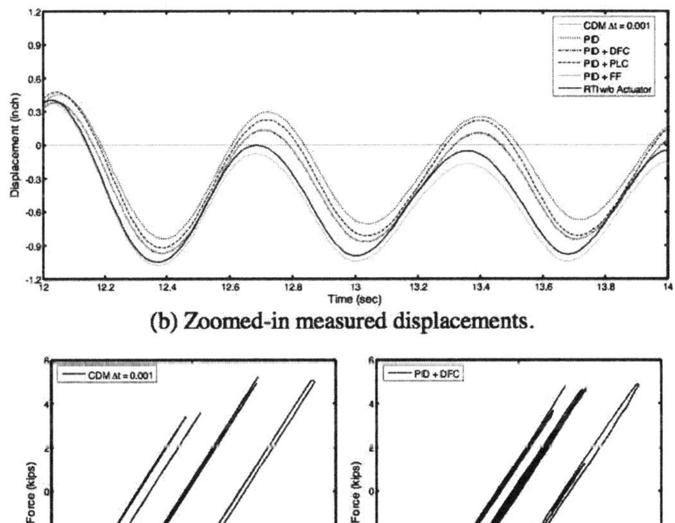

(b) Zoomed-in measured displacements.

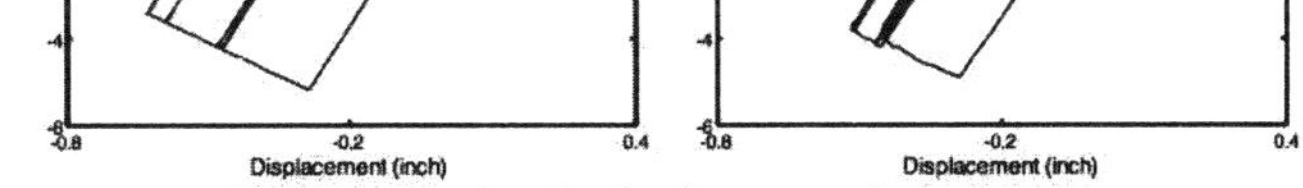

(c) Top-story load - displacement hystereses.

Figure 16. 2-DOF structure with strain softening (1 in. = 24.5 mm; 1 kip = 4.45 kN).

load-deformation relations is considered. The elastic stiffness coefficients are K_{11} = 36.48 x 10^3 kN/m (7 kips/in), $K_{12} = K_{21}$ = -2.98 x 10^3 kN/m (-17 kips/in), and K_{22} = 2.98 x 10^3 kN/m (17 kips/in). The mass at each DOF of the model is assumed to be 1.41 x 10^4 kg (82.3 $lb \cdot s^2/in$), while that of the physical structure is assumed to be 683 kg (3.9 $lb \cdot s^2/in$). Damping is assumed to be 5% of the critical for each mode. In modeling the physical specimen, 5% damping is assumed for each mode to account for the additional damping C_t. The NS component of the 1940 El Centro ground motion is used with the peak ground acceleration scaled to 0.1g. The time interval for integration is 0.01 s with α, β, and γ set to -0.25, 0.391, and 0.75. Ten iterations are used for each integration time step.

To investigate the effectiveness of the error-compensation schemes, real-time simulations are performed with the DFC, PLC, and simple feed-forward schemes presented previously. By a trial-and-error method, it has been found that the optimal value of the proportional gain is 390 V/m (10 V/in). The integral, derivative, and delta pressure gains are selected to be 4V/m-s (0.1 V/in-s), 0, and 0, respectively. For the DFC and feed-forward compensation schemes, k_{DFC} and k_{ff} are set to 0.8 and 0.15, respectively, for both degrees of freedom. The values of T_{d} for the phase-lead compensator as determined from a sine-sweep test are 30.9 and 24.2 ms for the first and second degrees of freedom, respectively, and α_{d} is set to 0.1. In the sine-sweep test, two chirp signals that have the same frequency but different amplitudes are used for the first and second degrees of freedom. The ratio of the amplitudes of the two signals applied is the average ratio of the two displacements obtained from the "exact" numerical solution.

The "measured" displacement responses are compared with the "exact" numerical solutions, which are obtained using the CDM with an integration time step of 0.001 s. For the exact numerical solution, damping is set to be 6% of the critical to account for the additional damping introduced by the physical test structure. The initial stiffness of the structural model used in the iteration and response correction is set to be 5% larger than the actual initial stiffness.

Figure 15 compares the displacement responses at the first story for the strain-hardening case. It can be observed that the system provides a reasonably good performance, and both the DFC and feed-forward schemes are effective in reducing the phase-lag errors. However, the performance of the phase-lead compensator is not as good as the other two.

Next, a 2-DOF structure that has a softening-type load-displacement relation is considered. Figure 16 shows the displacement responses at the top floor obtained from real-time simulations. It can be seen that the accuracy of the displacement responses is not as good as that for the strain-hardening case. This can be attributed to the fact that the initial stiffness used in the iteration and response update is very different from the actual tangent stiffness of the structure in the softening regime, which results in a more significant damping effect and, therefore, reduces the displacement drift as shown in Figure 16. This is verified by the fact that the simulated response matches the "exact" solution well when the time delay due to the dynamics of the actuator is removed as shown by the curve labeled "RTI w/o actuator" in the figure. The small discrepancy that remains is probably due to the numerical convergence errors introduced in the iterative scheme.

9 Summary and Conclusions

Real-time hybrid testing extends the concept of partitioned structural analysis (substructuring) methods, and it is developed to complement numerical simulation with physical tests when a portion of a structural system being evaluated is difficult to simulate accurately with analytical models alone. This chapter has presented an overview of real-time hybrid test methods from a general perspective, and discussed schemes of structural partitioning and issues related to numerical computation, system dynamics, and system performance. Three different hybrid test approaches can be identified in terms of the structural partitioning scheme, numerical solution scheme, and the treatment of the physical substructure. One is the total approach, where all the degrees of freedom of the physical substructure are explicitly expressed in the equations of motion and the resulting equations for the analytical and physical substructures are solved in an integrated fashion. The second is the coupled subdomain approach, in which the analytical and physical substructures are treated as separate subdomains whose equations of motion can be solved with different numerical schemes. The third is the dynamic substructuring approach, in which the dynamic properties of the physical substructure are not modeled explicitly but reflected in the testing. Each approach has its own merits that may best suit one type of problem or the other. A specific implementation, including the hardware configuration, numerical integration scheme, and delay compensation methods, has been discussed in detail. Results of experimental validation and numerical simulation study have been presented to illustrate the effectiveness of several delay compensation schemes and the subtleties of real-time testing.

The possibility of conducting reliable real-time hybrid tests using an unconditionally stable implicit integration scheme has been demonstrated in this chapter. However, the system dynamics, due to the mass and damping of the physical test structure and the loading apparatus, is an important consideration that may affect the accuracy of the test results. The significance of this can be identified with system transfer functions and simulation models. The effectiveness of delay compensation schemes for the control of three or more actuators remains to be studied.

Bibliography

Bayer, V., Dorka, U.E., Fllekrug, U., and Gschwilm, J. On real-time pseudodyanmic sub-structure testing: algorithm, numerical and experimental results. *Aerospace Science and Technology* 9: 223–232, 2005.

Blakeborough, A., Williams, M.S., Darby, A.P., and Williams, D.M. The development of real-time substructure testing. *Philosophical Transactions of the Royal Society of London* 359: 1869–1891, 2001.

Bonelli, A., Bursi, O.S., He, L., Magonette, G., and Pegon, P. Convergence analysis of a parallel interfield method for heterogeneous simulations with substructuring. *International Journal for Numerical Methods in Engineering*: In Press.

Bursi, O.S., Gonzalez-Buelga, A., Vulcan, L., Neild, S.A., and Wagg, D.J. Novel coupling Rosenbrock-based algorithms for real-time dynamic substructure testing. *Earthquake Engineering and Structural Dynamics*: In Press.

Chang, S.Y. 2002. Explicit pseudodynamic algorithm with unconditional stability. *Journal of Engineering Mechanics* 128(9): 935–947.

Dermitzakis, S.N. and Mahin, S.A. Development of substructuring techniques for on-line computer controlled testing. *UBC/EERC-85/04*, 1985. Earthquake Engineering Research Center, University of California, Berkeley, California.

Dimig, J., Shield, C., French, C., Bailey, F., and Clark, A. Effective force testing: a method of seismic simulation for structural testing. *Journal of Structural Engineering* 125(9): 1028–1037, 1999.

Dorf, R.C. and Bishop, R.H. *Modern Control Systems*, 9th ed., 2000. Prentice Hall, NJ.

Goodwin, G.C., Graebe, S.F., and Salgado, M.E. *Control System Design*, 2001. Prentice Hall, NJ.

Horiuchi, T., Masahiko, I., and Konno, T. Development of a real-time hybrid experimental system using a shaking table. In *Proceedings of the 12th World Conference on Earthquake Engineering*: Paper No. 0843, 2000.

Hughes, T.J.R. Analysis for transient algorithms with particular reference to stability behavior. In T. Belyschko and T.J.R Hughes (eds.), *Computational Methods for Transient Analysis.* 1983.

Jung, R.-Y. and Shing, P.B. Performance evaluation of a real-time pseudodynamic test system. *Earthquake Engineering and Structural Dynamics* 35(7): 789–810, 2006.

Jung, R.-Y. and Shing, P.B. Performance of a real-time pseudodynamic test system considering nonlinear structural response. *Earthquake Engineering and Structural Dynamics* 36(12): 1785–1809, 2007.

Magonette, G., Pegon, P., Molina, F.J., and Buchet, P. Development of fast continuous pseudodynamic substructuring tests. In *Proceedings of Second World Conference on Structural Control*, 1998.

McKenna, F. and Fenves, G.L. An object-oriented software design for parallel structural analysis. In *Proceedings of the SEI/ASCE Structures Congress*, 2000.

Nakashima, M., Kaminosono, T., and Ishida, M. Integration techniques for substructure pseudodynamic test. In *Proceedings 4th US National Conference on Earthquake Engineering* II: 515–524,1990.

Nakashima, M., Kato, H., and Takaoka, E. Development of real-time pseudodynamic testing. *Earthquake Engineering and Structural Dynamics* 21: 79–92, 1992.

Nakashima, M. and Masaoka, N. Real-time on-line test for MDOF systems. *Earthquake Engineering and Structural Dynamics* 28: 393–420, 1999.

Pegon, P. (2008). Continuous PsD Testing with substructuring. in *Modern Testing Techniques for Structural Systems, Dynamics and Control*, O. Bursi and D. Wagg (Eds.), CISM courses and Lectures No. 383 Springer-Verlag Wein-New York.

Shing, P.B. and Mahin, S.A. Computational aspects of a seismic performance test method using on-line computer control. *Earthquake Engineering and Structural Dynamics* 13: 507–526, 1985.

Shing, P.B. and Mahin, S.A. Cumulative experimental errors in pseudodynamic tests. *Earthquake Engineering and Structural Dynamics* 15: 409–424, 1987.

Shing, P.B. and Mahin, S.A. Experimental error effects in pseudodynamic testing, *Journal of Engineering Mechanics* 116(4): 805–821, 1990.

Shing, P.B., Vannan, M.T., and Carter, E. Implicit time integration for pseudodynamic tests. *Earthquake Engineering and Structural Dynamics* 20(6): 551-576, 1991.

Shing, P.B. and Vannan, M.T. Implicit time integration for pseudodynamic tests: convergence and energy dissipation. *Earthquake Engineering and Structural Dynamics* 20(6): 809–819, 1991.

Shing, P.B., Bursi, O.S., and Vannan, M.T. Pseudodynamic tests of a concentrically braced frame using substructuring techniques. *Journal of Constructional Steel Research*, 29: 121-148, 1994.

Shing, P.B., Wei, Z., Jung, R., and Stauffer, E. NEES fast hybrid test system at the University of Colorado. In *Proceedings of the 13th World Conference in Earthquake Engineering*, 2004.

Sivaselvan, M.V., Reinhorn, A., Liang, Z., and Shao, X. 2004. Real-time dynamic hybrid testing of structural systems. In *Proceedings 13th World Conference in Earthquake Engineering*, 2004.

Takanashi, K. *et. al.*, Non-linear earthquake response analysis of structures by a computer actuator on-line system, part 1 – details of the system. *Transactions of Architectural Institute of Japan* 229: 77–83, 1975.

Wagg, D., Neild, S., and Gawthrop, P. Real-time testing with dynamic substructuring. in *Modern Testing Techniques for Structural Systems, Dynamics and Control*, O. Bursi and D. Wagg (Eds.), CISM Courses and Lectures no. 383, 2008. Springer Verlag, Wien-New York.

Wallace, M.I., Sieber, J., Neild, S.A., Wagg, D.J., and Krauskopf. Stability analysis of real-time dynamic substructuring using delay differential equation models. *Earthquake Engineering and Structural Dynamics* 34(15): 1817–1832, 2005.

Wei, Z. Fast hybrid test system for substructure evaluation. *Doctoral Thesis*, 2005. University of Colorado at Boulder, Boulder, Colorado.

Williams, D.M., Williams, M.S., and Blakeborough, A. Numerical modeling of a servo-hydraulic testing system for structures. *Journal of Engineering Mechanics* 127(8):816–827, 2001.

Wu, B., Xu, G., Wang, Q., and Williams, M.S. Operator-splitting method for real-time substructure testing. *Earthquake Engineering and Structural Dynamics* 35(3): 293–314, 2006.

Wu, B., Wang, Q., Shing, P.B., and Ou, J. Equivalent force control method for generalized real-time substructure testing with implicit integration. *Earthquake Engineering and Structural Dynamics* 35(9): 1127–1149, 2007.

Zhao J., *et. al.* Considerations for the development of real-time dynamic testing using servo-hydraulic actuation. *Earthquake Engineering and Structural Dynamics* 32: 1773–1794, 2003.

Real-Time Testing With Dynamic Substructuring

David Wagg[†,*] Simon Neild[†] and Peter Gawthrop[‡]
[†] Department of Mechanical Engineering, University of Bristol, Bristol, UK
[‡] Department of Mechanical Engineering, University of Glasgow, Glasgow, UK

Abstract In this chapter the real-time testing technique which has become known as *real-time dynamic substructuring* is discussed. A control systems approach is taken to develop stability and robustness criteria for real-time dynamic substructuring tests. In particular we discuss how delay and uncertainty effects in the experimental apparatus can lead to loss of accuracy, or worse, system instability. Models which take account of the delay effects are developed using both delay differential equations and transfer functions. From these models delay compensation schemes can be constructed either using inverse transfer functions or forward prediction methods. Three methods for improving robustness are described for use in combination with a delay compensator. Throughout the chapter experimental results are presented, and in the final part results from an industrial example of substructuring a helicopter lag damper are discussed in detail.

1 Introduction

The behaviour of engineering systems under extreme dynamic loading, such as earthquake ground motion or extreme weather conditions, is highly unpredictable and consequently difficult to model. Add to this the current trend for increasingly complex mechanical and structural systems, and accurate predictive modelling of dynamic behaviour becomes extremely difficult. This is a serious problem, for example, when designing complex

*David Wagg is supported by an EPSRC Advanced Research Fellowship. The authors would also like to acknowledge the work of Max Wallace and Alicia Gonzalez-Buelga, who produced many of the experimental results shown in this chapter. MW was supported for his PhD studies by an EPSRC DTA and AGB is supported by EPSRC grant (GR/S49780). Peter Gawthrop is a Visiting Reseach Fellow at the University of Bristol.

infrastructure in regions of high seismic activity. In fact large scale engineering structures, such as bridges, buildings and some aircraft, present a particular problem in terms of experimental testing. Large scale experimental testing facilities such as shaking tables and reaction walls have been used for many years to carry out experimental testing of civil engineering structures (Severn, 1995). However, size and weight limitations mean that full scale dynamic testing is impossible for most large structures — a problem which also affects wind tunnel testing of both civil and aerospace structures (Bushnell, 2006). The most common alternative approach is to use a scaled down experimental test specimen to fit the test facility. However, this has the significant disadvantage that complex experimental test results have to be interpreted with the additional problem of scaling between the model and the full size structure.

For numerically simulating modular structures or those with a high degree of self similarity the technique of *substructuring* (Prezemieniecki, 1968), which involves dividing the structure into multiple substructures, has been developed to reduce the computational workload. The concept of subdividing a structure can also be used to formulate a model where one (or more) of the substructures is a physical test specimen. In this approach the model of the overall structure, or *emulated system*, is created by a combination of numerical and experimental parts.

This type of combined modelling is particularly suitable for structures where a design *critical element* can be identified. Critical elements are those which are most difficult to model numerically, or most likely to fail during dynamic loading, and as such are of primary interest to the design engineer — in other words, parts of the structure which exhibit nonlinear or unpredictable behaviour. So, the critical element is included as a full scale physical test specimen while the remainder of the structure is modelled numerically. There are two main advantages of this method over scale model testing. The first is that the critical element is tested at full size and the problems associated with scaling are eliminated. The second is that parameters in the numerical model can be varied and/or many different models can be coupled to the critical component. As this approach is limited to testing only part of the structure, it can be viewed as a form of component testing.

Initially the numerical–experimental substructuring technique was used for testing structural components using a *pseudodynamic* approach in which dilated time scales are used, so that loading is applied to the structure quasi-statically (Shing and Mahin, 1987; Nakashima et al., 1992; Donea et al., 1996; Bursi and Shing, 1996; Nakashima, 2001; Bonelli and Bursi, 2004; Pinto et al., 2004). These type of tests are often performed using reaction wall facilities, where the reaction forces induced in the structure can be measured accurately, but real time inertia and velocity dependent damping forces have to be estimated numerically. Extending the concept of pseudodynamic testing to real time for testing velocity dependent components was first achieved by Nakashima et al. (1992). Performing tests in real time removed the limitation that inertia and damping forces need to be estimated, potentially leading to more realistic modelling of the dynamic response of the structure to dynamic loading. However, implementing real time tests is significantly more difficult than applying loads quasi-statically and one of the earliest successful implementations is described by Horiuchi et al. (1999). The challenge is to implement the the complete process in real-time and ensure that the RTDS test is both *stable* and *robust*, so that useful test results can be obtained.

In this chapter we will discuss the control and implementation or real-time dynamic substructuring for mechanical and structural engineering applications. In particular, delay compensation, stability and robustness for RTDS will be the topics focussed on. The topic of numerical methods for substructuring is treated in detail by many other authors, primarily for psuedodynamic testing (Shing and Mahin, 1987; Shing et al., 1991; Thewalt and Roman, 1994; Chang, 1997; Chang et al., 1998; Bursi and Shing, 1996; Chang, 2000; Combescure and Pegon, 1997; Chang, 2002; Bonelli and Bursi, 2004; Chang, 2005), but also for RTDS (Darby et al., 1999, 2001; Wu et al., 2006; Bursi et al., 2007; Wang et al., 2006), and will not be considered in detail here.

1.1 Real-time dynamic substructuring

Real-time dynamic substructuring (RTDS) is a form of hybrid numerical-experimental testing suited to testing large structures or complex systems containing a critical component of interest. The technique allows the critical component to be tested experimentally while the remainder of the system is modelled numerically. The coupling between the physical and numerical parts is achieved using real-time control of actuators connected to the physical substructure. Therefore, RTDS allows observation, at full scale, of the behaviour of a critical element, as if it were *in* the structure, under a prescribed dynamic loading.

The substructuring technique was originally intended to be applied to situations where numerical models of the experimental part were unreliable, such as failure of concrete columns under earthquake loading. However, the technique has now been developed for a broader range of applications, and in fact can now be viewed as a form of advanced real-time component testing. The key concept underlying these methods is that only the nonlinear component of interest is tested experimentally, while the remaining part of the structure is computed numerically. In simple systems — such as those studied in this chapter— we can map the linear and nonlinear parts directly onto the numerical and experimental division.

In structural dynamics testing, this approach allows the single nonlinear element of interest to be tested experimentally, often at full scale. It also offers the potential for parameters in the computer model to be varied to undertake parametric testing. So far RTDS tests have also been carried out on a range of (mainly small scale) systems by several different researchers (Horiuchi et al., 1999; Darby et al., 1999; Wagg and Stoten, 2001; Darby et al., 2001; Gawthrop et al., 2005; Wallace et al., 2005b,a; Bayer et al., 2005; Bonnet et al., 2007). It is important to note that RTDS is very similar *in concept* to Hardware-in-the-loop (HiL) (Maclay, 1997) which has been been extended to test components other than control electronics (see for example Zhu et al. (2005); Misselhorn et al. (2006)). However a key difference is the effect of actuator and more general *transfer system* dynamics in RTDS, which is described in detail below.

1.2 Hardware-in-the-loop (HiL)

Hardware-in-the-loop (HiL) is a form of component testing where physical components of the system communicate with software models which simulate the behaviour of the rest of the system (Brendecke and Kucukay, 2002; Faithfull et al., 2001; Zhang and Alleyne, 2005) — a comprehensive review is given by Bacic (2006). Typically the hardware components being tested are control systems and the method has particular applications in the automotive industry (Hong et al., 2002; Rulka and Pankiewicz, 2005; Misselhorn et al., 2006) and a range of other applications (de Carufel et al., 2000; Mansoor et al., 2003; Ferreira et al., 2004b,a; Ganguli et al., 2005; Jezernik, 2005; Lambrechts et al., 2005). In a typical hardware-in-the-loop test, the hardware component consists of a box of electronic components which can communicate with the software models via electrical signals exchanged using a data acquisition and control system such as dSpace. Extending the HiL technique to test mechanical components has been an area of interest for some time, for example, for use in suspension development, see Misselhorn et al. (2006) and references therein. The main difficulty is that connecting a mechanical component to a software model requires the transfer of forces and velocities, and to achieve this an additional dynamic *transfer system* (Wagg and Stoten, 2001) must be included in the loop. Typically the transfer system is a set of actuators, which will have dynamic characteristics that must be compensated for if the test is to be carried out in real time.

Therefore, real time dynamic substructuring is essentially an actuator based HiL technique, which so far has primarily been considered for civil engineering systems. For RTDS instability is a frequent problem because the systems being modelled usually have lightly damped resonant behaviour, and any small delays due to the transfer system dynamics have the effect of negative damping. In contrast, in applications such as the automotive suspension systems studied by Misselhorn et al. (2006) the damping levels are significantly higher than in most RTDS tests, such that phase margin instabilities can be avoided. In fact the approach is to use PID control[1], and operate in a frequency range where actuator phase lag is seen to be acceptable. However, for mechanical components with lower damping, delay compensation techniques such as those developed for RTDS will be required to make the system robust. This will also apply to applications where electro-mechanical devices or complex circuitry are used as transfer systems (Driscoll et al., 2005; Zhu et al., 2005) and for the development and techniques such as model-in-the-loop (Zhu et al., 2005; Plummer, 2006) and engine-in-the-loop testing (Fathy et al., 2006) which are further extensions of the HiL technique.

2 The structure of a real-time dynamic substructuring test

2.1 Basic definitions

To carry out a substructuring test, the numerical model and the physical substructure are run in parallel and interact in real-time to emulate the dynamic behaviour of the complete structure. This interaction is achieved through the exchange of information at the interface between the numerical model and the substructure. Firstly, the displacements at the interface are calculated using the numerical model and imposed via actuation devices on the physical substructure. Secondly, the forces required to impose these displacements on the physical substructures are measured and fed back to the numerical model where they are included at the interface. A controller is used to minimise the effects of dynamics introduced by the actuators. Most dynamic test methods use a proprietary controller, which typically performs PID control, to reduce the effect of actuator dynamics to an acceptable level. However to achieve accurate substructuring more sophisticated compensation is required to further reduce the effect of the actuator dynamics.

[1]Proportional-Integral-Derivative. See any standard test on linear control theory

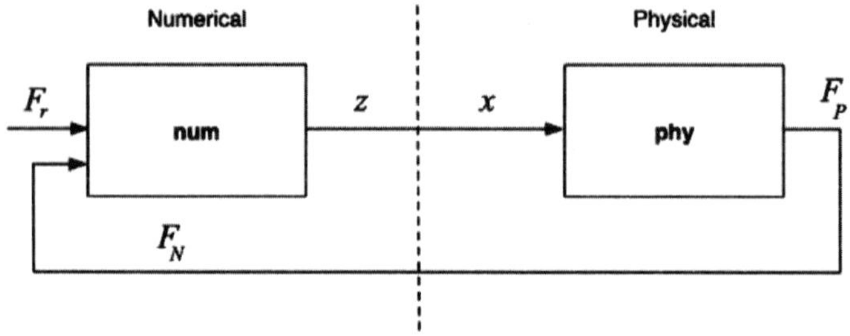

(a) Ideal substructured system (and HiL case)

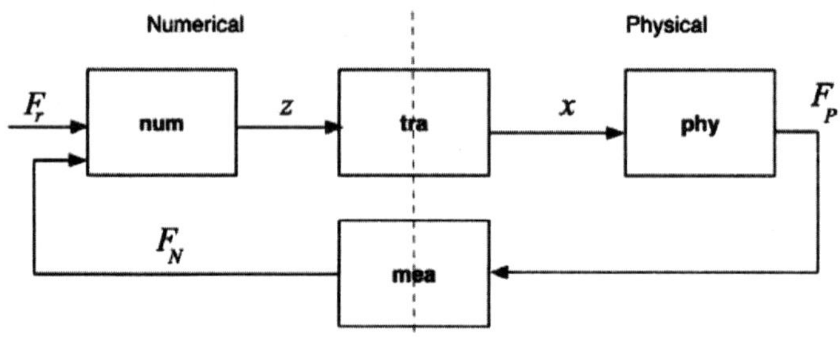

(b) Substructured system with a transfer system

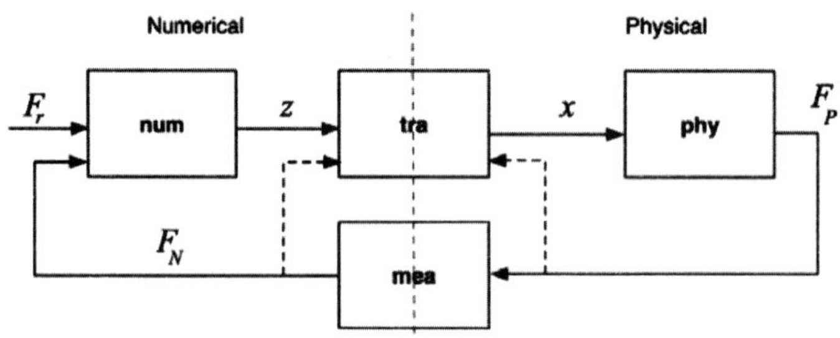

(c) Substructured system with force feedback to transfer system

Figure 1. Substructuring: basic definitions using a block diagram approach

Figure 1 shows three block diagram representations of substructuring[2]. Figure 1(a) shows the ideal case where a numerical substructure (**num**) is coupled directly to a physical substructure (**phy**) and there are no transfer system dynamics and hence no need for a controller. The detail of the two substructure blocks is: for **phy** the output F_p and input x are a collocated force and displacement pair connecting the **phy** to the numerical part of the substructured system. Similarly for **num** the input F_N and output z are the collocated force and displacement pair connecting **num** to the physical part of the substructured system. The signal F_r represents the net effect of external forces in the numerical part of the model[3].

In this ideal situation, $F_N = F_P$ and $x = z$ and the dynamic behaviour of the substructured system exactly replicated that of the emulated system[4]. However, for structural or mechanical systems, the physical system input x has to be generated by a *transfer system* which has the control objective of setting $x \approx z$. The physical force F_P is measured by a *sensor system* which also has it's own dynamics. In practice the ideal sensor system has the relationship that $F_P \to F_N$ as $x \to z$, in which case we can assume $F_P = F_N = F$.

Figure 1(b) gives a block diagram representation of a substructuring system with a transfer system providing the interface between the numerical and experimental components. In addition to the two blocks of Figure 1(a) **tra** represents the controlled transfer system including both an inner- and outer-loop control system (details below). The measurement sensor system, which includes the force transducer and associated power supplies is represented by **mea** — this is assumed not to interact with **phy**. We note that physical part of **tra** usually consists of the actuator and inner-loop control, while the numerical (augmented) part of **tra** would typically include elements such as the outer-loop controller and robustness compensator — described in section 5.

Finally we note that the physical part of **tra** is affected by F_P, especially if **tra** is operating close to its limits (discussed below). In addition the measured version of F_P (F_N) can be used in the control system associated with **tra**. These two interactions are shown as the dotted connections in Figure 1(c), and are often ignored when analysing substructuring systems.

[2]An alternative bond graph representation is given by Gawthrop et al. (2005).

[3]By interchanging the force and displacement signals, the causality can be reversed to give a situation where the actuator is controlled in force rather than displacement.

[4]This is the case in Hardware-in-the-loop testing, where because of the structure of the system being tested, transfer systems are unnecessary, only voltage signals are passed between substructures.

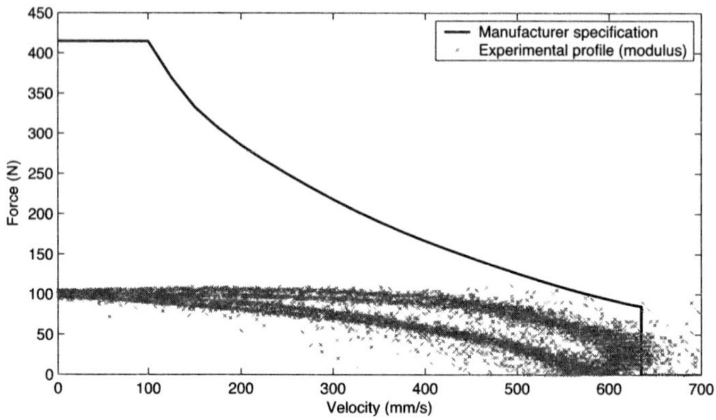

Figure 2. Actuator performance envelope compared to RTDS test data. Reproduced with kind permission of the Royal Society Wallace et al. (2005b).

2.2 Transfer systems

It is important to notice from the discussion above that the transfer system consists of both the hardware and software components required to transfer force and velocity between the numerical and experimental substructures. The hardware components of transfer systems are typically actuators (electric or hydraulic), but can also be in the form of a more complex test facility such as a shaking table. The dynamic capability of the transfer system hardware will impose a restriction on what forces and velocities can be transferred, and this is one of the key factors in designing a successful substructuring experiment.

As a starting point we consider the case for a single actuator for which a performance envelope can be defined in terms of a force-velocity graph. An example is shown in Figure 2 where the performance envelope for an electrical actuator (as specified by the manufacturer) is shown in comparison with data recorded from the actuator during a RTDS test. In this case we can see that during the RTDS test the actuator is operating right up to, and just beyond, the specified maximum velocity performance envelope. For small scale applications, actuators can be selected with appropriate performance characteristics as part of the design process for a RTDS test. However, for large scale applications it is not normally possible to change the test facilities or actuators. As a result, one of the major challenges in this area is to optimise the performance of such facilities in order to be able

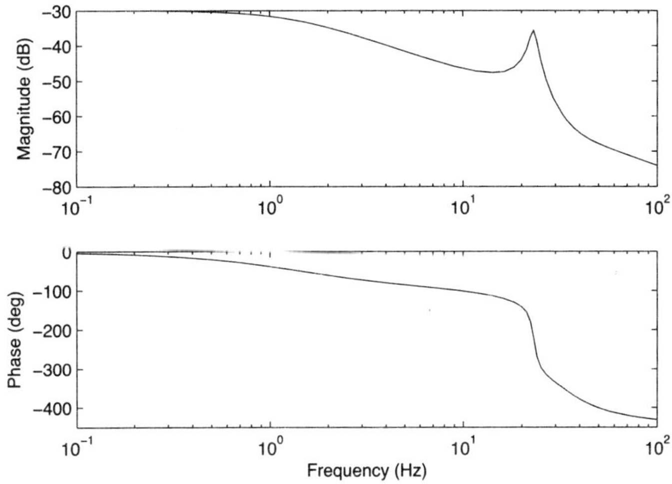

Figure 3. System identification for the University of Bristol shaking table. A 4th order representation of the table capturing the resonance peak due to oil column resonance.

to carry out as wide a range of tests as possible.

The critical dependence of RTDS on the transfer system characteristics means that accurate system identification of the transfer system(s) is the first step towards designing a RTDS test. An example is shown in Figure 3, where a system identification has been carried out for the University of Bristol shaking table[5]. In this example there is a clear resonance peak at 25Hz due to oil column resonance — a phenomenon which occurs in many shaking table systems. This resonance is an example of unwanted transfer system dynamics. All mechanical transfer systems will introduce some form of undesirable dynamics into the system. The most immediately noticeable form of unwanted dynamics is phase delay, but amplitude errors and other undesirable effects such as resonances can commonly occur. For accurate test results the unwanted dynamics must be compensated for and ideally eliminated from the closed loop system. A large number of compensation techniques have been proposed to achieve this, see for example Horiuchi et al. (1999); Darby et al. (2002); Gawthrop et al. (2005); Wallace et al. (2005b) and references therein[6].

[5]This table was reconditioned in 2004 with new actuators and now does not have the oil-column resonance shown in Figure 3.

[6]We also note that related delay compensation techniques have been developed in

The other part of the interface between numerical and experimental substructures is the measurement system. This is usually a force sensor(s), and the main effect of the sensor is to introduce a considerable high frequency noise content to the signal being fed back into the numerical model. This effect can normally be mitigated by appropriate filtering, which in turn will introduce an additional component of phase delay into the loop. This phase delay is measurable and can be compensated for as part of the delay compensation process.

2.3 The control problem

When implementing a RTDS test there are two main differences from a conventional control problem. First there is an additional force feedback from the plant into the numerical model (which changes the control demand). Second, the plant is highly nonlinear and/or difficult to model in some way, so linear analysis cannot be applied directly in most cases. In addition, the RTDS system is often required to model lightly damped structural systems which have significant resonance peaks close the the stability boundary. As a result, robustness is an essential consideration for RTDS testing.

A block diagram representation of the control implementation for a substructuring system using P-control and delay compensation is shown in Figure 4. The external excitation forces, F_r, and feedback force, F, from the outer control loop are used by the numerical model to produce the interface demand signal, z. This is then fed into the transfer system, which then tries to apply z to the experimental substructure such that, x, the actuator output follows, z, with the result that $x = z$ in the ideal case. To achieve this a delay compensation is applied to z, based on the inner loop control error, e, and a model of the actuator dynamics. The delay compensated signal, z', is then used as the setpoint for the inner control loop, with k_p as the proportional gain.

Several different delay compensation techniques can be used in combination with the P-controller, these are described below, and the combined system has been shown to be a highly effective and accurate method for eliminating actuator delay, see Wallace et al. (2005b); Gonzalez-Buelga et al. (2007) and references therein. The use of a relatively simple inner control loop helps to keep the system robust, and is easy to implement as most actuation systems come with a proprietary in-built P-controller[7]. This means

relation to active control systems for structural control; McGreevy et al. (1988); Soong (1990); Agrawal and Yang (2000); Chu et al. (2005); Preumont (2002).

[7]Usually in the form of a PID controller, see Goodwin et al. (2000) for details.

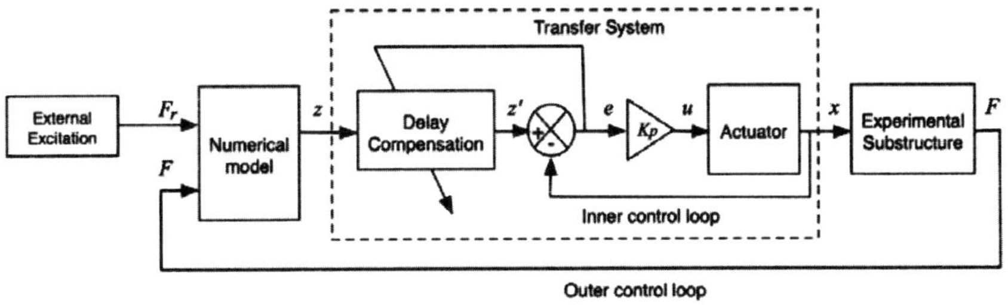

Figure 4. Block diagram of the control implementation for a substructuring system using P-control and delay compensation.

that the control gains for the proprietary controller can be selected to give the required inner loop control, with the delay compensation implemented using a digital computer. Typically the computer would also be used to run the numerical model.

There are two important factors here. First the implementation outside the inner loop control is digital, and secondly the process must be implemented in (hard) real-time. This means that the time taken to compute the numerical model must fit within some real-time constraints. Here we will assume that the numerical computation step must fit within the real-time control step. This is a single step approach. Other approaches which use a different time base for the control and numerical computation have been considered by a range of authors, see for example Combescure and Pegon (1997); Nakashima and Masaoka (1999); Darby et al. (1999, 2001, 2002); Pan et al. (2005) and references therein.

3 Delay compensation

The aim of a delay compensation strategy is to eliminate the effect of the transfer system delay from the combined experimental-numerical model. There are two main categories of compensator, those that provide (time) delay (e^{-sT}) compensation and those that provide phase lag (e.g. $\frac{1}{1+sT}$) compensation. The most straightforward time delay compensator is obtained by assuming that the dynamics of the transfer system can be approximated by a pure delay. The delay compensator will then simply shift the numerical model displacement forward by a fixed number of time-steps (see for example Horiuchi et al. (1999); Blakeborough et al. (2001)). This basic approach can be extended to vary the amount of delay compensation, based on the error between the actuator displacement and the desired nu-

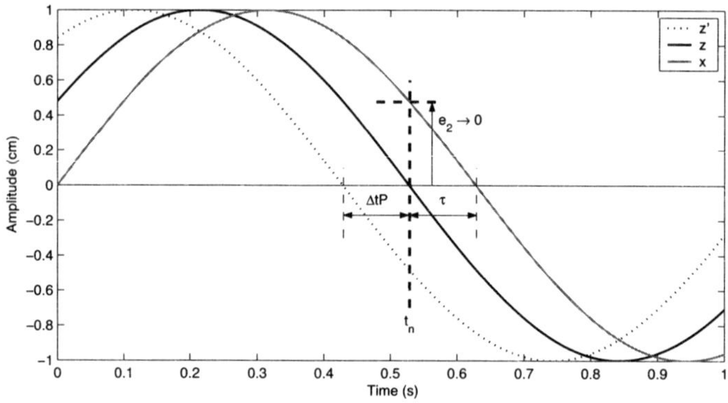

Figure 5. Delay compensation via forward prediction. Reproduced with kind permission of the Royal Society Wallace et al. (2005b).

merical model displacement (Darby et al., 2002). By introducing adaption the assumption of pure delay can be relaxed, and effects such as amplitude correction can be included Wallace et al. (2005b). The most straight forward form of lag compensation is to use the inverse of an estimated transfer function model of the transfer system, see for example Gawthrop et al. (2005) also Sivaselvan et al. (2004); Jung and Shing (2006); Wallace et al. (2007)[8]. However, for a large number of cases the transfer system will not be (causally) invertible, for example those including a pure time delay.

It is also possible to use a combined lag and time delay compensation approach. In this case, the lag compensator assumes a simple transfer system model (inverted) and then adaptive forward prediction is used as a "fine tuning". This works well as most RTDS tests are not operating in the linear steady state, which is the implicit assumption behind the phase lag transfer function model, and must be able to cope with transient nonlinearity.

3.1 Forward prediction

The concept of using forward prediction for delay compensation is shown in Figure 5. Here the interface reference signal, z, is a single frequency sine wave and without any delay compensation the effect of the actuator would be to delay the signal by a fixed delay, τ. The result is to create an error, e_2,

[8]Both Smith predictor (Sivaselvan et al., 2004) and adaptive control Lim et al. (2004) have been proposed as lag compensation strategies. However, both methods suffer from significant limitations which restrict their usefulness in experimental implementation.

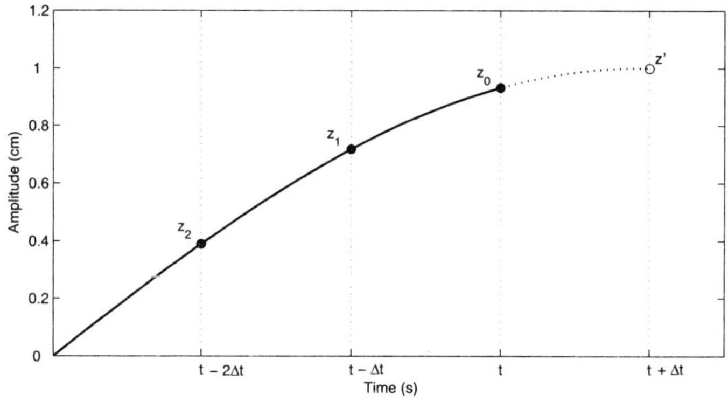

Figure 6. Single-step forward prediction. Reproduced with kind permission of the Royal Society Wallace et al. (2005b).

between the desired signal, z, and the actuator output x. To eliminate the delay effect, z is shifted forward by P time steps of Δt such that $P\Delta t = \tau$.

Single step forward prediction

Single step forward prediction is based on using predefined coefficients, a_i, for an Nth order polynomial fit of n control points such that

$$z' = \sum_{i=0}^{N} \bar{a}_i \; z_i, \tag{3.1}$$

where z_0 is the current numerical model point and z_i are previous points. Figure 6 shows the new point z' which is obtained by extrapolating the polynomial function over the previous three points with $n = 3$ and $N = 2$. For a 2nd order polynomial fit we use the following constants $\bar{a}_0 = 3$, $\bar{a}_1 = -3$ and $\bar{a}_2 = 1$, the derivation of which is described below. To predict further ahead than a single time step Equation 3.1 is applied recusively Darby et al. (2001). For example to predict two time steps forward

$$\begin{aligned} z'^{(2\Delta t)} &= 3z'^{(\Delta t)} - 3z_0 + z_1, \\ &= 6z_0 - 8z_1 + 3z_2, \end{aligned} \tag{3.2}$$

where

$$z'^{(\Delta t)} = 3z_0 - 3z_1 + z_2. \tag{3.3}$$

To derive these relationships we first consider a polynomial of order N expressed in terms of the time parameters as,

$$z_i = a_0 + a_1 t_i + \ldots + a_N t_i^N. \tag{3.4}$$

In order to predict forward, we first need to compute the a_i coefficients. Starting at the current point, z_0, and including $n-1$ previous data points we can write

$$\begin{bmatrix} z_0 \\ z_1 \\ \vdots \\ z_{n-1} \end{bmatrix} = \begin{bmatrix} 1 & t_0 & t_0^2 & \ldots & t_0^N \\ 1 & t_1 & t_1^2 & \ldots & t_1^N \\ \vdots & \vdots & \ddots & \vdots & \\ 1 & t_{n-1} & t_{n-1}^2 & \ldots & t_{n-1}^N \end{bmatrix} \begin{bmatrix} a_0 \\ a_1 \\ \vdots \\ a_N \end{bmatrix}, \tag{3.5}$$

or, in vector notation

$$z = Xa, \tag{3.6}$$

which can be solved by premultiplying by the matrix transpose, such that $X^T z = X^T X a$. The matrix can then be inverted directly to give the coefficient values,

$$a = (X^T X)^{-1} X^T z. \tag{3.7}$$

For the case shown in Figure 6, $n = 3$ and $N = 2$, gives

$$\begin{bmatrix} z_0 \\ z_1 \\ z_2 \end{bmatrix} = \begin{bmatrix} 1 & t_0 & t_0^2 \\ 1 & t_1 & t_1^2 \\ 1 & t_2 & t_2^2 \end{bmatrix} \begin{bmatrix} a_0 \\ a_1 \\ a_2 \end{bmatrix}, \tag{3.8}$$

where, t_0 is taken as time zero. Therefore, for a sample time step size Δt as shown in Figure 6,

$$\begin{bmatrix} z_0 \\ z_1 \\ z_2 \end{bmatrix} = \begin{bmatrix} 1 & 0 & 0 \\ 1 & -\Delta t & \Delta t^2 \\ 1 & -2\Delta t & 4\Delta t^2 \end{bmatrix} \begin{bmatrix} a_0 \\ a_1 \\ a_2 \end{bmatrix}. \tag{3.9}$$

For the predicted point z', one time step ahead we now have

$$z' = X_P \, a, \tag{3.10}$$

where, X_P is the forward prediction vector

$$X_P = [1 \quad P\Delta t \quad \ldots \quad P^N \Delta t^N], \tag{3.11}$$

and P is the number of time steps to be predicted forward. As X is a square matrix in this case, $(X^T X)^{-1} X^T = X^{-1}$, therefore,

$$z' = X_P X^{-1} z. \tag{3.12}$$

Evaluating $X_P X^{-1}$ gives an expression which is independent of the sample time step size Δt. Substituting into Equation 3.12 for the case when $P = 1$

$$z'^{(\Delta t)} = 3z_0 - 3z_1 + z_2, \tag{3.13}$$

matching the coefficients of the one step method in Equation 3.3. For $P = 2$

$$z'^{(2\Delta t)} = 6z_0 - 8z_1 + 3z_2, \tag{3.14}$$

matching the coefficients found in Equation 3.2 but in a single operation. Therefore, the coefficients $\bar{a}_i$ from Equation 3.1 are actually the pre-multiplication of the forward prediction vector X_P for the special case of $P = 1$, such that

$$\bar{a} = X_P[(X^T X)^{-1} X^T]. \tag{3.15}$$

Using this approach, the coefficients $\bar{a}$ can be predefined and thus fix the level of forward prediction.

Adaptive forward prediction

To implement online forward prediction a buffer of n data points of the interface reference signal, z, are used. These values are used in a least squares polynomial fitting sub routine to calculate the Nth order polynomial fit and thus find the coefficient vector a from Equation 3.7 for that time step The current coefficients are then used to compute the predicted point z' via Equation 3.10 using the forward prediction vector X_P, such that

$$z' = k_a \sum_{i=0}^{N} \bar{a}_i z_i = k_a \sum_{i=0}^{N} \bar{a}_i P^i \Delta t^i. \tag{3.16}$$

The forward predicted point z' can be multiplied by a gain k_a to remove any amplitude error and increase accuracy.

The basic forward prediction algorithm can be used to effectively remove the transfer system delay by tuning the magnitude of the forward prediction, P, and the amplitude gain, k_a. To remove the need for this tuning and to allow the algorithm to achieve higher levels of accuracy for transient plant conditions a feedback control can be added. To implement this the magnitude of the forward prediction τ is now governed by,

$$\tau = \Delta t(P + \rho), \tag{3.17}$$

where, Δt is the sample time step size, P is the *fixed* initial number of time steps to be forward predicted and ρ is the *adaptive* number of time steps to

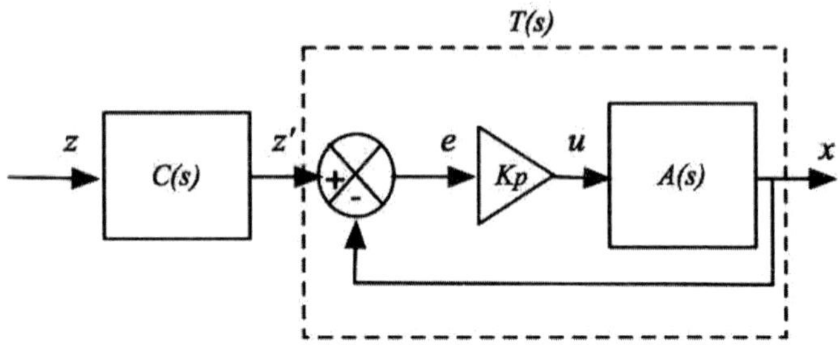

Figure 7. Transfer system delay compensation

be forward predicted. Similarly, the amplitude accuracy is can be defined as,

$$z' = z'(k_a + \sigma), \tag{3.18}$$

where, k_a is the *fixed* amplitude gain and σ is an *adaptive* amplitude variable which together control the amount of the predicted reference signal to be used. The adaptive coefficients are computed using

$$\rho_{n+1} = \rho_n \ \pm \ \alpha e^{\gamma} \ , \tag{3.19}$$

and

$$\sigma_{n+1} = \sigma_n \ \pm \ \beta e^{\gamma} \ , \tag{3.20}$$

where e is the error, $\gamma \geq 1$ sets the convergence and α and β are the adaptive gains. The sign of the error determines the plus or minus sign — further details can be found in Wallace et al. (2005b).

3.2 Lag compensation

To implement lag compensation it is assumed that the transfer system can be modelled as a frequency dependent transfer function, $T(s)$, where s is the Laplace variable. We note that this type of model assumes that the system is linear and that initial conditions can be ignored. In addition, $T(s)$ must be a stable transfer system with no zero dynamics. Further assumptions may be required in order to ensure that the model of the transfer system is proper and has a state-space realisation.

A block diagram representation of the transfer system is shown in Figure 7. The delay compensation function is represented by $C(s)$ and the actuator by $A(s)$. The simplest approach is set such that $C(s) = T(s)^{-1}$. To achieve this we set

$$z' = T^{-1}z = \frac{1 + K_pA(s)}{K_pA(s)}z \tag{3.21}$$

which give the relationship $x = z$. The simplest first order lag compensator has a transfer function of the form $T(s) = \frac{1}{1+sT}$.

An alternative approach is to use a phase advance compensator of the form

$$C(s) = \frac{\alpha s + \omega_c}{\frac{1}{\alpha}s + \omega_c}, \tag{3.22}$$

where the parameter $\alpha \geq 1$ and ω_c is the critical frequency — the frequency at which the magnitude of the closed loop gain equals one. Typically, for this application, $1 < \alpha < 2$ and $\alpha = 1$ corresponds to a unit transfer function which has no effect. The maximum phase advance rises to about $\frac{\pi}{4}$rad $= 45^o$ which occurs when $\alpha \approx 10$. (Goodwin et al., 2000, sec. 6.6). Other standard control techniques for phase compensation can be also be applied, although it should be noted that the physical substructure is nonlinear, so closed loop analysis should be applied with caution.

Finally we note that both transfer functions for a fixed delay e^{-sT} and a first order lag $\frac{1}{1+sT}$ have the same first-order Taylor expansion $(1 - sT)$. As a result there is not much practical difference between the approaches in the case of small delays.

4 Stability and robustness

So far our discussion regarding delay compensation has excluded the issue of system stability and robustness. The effect of delays in the transfer system can have highly destabilising effects, especially for systems with very light damping. This is due to the fact that delay can act as a form of negative damping (Horiuchi et al., 1999), reducing the combined system damping to zero and therefore instability.

In this section we first consider the stability of the system by modelling the transfer system dynamics as a fixed delay using delay differential equations (DDEs). We then turn to a frequency domain analysis, based on linear control theory, by assuming the physical substructure has an approximate linear model. This leads to a discussion of the system robustness, where we demonstrate how robustness can be increased by adjusting the system parameters.

4.1 Delay differential equations

Delay differential equations include the current state of the system and the state of the system some fixed time τ ago (Stépan, 1989; Diekmann et al., 1995). Typically, the onset of instability can be observed as oscillations which occur as delay increases. This corresponds to a pair of complex conjugate eigenvalues with zero real part in the DDE model. As a result it is possible to determine the critical delay time τ_c, above which the system is unstable. This can be done either by considering the characteristic equation for the eigenvalues of the DDE (Stépan, 1989; Diekmann et al., 1995) or with the numerical tool DDE-BIFTOOL (Engelborghs et al., 2002) for the stability analysis of DDEs.

If the delay in the transfer system is less than τ_c the substructured system is stable[9]. However, there are a significant number of systems for which the delay of the transfer system is larger than the critical delay, so that without delay compensation, the substructured system is unstable. The DDE stability analysis gives the generic stability properties of the substructured system and can be carried out without choosing a specific type of controller. As such it is a useful precursor to control system design, as τ_c can be related to a phase margin value(s) for the system, as we will show.

Single-degree-of-freedom linear oscillator example

A single-degree-of-freedom (SDOF) mass-spring oscillator system with one excitation input, r, is shown schematically in Figure 8(a). This is the emulated system, with displacement z^*, where $(.)^*$ indicates that these dynamics represent the complete dynamics. In this example, the spring k_s is the physical substructure[10], which combined with a numerical model and transfer system gives a substructured model of the system shown in Figure 8(a). The governing equation of the numerical model is

$$m\ddot{z}^* + c(\dot{z}^* - \dot{r}) + k(z^* - r) = F, \tag{4.1}$$

where the feedback force $F = -k_s x$ (from Figure 8(b)). The key assumption is that the transfer system introduces a delay such that $x(t) = z(t-\tau)$, $\tau > 0$. The overall substructured system (from Figure 8(a)) is then governed by the delay differential equation

$$m\ddot{z} + c\dot{z} + kz + k_s z(t-\tau) = c\dot{r} + kr\,. \tag{4.2}$$

The characteristic equation can be solved by setting the input to zero and assuming solutions of the form $z = Ae^{\lambda\hat{t}}$ which gives

$$\lambda^2 + 2\zeta\lambda + 1 + pe^{-\lambda\hat{\tau}} = 0, \tag{4.3}$$

[9]Note that we are assuming that no other instability mechanisms are operating.

[10]This is taken as linear to simplify the analysis, a nonlinear pendulum example is considered next.

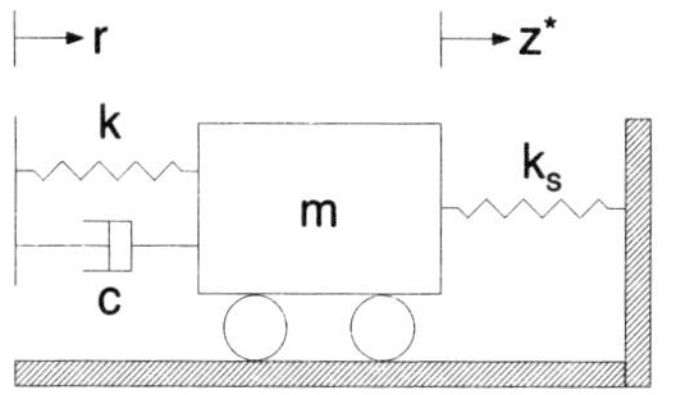

(a) Emulated system

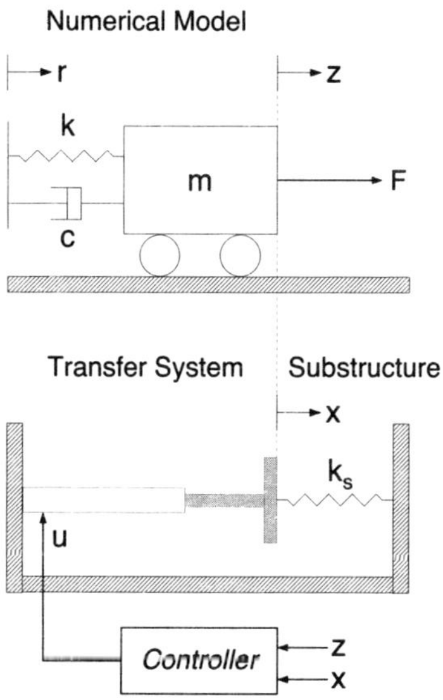

(b) Substructured system

Figure 8. Schematic representation of the SDOF oscillator example where complex roots λ_i are the system eigenvalues, $\hat{t} = \omega_n t$, $\hat{\tau} = \omega_n \tau$, $\omega_n = \sqrt{k/m}$, $p = k_s/k$ and $\zeta = c/(2\sqrt{mk})$. If the delay is small, then we can make the approximation $e^{\lambda\hat{\tau}} \approx (1 - \lambda\hat{\tau})$ which gives a critical delay value of

$$\tau_c = \frac{2\zeta}{p\omega_n} = \frac{c}{k_s}. \tag{4.4}$$

This expression shows that lightly damped stiff structures, such as those commonly encountered in structural engineering, will have a small τ_c and, consequently, it will be more difficult to maintain stability of the substructuring test. Equation 4.4 also clearly demonstrates how the delay can be interpreted as negative damping, with as equivalent negative damping term of $c_{neg} = -k_s\tau$.

For this example explicit relationships for the stability boundaries can be derived to construct a stability chart in terms of p and τ Wallace et al. (2005a). This is shown in Figure 9(a1) where solid lines represent the stability boundaries, and the shaded region is stable. The dashed curve is the stability boundary obtained from the $e^{\lambda\hat{\tau}} = (1 - \lambda\hat{\tau})$ approximation, which only holds for small values of delay. Figure 9(a2) shows an enlargement of the region where $p = 1$, and $\hat{\tau}_c$ can be read off as $\hat{\tau}_c = 0.04627$, from which the critical time delay is $\tau_c = 1.335\,\text{ms}$.

Figures 9(a1) and 9(b1) also highlight that there are regions where the system is stable regardless of the delay $\hat{\tau}$. We note that the vast majority of structures, especially in the civil engineering field, are lightly damped such that operating in the region of stability for all $\hat{\tau}$ would be extremely unlikely. However, when substructuring mechanical components, such as individual damper units, other parts of the stable region could be accessible.

Coupled pendulum-oscillator example

We now consider the coupled pendulum-oscillator system shown schematically in Figure 10. This system consists of a pendulum mounted with it's pivot point in the center of the mass, M, belonging to the mass-spring-damper. The pendulum mass, m is assumed to act at a single point, a distance l from the pivot point. The equations of motion for the complete, emulated, system are given by

$$(M + m)\ddot{z}^* + C\dot{z}^* + Kz^* + ml[\ddot{\theta}\sin\theta + \dot{\theta}^2\cos\theta] = F_r, \tag{4.5}$$

$$ml^2\ddot{\theta} + b\dot{\theta} + mgl\sin(\theta) + ml\ddot{y}\sin(\theta) = 0, \tag{4.6}$$

where z^* is the displacement of the mass M, θ is the angular displacement of the pendulum which has mass m and length l, C and K are the damping and

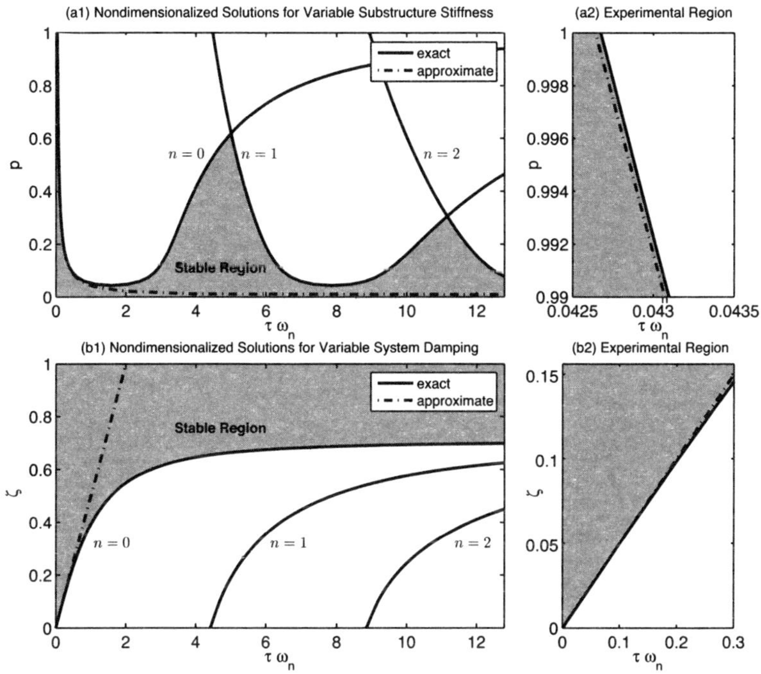

Figure 9. Non-dimensionalized Hopf stability boundaries of the SDOF example. Fixed parameters: $\zeta = 0.0213$ in panels (a) and $p = 1$ in panels (b). Individual solution branches computed for $n = 0, 1, 2, ...$ as detailed in Wallace et al. (2005a).

stiffness of the mass-spring-damper respectively, b is the viscous damping of the pendulum, g is acceleration due to gravity and F_r is the external exciting force.

During testing (Gonzalez-Buelga et al., 2007) it was found that the platform on which the pendulum pivot is mounted had a significant effect of the substructuring test results. Therefore, to accurately capture this behaviour the mass M is split into two components M_1 and M_2, such that $M = M_1 + M_2$. Now the numerical mass component is M_1 and mass M_2 represents the physical mass of the mounting platform. A similar effect occurs with the viscous damping parameter C, so this too is divided such that $C = C_1 + C_2$, where C_2 corresponds to the physical damping in the mounting platform. A schematic representation of the complete substructuring testing process is shown in Figure 11.

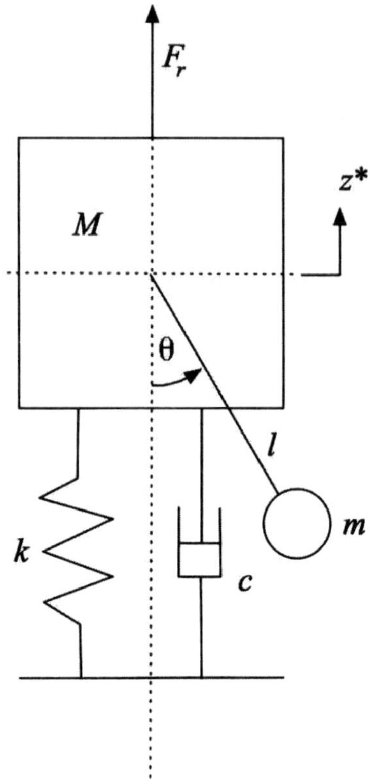

Figure 10. Schematic representation of the complete pendulum-mass-spring-damper system.

Including the mounting platform effects, the dynamics of the numerical model as shown in Figure 11 can be written as

$$M_1\ddot{z} + C_1\dot{z} + Kz = F_d, \tag{4.7}$$

where $F_d = F_r - F$, and z represents the numerical model coordinate in this case. The substructuring force, F, can be expressed as a combination of the testing platform and pendulum dynamics, such that

$$F = (M_2 + m)\ddot{x} + C_2\dot{x} + ml[\ddot{\theta}_x \sin\theta_x + \dot{\theta}_x^{\,2} \cos\theta_x]. \tag{4.8}$$

where $\ddot{x}$ is the actual acceleration acting on the pivot point and θ_x, $\dot{\theta}_x$ and $\ddot{\theta}_x$ are the measured angle, angular velocity and angular acceleration of the pendulum respectively. When considering the the semi-trivial solution $\theta = \dot{\theta} = 0$, equation (4.8) simplifies to

$$F = (M_2 + m)\ddot{x} + C_2\dot{x}. \tag{4.9}$$

Physically this corresponds to the case when the pendulum is not moving while the pivot point is subject to an oscillatory motion.

Using the same approach as above, we model the actuator dynamics as a fixed delay τ, such that $x(t) = z(t - \tau)$. Then by combining equations 4.7 and 4.9 we obtain

$$M_1\ddot{z} + C_1\dot{z} + Kz + m_s\ddot{z}(t - \tau) + C_2\dot{z}(t - \tau) = F_r, \tag{4.10}$$

where $m_s = M_2 + m$. From this expression the characteristic equation (or complementary function) of the system is found by substituting $z = A\mathrm{e}^{-i\omega t}$ so that

$$\begin{aligned} -\omega^2\left[M_1 + m_s\cos(\omega\tau)\right] + i\omega[C_1 + m_s\omega\sin(\omega\tau) + C_2\cos(\omega\tau)] \\ +[K + C_2\omega\sin(\omega\tau)] = 0. \end{aligned} \tag{4.11}$$

Equation (4.11) indicates how actuator delay can potentially cause changes in the effective values of mass, damping and stiffness. The DDE analysis of this system is discussed in detail by Kyrychko et al. (2005).

This example is shown in Figure 12 where we demonstrate the effects on the stability boundary as the parameter m_s is varied for two different C_2 cases — note $p = m_s/M_1$. The solid line in Figure 12 corresponds to the stability chart calculated for the case when $C_2 = 0$ with $\omega_n = 10$rad/s and $\zeta = 0.05$. The theoretical stability boundary which occurs when $C_2/C_1 = 1$ is plotted as a dashed line (Kyrychko et al., 2005). A very close agreement exists between the experimental results and the case when $C_2/C_1 = 1$ — indicating why the effect of damping feedback, C_2 is significant. This example demonstrates that mass and damping effects in the experimental apparatus used to mount the experimental substructure need to be included as part of the transfer system.

4.2 Linear control theory

A clear distinction of RTDS from related linear control techniques, such as HiL, is that the physical substructure (**phy**) is strongly nonlinear. As a result linear control theory cannot be applied in general. However, it is useful when studying system robustness to analyse the closed loop system with (**phy**) approximated by a linear transfer function with unmodelled nonlinear dynamics. Therefore we assume that (**phy**) can be approximated

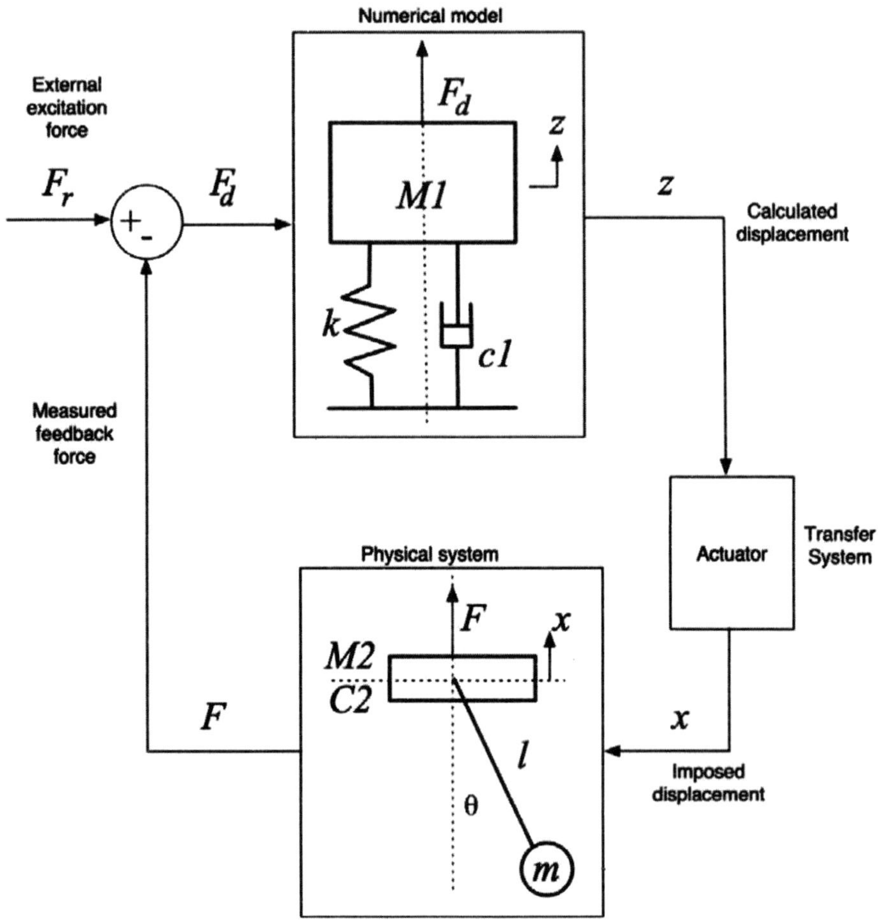

Figure 11. A schematic representation of the real time dynamic substructure testing method for the pendulum-oscillator system

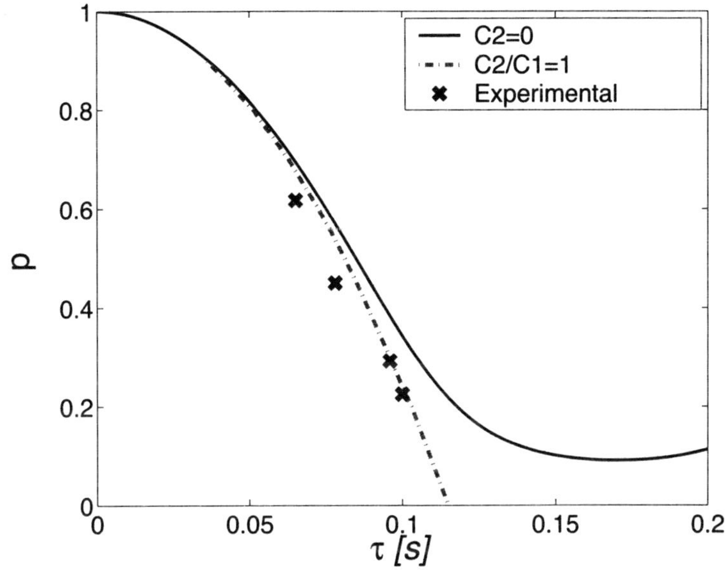

Figure 12. Effects of the viscous delayed feedback in the stability chart showing experimental versus theoretical results $p = m_s/M_1$. The experimental points are compared with the two C_2 cases. Numerical model parameters: $M_1 = 4$kg, $C_1 = 10$kg/s and $K = 3000$N/m, $C_2/C_1 = 1$.

by $P(s)$, when the actual behaviour is governed by $P(s) + f(\cdot)$, where $f(\cdot)$ are the neglected nonlinear dynamics.

Then for the ideal case shown in Figure 1(a), with the relations $F_p = P(s)x$, $z = N(s)(N_r F_r - F_N)$, $z = x$ and $F_N = F_p$ we obtain a closed loop transfer function of the form

$$F_P = \frac{L_0(s)}{1 + L_0(s)} N_r(s) F_r, \tag{4.12}$$

where $L_0(s) = P(s)N(s)$ and is defined as the ideal, or *n*ominal loop gain and N_r is the part of the numerical model which relates to F_r. From this we define the transfer function relating to the ideal substructured system (i.e. the emulated system), as

$$D_0(s) = \frac{L_0(s)}{1 + L_0(s)}. \tag{4.13}$$

For the substructuring case shown in Figure 1(b) the additional relationship of $x = T(s)z$ and $F_N = M(s)F_p$ must be added, leading to

$$F_P = \frac{L_0(s)}{1 + \Lambda(s)L_0(s)}\Lambda_r(s)N_r(s)F_r, \tag{4.14}$$

where $\Lambda(s) = T(s)M(s)$ and $\Lambda_r(s) = T(s)$. By comparing equations 4.12 and 4.14 we can see that for the substructured system to match the ideal (or nominal) case we require that $T(s) = M(s) = I$. The case shown in Figure 1(c) is discussed in Gawthrop et al. (2007), see also Jung and Shing (2006) for a discussion of transfer function modelling for real-time hybrid testing.

Phase margin

It is now possible to apply a range of standard control system design techniques, with the proviso that we are ignoring the nonlinear dynamics in **phy**. We define the loop gain, $L(s)$, for the substructuring case as $L(s) = \Lambda(s)L_0(s)$ (where $L_0(s)$ can be considered as the nominal loop gain). From the Nyquist stability criterion (Goodwin et al., 2000) we know that $|L(i\omega)| = 1$ and $\angle L(i\omega) = -\pi$ at the point of instability[11]. So the critical frequency, ω_c, is defined as the solution of $|L(i\omega_c)| = 1$. Then the *relative stability* of the system can be found by computing the *phase margin* ϕ_m which is given by

$$\phi_m = \pi - \angle L(i\omega_c). \tag{4.15}$$

The phase margin provides a measure of how near the system is to instability, with $\phi_m = 0$ denoting the instability boundary[12]. Now we can examine the effect of the transfer system dynamics by comparing the nominal system ($\Lambda(s) = 1$, e.g. no transfer system effects) to cases where a the transfer system has an effect, corresponding to $\Lambda(s) \neq 1$. As the phase margin decreases the system gets closer to the instability boundary, and therefore the phase margin gives an indication of how robust the system is.

In the case when the transfer system dynamics comprise a pure delay ($\Lambda(s) = e^{-s\tau}$) then the *critical delay*, τ_c, is the time delay which would give a *phase lag* of ϕ_m and is given by

$$\tau_c = \frac{\phi_m}{\omega_c}. \tag{4.16}$$

[11] Here $\angle L(i\omega_c)$ denotes the argument of the complex function $L(i\omega_c)$.

[12] In other words, ϕ_m can be thought of as the angle in the Nyquist plot required to reach the -1 point.

This gives an alternative method for computing the critical delay, in addition to Delay Differential Equation (DDE) models. Equation 4.16 can be used in many cases (even if only as a linear approximation) to give an estimate of τ_c.

Coupled pendulum-oscillator example

As an example, we return to the coupled pendulum-oscillator example which has been described above. For this system, the dynamics of the numerical model displacement are written using Laplace transform as

$$Z(s) = \frac{N(s)F_r(s)}{1 + P(s)N(s)T(s)}, \tag{4.17}$$

where $P(s) = m_s s^2 + C_2 s$, is the linear part of the pendulum dynamics, $N(s) = \dfrac{1}{M_1 s^2 + C_1 s + K}$ is the numerical model and $T(s) = e^{-s\tau}$ represents the delay due to the transfer system.

The transfer function between input $F_r(s)$ and output $Z(s)$ can now be studied as a typical closed loop system. The stability of the system can be examined via the characteristic equation given by

$$1 + P(s)N(s)T(s) = 0 \tag{4.18}$$

which corresponds to the complementary function given by equation 4.11 with $s = i\omega$. In this example $\Lambda(s) = \Lambda_r(s) = T(s)$, $L_0(s) = P(s)N(s)$ and $L(s) - P(s)N(s)T(s)$. By plotting the Bode diagram of $L_0(s)$ we can estimate the phase margin in the emulated system. As the Bode diagram uses decibels, $|L(i\omega_c)| = 1$ corresponds to the amplitude curve crossing the 0dB gain line. From this we can find the critical frequencies, ω_1 and ω_2. The corresponding phase margins are found, by measuring the difference between the phase curve and -180 degrees (i.e. $-\pi$) at each critical frequency. The phase margin relates directly to the maximum allowable delay (via $T(s)$) before the system becomes unstable. An example is shown in Figure 13, from which we see that, in this case, the maximum phase delay before stability loss is approximately 0.223 seconds. Using this approach for a particular set of parameter values the phase margin can be calculated and from this the maximum delay and its associated instability frequency can be found using the Bode diagram.

4.3 Robustness

For substructuring systems in the form developed in the preceding section we can apply more general, *robust stability* methods (see for exam-

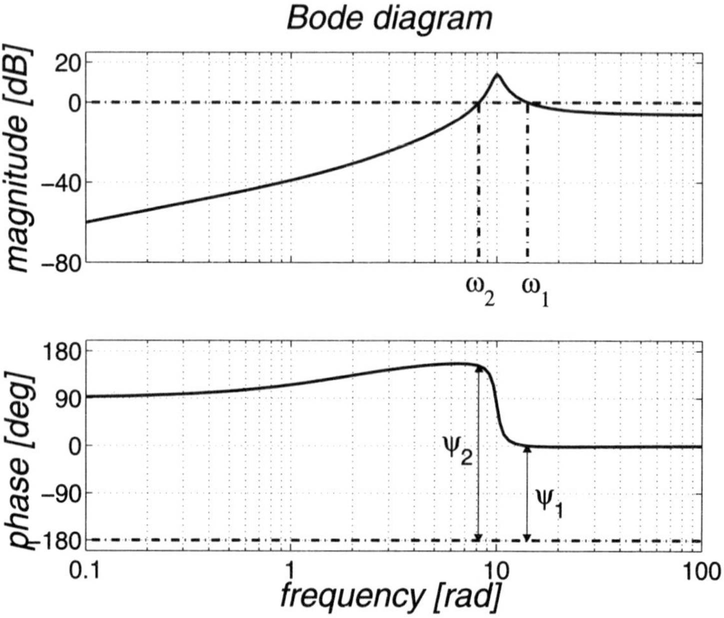

Figure 13. Phase margin calculation. The upper plot shows amplitude $|L_0|$ and the lower plot show phase $\angle L_0$. The phase margin is the difference in phase between the phase curve and -180 degree line at the point corresponding to the frequencies that give 0dB gain. Parameters values for the example: $M_1 = 20$kg, $C_1 = 20$kg/s, $K = 2000$N/m, $m = 10$kg, $C_2 = 20$kg/s. Phase margins: $\omega_1 = 14.1$rad/s, $\psi_1 = 180$degrees, $\omega_2 = 8.16$rad/s, $\psi_2 = 332.5$degrees. Critical delay values are $\tau_1 = \frac{\psi_1 \pi}{\omega_1 180} = 0.223$s, and $\tau_2 = \frac{\psi_2 \pi}{\omega_2 180} = 0.711$s. $\tau_1 < \tau_2$ in this case.

ple Goodwin et al. (2000); Preumont (2002)). We start with the assumption that both $L_0(s)$ and $\Lambda(s)$ are stable. We know that in the ideal case $\Lambda(s) = 1$, so we define the uncertainty in the system as the difference between the actual $\Lambda(s)$ value and the ideal value, so that the uncertainty function $\Delta(s) = \Lambda(s) - 1$. This effect can be visualised as an 'uncertainty circle' of radius $|L_0(s)||\Delta(s)|$ centred at any point on the Nyquist diagram for the nominal system $L_0(s)$. Then the key idea is that the distance to instability $|1 + L_0(s)|$ should always be greater than the uncertainty $|L_0(s)||\Delta(s)|$. Then the closed-loop system will be robustly stable if

$$|1 + L_0(i\omega)| \geq |L_0(i\omega)||\Delta(i\omega)| \quad \forall\omega, \tag{4.19}$$

or alternatively

$$|D_0(i\omega)|\,|\Delta(i\omega)| \leq 1 \quad \forall\omega. \tag{4.20}$$

Single-degree-of-freedom example

We now repeat the example shown schematically in Figure 8. The system has a numerical substructure consisting of a mass of $m$kg, a spring with stiffness kN/m and damper with constant cNs/m and a physical substructure consisting of a spring with stiffness k_sN/m. In this case,

$$L_0(s) = \frac{k_s}{ms^2 + cs + k}, \tag{4.21}$$

$$N_r(s) = cs + k. \tag{4.22}$$

Defining, as before, the natural frequency of the numerical subsystem as $\omega_n = \sqrt{\frac{k}{m}}$, the corresponding damping ratio $\zeta = \frac{c}{2m\omega_n}$ and $p = \frac{k_s}{k}$ we can write

$$L_0(s) = \frac{p\omega_n^2}{s^2 + 2\zeta\omega_n s + \omega_n^2}, \tag{4.23}$$

$$N_r(s) = m(2\zeta\omega_n s + \omega_n^2). \tag{4.24}$$

Defining $\hat{s} = \frac{s}{\omega_n}$, this can be rewritten in normalised form as

$$L_0(s) = \frac{p}{\hat{s}^2 + 2\zeta\hat{s} + 1}, \tag{4.25}$$

$$N_r(s) = k(2\zeta\hat{s} + 1). \tag{4.26}$$

Defining $\hat{\omega} = \frac{\omega}{\omega_n}$ and multiplying equation (4.25) by the complex conjugate $\bar{L}(i\omega)$ we can evaluate the critical frequency corresponding to the roots of $|L(i\omega_c)|^2 = 1$ as the solution of

$$(1 - \hat{\omega}^2)^2 + 4\zeta^2\hat{\omega}^2 - p^2 = \hat{\omega}^4 + (4\zeta^2 - 2)\hat{\omega}^2 + (1 - p^2) = 0. \tag{4.27}$$

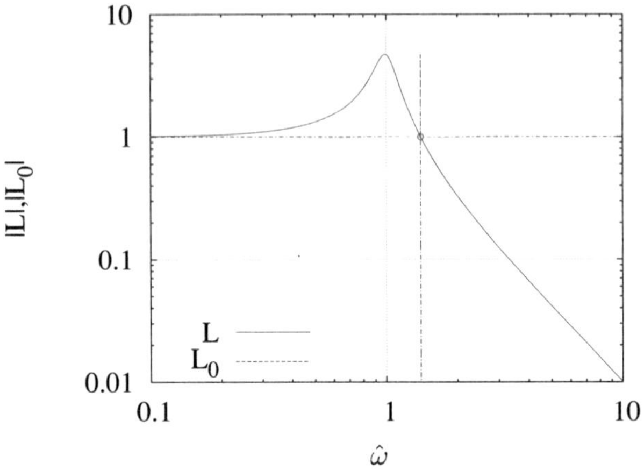

(a)

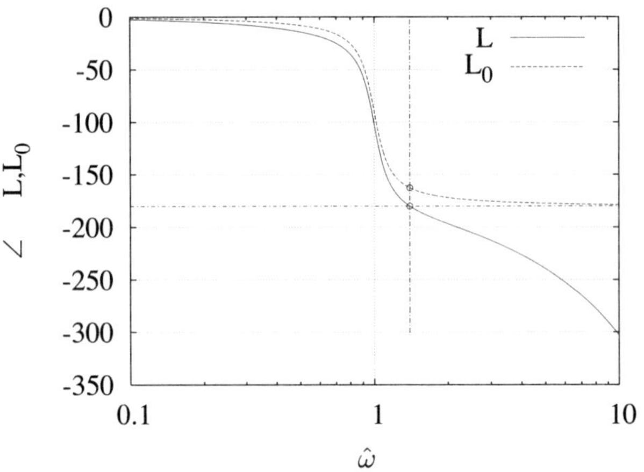

(b)

Figure 14. Robustness. (a) shows the log magnitude of the nominal $\log |L_0|$ and actual $\log |L|$ plotted against log normalised frequency $\log \hat{\omega}$; because there is only a phase error (pure delay), the curves are the same; the critical frequency $\hat{\omega}_c$ is marked by a vertical line and a unit gain by a horizontal line. (b) shows the corresponding phases together with a vertical line at the critical frequency $\hat{\omega}_c$ and a horizontal line at $-180°$. The phase margin is the vertical distance between $-180°$ and the corresponding phase curve $\angle L_0(i\hat{\omega})$ at $\hat{\omega} = \hat{\omega}_c$. In this case, the actual system, L, is on the boundary of stability.

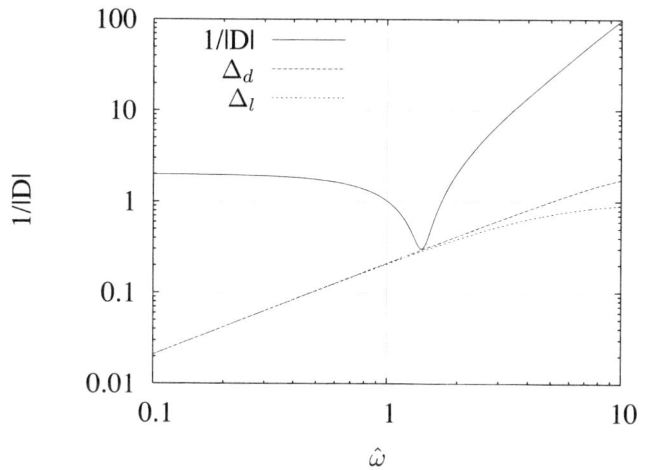

Figure 15. Robustness: This plot shows the inverse magnitude of $D(i\hat{\omega})$ against $i\hat{\omega}$ on a logarithmic scale. For comparison two possible uncertainty transfer functions $e^{-i\hat{\omega}\hat{\tau}}$ and $\frac{1}{1+i\hat{\omega}\hat{\tau}}$ are plotted for $\hat{\tau} = 0.21$.

Equation (4.27) is quadratic in $\hat{\omega}^2$. It has real solutions if

$$(4\zeta^2 - 2)^2 > (1 - p^2). \tag{4.28}$$

There are two cases: if $p \geq 1$ then Equation (4.28) is satisfied independent of ζ, otherwise the condition depends on the value of ζ.

Figure 14(a) shows $\log |L_0(i\hat{\omega})|$ plotted against $\hat{\omega}$ for $p = 1$ and $\zeta = 0.107$. In this case, $\hat{\omega}_c = 1.398$, corresponding to the frequency at which $|L_0(i\hat{\omega})| = 1$. Figure 14(b) gives the corresponding phase (in degrees) indicating a phase margin of $17.34° = 0.3027$rad (dashed line). Using equation 4.16, this gives a critical delay of $\hat{\tau}_c = 0.2165$, in this case. Also shown in Figures 14(a) and 14(b) are the cases for critical delay when $L = e^{-i\hat{\omega}\hat{\tau}} L_o$. As would be expected, the case for critical delay is on the stability boundary; $L(i\hat{\omega}_c) = -1$ corresponding to $|L(i\hat{\omega}_c)| = 1$ and $\angle L(i\hat{\omega}_c) = -180°$.

From equation (4.20), the system will have robust stability if $|\Delta(i\omega)| \leq \frac{1}{|D(i\hat{\omega})|}$. So, by plotting both $\frac{1}{|D(i\hat{\omega})|}$ and $|\Delta(i\omega)|$ on the same diagram we can assess the system robustness. For example, Figure 15 shows $\frac{1}{|D(i\hat{\omega})|}$ when $p = 1$ and $\zeta = 0.107$. On the same diagram, $|\Delta(i\omega)|$ is plotted for two cases:

$$\Lambda(s) = T(s) = \begin{cases} e^{-\hat{\tau}\hat{s}}, & \text{(shown as } \Delta_d \text{ in Figure 15)} \\ \frac{1}{1+\hat{\tau}\hat{s}}, & \text{(shown as } \Delta_l \text{ in Figure 15)} \end{cases} \tag{4.29}$$

where $\hat{\tau} = 0.21$. In this example, stability is predicted in for both Δ_d and Δ_l. However, the minimum of $\frac{1}{|D(i\hat{\omega})|}$ is at the stability boundary $|\Delta(i\omega)| = \frac{1}{|D(i\hat{\omega})|}$, so any increase in delay would lead to instability. It is worth noting that both forms of $\Lambda(s)$, delay and lag, of Equation (4.29) give similar results in this case indicating that phase error is more important that amplitude error.

The minimum value of $\frac{1}{|D(i\hat{\omega})|}$ occurs at $\hat{\omega}^2 \approx (1+p)$ with a value of approximately $\frac{2\zeta\sqrt{1+p}}{p}$. Noting that the maximum value of $\Delta = 1 - e^{-\hat{\tau}\hat{s}}$ is 2, it follows that the substructured system will be stable for *any* delay τ if

$$\zeta > \frac{p}{\sqrt{1+p}}. \tag{4.30}$$

Figure 9(b1) which shows the the boundary implied by Equation (4.30) for the DDE formulation.

5 Robust Transfer System Design

In this section we describe a methodology for the *design* of the transfer system to achieve robust stability. The use of linear theory — and particularly the assumptions that the experimental substructure and transfer systems are approximated by linear transfer functions are a limitation of this analysis. However, these results can be applied to substructuring tests with nonlinear elements. This is because a robust linear system can cope with a significant amount of nonlinear 'disturbance'.

Using the linear analysis discussed in the previous section, a suitable 4 stage controller design strategy for each transfer system (i.e. actuator) is:

1. Design an inner-loop controller around the actuator.
2. Use system identification to estimate the transfer function of the actuator plus inner-loop controller to give a *nominal* model and an *uncertainty* model for the transfer system.
3. Use the *nominal* model from step 2 to design an outer-loop *transfer system cancellation* controller.
4. Use the *uncertainty* model from step 2 to design a robustness compensator.

The adaptive nature of some of the outer-loop controllers proposed in the literature (see for example (Blakeborough et al., 2001; Wallace et al., 2004a; Neild et al., 2005)) allow for the compensation of the transfer system dynamics despite uncertainty in the nominal model derived in step 2. Although they incorporate some level of robustness due to this adaptation, they do not explicitly include the robustness compensator proposed in step 4. This is the key reason why the linear results can be applied so readily to systems where the experimental substructure is nonlinear. The adaptive nature of the outer-loop controller plus the robustness compensation allows the system to cope with a significant degree of nonlinear 'disturbance'.

It is worth emphasising that the analysis applies primarily to the design of the transfer system. The object is to design a stable robust control strategy to eliminate (or at least mitigate) the effect of uncertainty and non-linearity from the transfer system. In other words we would like to make the transfer system dynamics as close to linear as possible. The only time we actually require a definition of the transfer function of **phy** is in order to apply the robustness compensation technique based on physical model emulation, which is just one of three approaches described here. The three methods are:

1. Damping-ratio compensation: ζ-robustness
2. Phase-advance compensation: α-robustness
3. Physical model emulation: γ-robustness

All three approaches have a single parameter which provides a trade off between performance and robustness and are considered in the following sections. Methods 1 and 3 have been developed specifically with substructuring in mind; method 2 is a standard control system technique (Goodwin et al., 2000, section 6.6) applied to substructuring.

Damping-ratio compensation: ζ-robustness

From Figure 15, it is clear that it is the magnitude of the resonant peak of the closed-loop transfer function $|D_0(i\omega_c)|$ that restricts the maximum allowed value of uncertainty $|\Lambda(i\omega_c)|$. As $|D_0(i\omega_c)|$ decreases with increasing damping, a simple way of trading robustness for stability is to increase the damping coefficients of the numerical model above their correct values. This method has the advantage of requiring no knowledge about the system properties but has the disadvantage of distorting the nominal closed loop system.

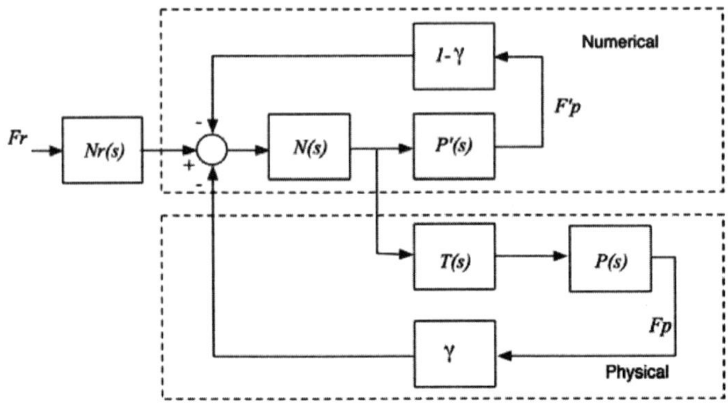

Figure 16. γ-robustness

Phase-advance compensation: α-robustness

This technique has already been described in section 3, and the phase advance is introduced via equation 3.22. This method has the advantage of requiring only knowledge of ω_c, but has the disadvantage of distorting the nominal closed loop system. The same effect can be obtained with forward prediction techniques, and an *overcompensation* method of this type is described in detail in Wallace et al. (2005a).

Physical model emulation: γ-robustness

In this approach the **phy** subsystem is numerically simulated in addition to the numerical simulation of the **num** subsystem. As indicated in block diagram form in Figure 16, the output of **num** is fed into both the simulation of **phy**, denoted $P'(s)$, and the physical subsystem **phy**. In comparison to Figure 1 (b), there are now *two* feedback loops: γ times the output of the physical system and $(1-\gamma)$ times $F_p^{'}$, which is the output of the simulated physical system. There are two limiting cases, $\gamma = 0$ gives a purely numerical simulation such that $T(s)$ is not part of the feedback , whereas $\gamma = 1$ gives the hybrid numerical/physical simulation discussed in the preceding sections — equation 4.14 with $\Lambda(s) = \Lambda_r(s) = T(s)$. When $0 < \gamma \leq 1$, the system can be represented by

$$F_p = \frac{L_O(s)}{1+\gamma\Lambda(s)L_O(s)}\Lambda(s)N_r(s)F_r - \frac{L_O(s)\Lambda(s)(1-\gamma)F_p^{'}}{1+\gamma\Lambda(s)L_O(s)}. \tag{5.1}$$

This method has the disadvantage of requiring an accurate model of the physical system transfer function $P(s)$ but has the advantage of *not* distorting the nominal closed loop system.

Example (continued)

Figure 17(a) is the same as Figure 15 except that the results of each of the three compensators are shown as well. In each case, the stability margin is significantly increased at the resonant frequency. For all three compensators the critical delay is $\hat{\tau} \approx 0.35$, which is more than 1.5 times larger than the uncompensated case.

Figure 17(b) shows the closed loop system when a system comprising of the nominal system with a pure time delay is compensated by the three robustness compensators. As expected, these three closed-loop responses differ slighlty from the nominal (D_0), but they offer a much better match than the uncompensated case which displays a large resonant peak indicating near-instability.

6 Example application: EH101 Lag damper

In this section we discuss substructuring results from laboratory tests of a lag damper for the EH101 helicopter manufactured by Agusta-Westland helicopters. Lag dampers are found on all fully-articulated helicopters, each one connected from the main rotor hub onto an individual blade. They perform a vital function with respect to the stability of the aircraft by controlling blade motion to avoid resonances — particularly on rotor start up, where the rotor frequency must pass through a resonant region before reaching operating frequency levels.

However, as a component of the rotor hub dynamic system, the dynamics of the damper influences the general vibration characteristics of the entire aircraft. This occurs because significant loads generated by the damper are transferred back into the airframe through the rotor hub at the damper attachment point. The fluid filled lag-dampers have highly nonlinear dynamic characteristics and the effect of this nonlinear behaviour on the combined rotor blade–lag-damper system is of significant interest in the design and manufacture of helicopters (Panda et al., 1996; Smith et al., 1996; Kunz, 1999; Hu and Wereley, 2005). Further details of the study and experimental setup up can be found in Wallace (2006).

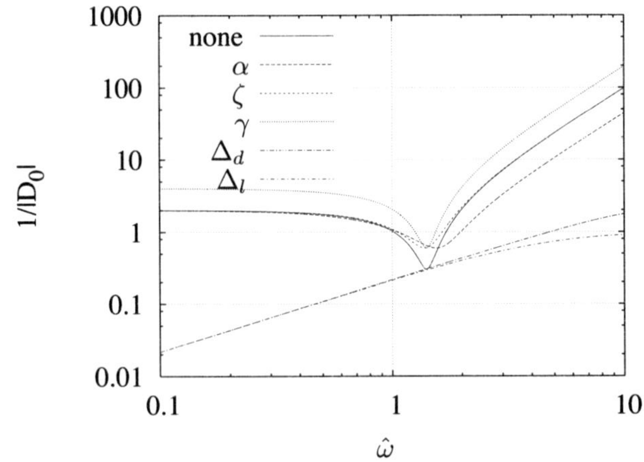

(a) Robustness analysis

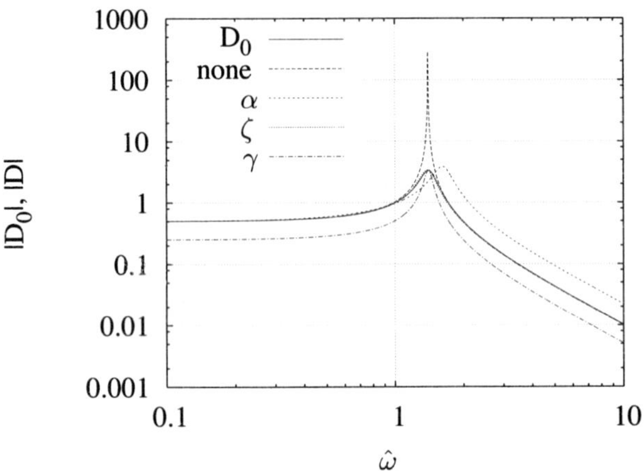

(b) Closed-loop response

Figure 17. Robustness Compensation. (a) The inverse magnitude of $D(i\hat{\omega})$ is plotted against $i\hat{\omega}$ on a logarithmic scale for the three robustness compensators with $\alpha = 1.5$, $\zeta_r = 2\zeta$ and $\gamma = 0.5$. For comparison two possible uncertainty transfer functions $e^{-i\hat{\omega}\hat{\tau}}$ and $\frac{1}{1+i\hat{\omega}\hat{\tau}}$ are plotted for $\hat{\tau} = 0.35$. (b) The lines marked α, ζ and γ give the corresponding closed-loop systems for each compensator, in the presence of $e^{-i\hat{\omega}\hat{\tau}}$. The case of no compensator, marked with none, is given for comparison

Figure 18. Close-up of the EH101 hub rotor system.

6.1 A lag damper model

Figure 18 shows the EH101 helicopter and detailed views of the lag damper in position between the helicopter rotor hub and rotor blade. The damping force is generated as fluid is forced through an orifice in the damper piston. The dampers also have relief (or blow-off) valves to keep the damper loads to an acceptable level. The force generated by the damper will act on the rotor blade in the opposite sense to the relative blade motion, damping the vibration of the blade in its lag degree of freedom i.e. parallel to the ground.

The vibration response of a single rotor blade can be adequately modelled using 8 modes represented by modal displacements ϕ_i where $i = 1 \ldots 8$. This low order modal model allows us to compute the blade response with minimal computational effort. The forcing effect of the damper is included on the right hand side of the forced response equation, thus giving the equation of motion for each mode as

$$\frac{1}{\Omega^2}\frac{d^2\phi_i}{dt^2} + \frac{2\upsilon_i^B\lambda_i^B}{\Omega}\frac{d\phi_i}{dt} + (\lambda_i^B)^2\phi_i = \frac{1}{I_i^B}(RF_i^{code} + LDF_i^{exp}), \qquad i = 1\ldots 8 \tag{6.1}$$

where λ_i^B is modal frequency, I_i^B modal inertia and υ_i^B modal damping of the blade with an angular velocity of Ω.

The 8 modal equations are forced by the terms RF_i^{code}, which represents the modal forcing from the main rotor and LDF_i^{exp} which is the experimentally measured lag damper force transformed to represent the effect of the lag damper force on each mode. This effective total modal forcing, which is part numerically defined (RF_i^{code}) and part experimentally measured (LDF_i^{exp}), drives the set of differential equations given by equation 6.1. The modal forcing provided by the damper is given for the ith mode as

$$LDF_i^{exp} = \frac{1}{\Omega^2}\sum_{j=1}^{7} T_i^{(j)}, \tag{6.2}$$

where the seven quantities $T_i^{(j)}$ are calculated for each mode using small angle approximations and constant vectors ω_{i_D}, $\bar{\upsilon}_{i_D}$ and t_{i_D} for flap, lag and twist — these are given in Wallace (2006).

Experimental testing set-up

Figure 19 shows the experimental test rig setup including the EH101 lag damper. The experimental test rig consists of two steel supports which are bolted directly into the "strong" floor of the laboratory with the central axis of the actuator and damper aligned. The base of the actuator is rigidly bolted into one support and then supported by a vertical stand. This stand has a height adjustment feature allowing for alignment and additionally ensures that the actuator does not vibrate during testing. A ±100kN Instron Dynacell dynamic load cell is then rigidly attached to the actuator piston. A yoke is required to connect the damper to the active part of the load cell.

As with the pendulum example discussed in section 4, the extra mass of the yoke then acts as if it were part of the experimental substructure component (the lag damper) which can distort the inertial response Kyrychko et al. (2005). To compensate for this effect, the load cell force reading would be altered such that $F_{cell} = F_{damper} + m_{yoke}a_{piston}$, where a denotes acceleration. The yoke can be seen in Figure 19 and is labelled as *added mass*. In this situation dynamic inertia compensation is essential to maintain a high level of accuracy during real-time dynamic substructure testing.

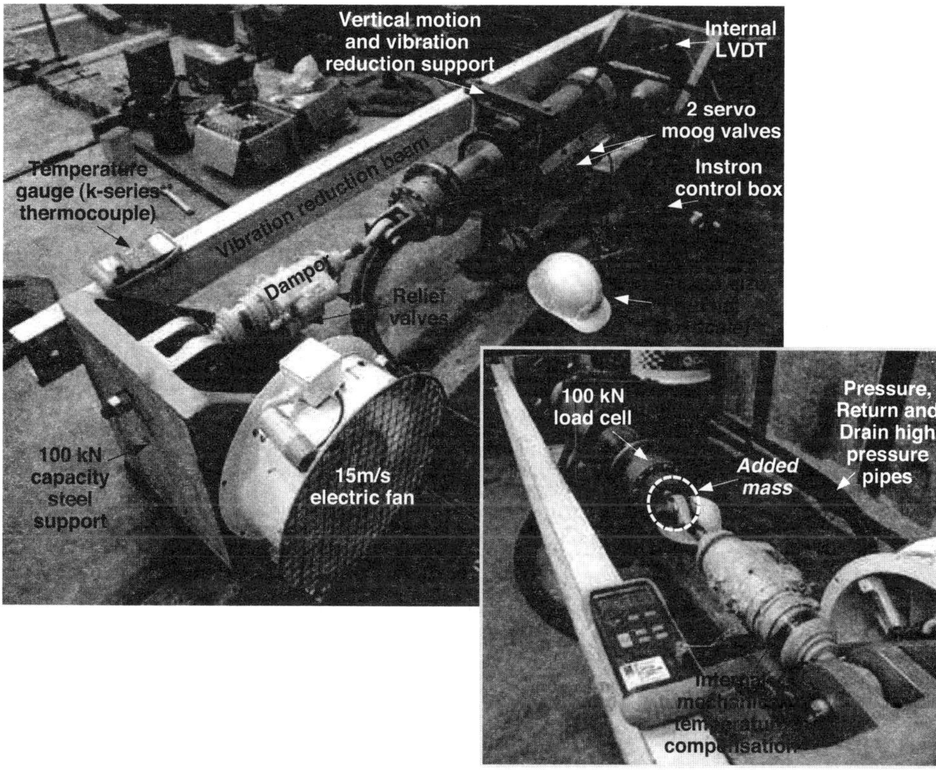

Figure 19. Experimental test rig setup for the EH101 lag damper; Note, the standard size “hard hat” for scale.

The control of the test rig is achieved using a P controller with adaptive forward prediction, as detailed in section 3. The control system consists of four constituent components:

1. *Control Hardware to drive the hydraulic actuator:* Two Instron 8800 digital servo-hydraulic controllers are used to drive the hydraulic actuator.
2. *Inner-loop PID controller:* A standalone PC is used for the *inner-loop* linear PID control of the transfer system (actuator).
3. *Outer-loop substructuring controller:* A second standalone PC is used for the substructuring algorithm and *outer-loop* control. The substructuring algorithm is designed in MATLAB/Simulink before being transferred to the hard real-time processor of the dSpace DSP.
4. *dSpace DSP:* The dSpace DS1103 R&D Controller Board is used to implement the substructuring algorithm experimentally in real-time. The substructuring algorithm is built into the processor and is connected to the Instron controller via an expansion board.

In practice, the inner-loop controller and Instron control tower are used to activate the system, achieve a high quality, repeatable response from the actuator (such that the transfer system has low uncertainty) and then to monitor the experimental signals during operation to ensure that everything is operating within the correct physical tolerances. The tests are then completely controlled from the outer-loop controller PC which simply changes the parameters values within the dSpace model, which in turn provides the demand signal to the activated inner-loop system in real-time.

Experimental real-time substructure test results

As an example of the type of numerical-experimental RTDS tests that were carried out the case of steady state flight at 84knots is considered. For this flight case the constants for equation 6.1 can be determined. The design of the test follows the four step process described in section 5:

Step 1: Proprietary control. In this step the low level (proprietary) controller which is part of the actuator hardware is tuned. The Instron 8800 control hardware contains a self-tuning algorithm, which was used to design a PID controller with gain values of $P = 32$ dB, $I = 1.2$ 1/s and $D = 0.8$ ms corresponding to an approximate damping ratio of 0.8.

Step 2: Transfer system identification. The resultant characteristic performance of the transfer system (actuator plus proprietary controller) was found to be highly repeatable with only low nonlinearity. A closed-loop transfer system identification was carried out using a sine sweep excitation (from 0-10Hz in 60s at ±5mm), from which the first order transfer function relating the experimental response of the actuator (x) to the sine sweep demand (r), was found to be

$$\frac{x(t)}{r(t)} \approx \frac{166.5}{\mathrm{s} + 169.3} = G_n(s). \quad (6.3)$$

This will be defined as the *nominal* model for the transfer function.

Step 3: Cancelling the transfer system dynamics In this step an outer-loop controller is designed which cancels the effect of the transfer system dynamics defined by equation 6.3. As equation 6.3 is a simple first order model, the most direct way to compensate for the actuator dynamics is to use a lag compensator which is exactly the inverse of equation 6.3. The displacement and velocity states can be taken directly from the numerical model of the blade so there are no issues relating to having an improper transfer function.

Step 4: Robustness. We cannot compute an explicit uncertainty model for the unmodelled dynamics (as done in Gawthrop et al. (2007)) in this case, but it is still important to consider the robustness of the substructuring process. As we are interested in steady-state vibration of the coupled blade–lag damper system, one of the most appropriate robustness techniques is that of a γ-compensator, described in section 5. The experimental test is initiated and run with $\gamma = 0$ while there is any transient behaviour. Then, using a linear transistion of $\gamma = 0 \rightarrow \gamma = 1$ in 5s, a fully numerical-experimental RTDS test is achieved.

6.2 Steady-state flight simulation

We now show results from RTDS tests for the steady state flight case of 84knots. Typical experimental results are shown in Figure 20 for one second of steady state system response when $\gamma = 1$. The numerical (idealized) force signal is shown in black whereas the actual force signal being fed back from the substructure is shown in grey — Figure 20(b).

It is clear from Figure 20 that the behaviour of the real damper is hysteretic as opposed to the piecewise linear idealized model. As a result, the experimental force signal provides more realistic information on the vibrational characteristics of the energy being transmitted back into the helicopter fuselage than the idealized model. This is because the substructuring tests contain the same modal frequency content as would be found from the same lag damper in service (i.e. on an actual helicopter) in steady-state flight at 84knots. This information can then be used to alter the characteristic dynamics of the lag damper by changing the tunable parameters

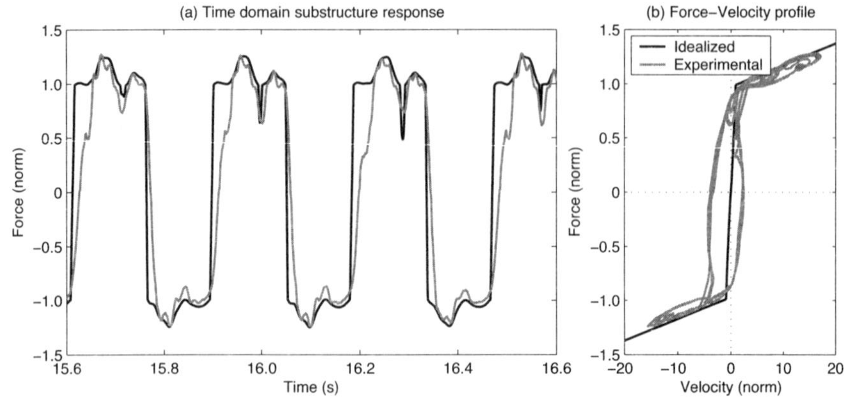

Figure 20. Force feedback during substructuring test.

(such as orifice size, bypass diameter, viscosity, relief valve arrangement and critical values etc.) to reduce damper loads at the critical harmonic frequencies.

7 Conclusions

In this chapter the control and implementation of real-time dynamic substructuring for mechanical and structural engineering applications has been discussed. The focus has been on small to medium scale applications, culminating with the example of an EH101 helicopter lag damper. A key difference with RTDS compared to HiL is that forces and displacment signals need to be transferred between the numerical model and physical substructure. As a result, the success of a RTDS test is critically dependant on the transfer system. To have confidence in the results from a RTDS test, accurate control of the transfer system is essential. Therefore, this chapter has focused on analysing the system from a dynamics and control perspective.

Methods for removing the effect of transfer system dynamics, using delay and amplitude compensation techniques have been discussed. For both HiL and pseudo-dynamic testing stability and robustness are issues which can be avoided, but for RTDS they are highly significant. In particular, for systems with low damping and high stiffness, only very small transfer system delays are necessary to destabilise the system. In such cases, delay compensation techniques are essential, just to stabilise the system.

Following on from the discussion on stability and robustness, a four stage robust transfer system design approach is described. Three methods of robustness compensation are discussed which can increase robustness at the expense of accuracy. Finally, the complete approach has been detailed for an industrial substructuring application — the EH101 helicopter lag damper.

7.1 Future research

One issue not discussed in this chapter is that of accuracy. Following early work on error propagation in pseudo-dynamic tests (Shing and Mahin, 1987), only a few other authors have considered this topic as a general issue (see Chang (2005); Combescure and Pegon (1997) and the references therein). Most studies centre on the improved accuracy of the numerical methods to compute **num**, (see for example Bursi and Shing (1996); Combescure and Pegon (1997); Chang et al. (1998); Algaard et al. (2001); Wu et al. (2006)). Accuracy of the *control* can be assessed by using synchronization subspace techniques as described by Wallace et al. (2005b). A combined analysis based on linear control and transfer functions has been considered by Jung and Shing (2006) and is mentioned briefly in this chapter.

The question which remains to be addressed is the relative importance of control accuracy compared to numerical integration accuracy. There is a related question of whether it is better to reduce the order of the numerical model, in order to have faster more efficient control. Clearly one can find examples where going faster is better and counter-examples where higher-order numerical models are desirable. The issue that has yet to be formalised is the exact trade-off between these cases so that the most appropriate hybrid simulation possible can be obtained.

Other areas of future work of interest to the development of substructuring are (i) robustness analysis applied to systems with strong nonlinearities, (ii) use of nonlinear control techniques, (iii) analysis of multiple transfer systems and systems with different and/or time varying critical delay values, (iv) investigation of alternative delay-compensation methods (for example, those based on emulators (Gawthrop et al., 1996)), and (v) causality in substructuring. The last topic is crucial in understanding (and designing) substructuring tests which will emulate the desired dynamics most closely.

Bibliography

A. K. Agrawal and J. N. Yang. Compensation of time-delay for control of civil engineering structures. *Earthquake Engineering & Structural Dynamics*, 29:37–62, 2000.

W. Algaard, A. Agar, and N. Bicanic. Enhanced integral form of the newmark time stepping scheme for pseudodynamic testing. *Engineering Computations*, 18(3-4):676–689, 2001.

M Bacic. Hardware-in-the-loop simulation: Taxonomic framework, survey and new research directions. University of Oxford Report No.2290/06, 2006.

V. Bayer, U. E. Dorka, U. Fullekrug, and J. Gschwilm. On real-time pseudo-dynamic sub-structure testing: algorithm, numerical and experimental results. *Aerospace Science & Technology*, 9(3):223–232, April 2005.

G. Benzoni. Challenges of new generation seismic testing facilities. *Experimental Techniques*, 25(2):20–23, 2001.

A. Blakeborough, M. S. Williams, A. P. Darby, and D. M. Williams. The development of real-time substructure testing. *Philosopical Transactions of the Royal Society of London A*, 359:1869–1891, 2001.

A. Bonelli and O.S. Bursi. Generalized-a methods for seismic structural testing. *Earthquake Engineering and Structural Dynamics*, 33(10):1067–1102, 2004.

P. A. Bonnet, C. N. Lim, M. S. Williams, A. Blakeborough, S. A. Neild, D. P. Stoten, and C. A. Taylor. Real-time hybrid experiments with newmark integration, mcsmd outer-loop control and multi-tasking strategies. *Earthquake Engineering & Structural Dynamics*, 36(1):119–141, January 2007.

T. Brendecke and F. Kucukay. Virtual real-time environment for automatic-transmission control units in the form of hardware-in-the-loop. *Internationa Journal of Vehicle Design*, 28(1-3):84–102, 2002.

O. S. Bursi, A. Gonzalez-Buelga, L. Vulcan, S. A. Neild, and D. J. Wagg. Novel coupling rosenbrock-based algorithms for real-time dynamic substructure testing. Earthquake Engineering and Structural Dynamics: In Press, 2007.

O.S. Bursi and P.B. Shing. Evaluation of some implicit time-stepping algorithms for pseudodynamic tests. *Earthquake Engineering and Structural Dynamics*, 25(4):333–355, 1996.

D. M. Bushnell. Scaling: Wind tunnel to flight. *Annual Review of Fluid Mechanics*, 38:111–128, 2006.

S. Y. Chang. Improved numerical dissipation for explicit methods in pseudodynamic tests. *Earthquake Engineering & Structural Dynamics*, 26(9): 917–929, September 1997.

S. Y. Chang. The gamma-function pseudodynamic algorithm. *Journal of Earthquake Engineering*, 4(3):303–320, July 2000.

S. Y. Chang. Explicit pseudodynamic algorithm with unconditional stability. *Journal of Engineering Mechanics ASCE*, 128(9):935–947, September 2002.

S. Y. Chang. Error propagation in implicit pseudodynamic testing of nonlinear systems. *Journal of Engineering Mechanics ASCE*, 131(12):1257–1269, December 2005.

S. Y. Chang, K. C. Tsai, and K. C. Chen. Improved time integration for pseudodynamic tests. *Earthquake Engineering & Structural Dynamics*, 27(7):711–730, July 1998.

S. Y. Chu, T. T. Soong, and A. M. Reinhorn. *Active, hybrid and semi-active structural control*. John Wiley:Chichester, England., 2005.

D. Combescure and P. Pegon. α-operator splitting time intergration technique for pseudodynamic testing error propergation analysis. *Soil Dynamics and Earthquake Engineering*, 16:427–443, 1997.

A. P. Darby, A. Blakeborough, and M. S. Williams. Real-time substructure tests using hydraulic actuator. *Journal of Engineering Mechanics*, 125 (10):1133–1139, 1999.

A. P. Darby, A. Blakeborough, and M. S. Williams. Improved control algorithm for real-time substructure testing. *Earthquake Engineering and Structural Dynamics*, 30:431–448, 2001.

A. P. Darby, M. S. Williams, and A. Blakeborough. Stability and delay compensation for real-time substructure testing. *ASCE Journal of Engineering Mechanics*, 128(12):1276–1284, 2002.

J. de Carufel, E. Martin, and J. C. Piedboeuf. Control strategies for hardware-in-the-loop simulation of flexible space robots. *IEE Proceedings-Control Theory and Applications*, 147(6):569–579, November 2000.

O. Diekmann, S. van Gils, S.M. Verduyn Lunel, and H.O. Walther. *Delay equations*, volume 110. Applied Mathematical Sciences, 1995.

J. Donea, P. Magonette, P. Negro, P. Pegon, A. Pinto, and G. Verzeletti. Pseudodynamic capabilities of the elsa laboratory for earthquake testing of large structures. *Earthquake Spectra*, 12(1):163–180, 1996.

S. Driscoll, J. D. Huggins, and W. J. Book. Electric motors coupled to hydraulic motors as actuators for hydraulic hardware-in-the-loop simulation. In *Proceedings of ASME International Mechanical Engineering*

Congress and Exposition, number IMECE2005-82124, Orlando, Florida USA, November 5-11 2005.

K. Engelborghs, T. Luzyanina, and D. Roose. Numerical bifurication analysis of delay differential equations using dde-biftool. *ACM Transactions on Mathematical Software*, 28(1):1–21, 2002.

P. T. Faithfull, R. J. Ball, and R. P. Jones. An investigation into the use of hardware-in-the-loop simulation with a scaled physical prototype as an aid to design. *Journal of Engineering Design*, 12(3):231–243, September 2001.

H. K. Fathy, R. Ahlawat, and J. L. Stein. Proper powertrain modeling for engine-in-the-loop simulation. In *Proceedings of the ASME International Engineering Congress and Exposition*, number IMECE2005-81592, 2006.

J. A. Ferreira, F. G. Almeida, M. R. Quintas, and J. P. E. de Oliveira. Hybrid models for hardware-in-the-loop simulation of hydraulic systems part 1: theory. *Proceedings of the Institution of Mechanical Engineers Part I-Journal of Systems and Control Engineering*, 218(I6):465–474, September 2004a.

J. A. Ferreira, F. G. Almeida, M. R. Quintas, and J. P. E. de Oliveira. Hybrid models for hardware-in-the-loop simulation of hydraulic systems part 2: experiments. *Proceedings of the Institution of Mechanical Engineers Part I-Journal of Systems and Control Engineering*, 218(I6):475–486, September 2004b.

A. Ganguli, A. Deraemaeker, M. Horodinca, and A. Preumont. Active damping of chatter in machine toolsdemonstration with a 'hardware-in-the-loop' simulator. *Proceedings of the Institution of Mechanical Engineers Part I-Journal of Systems and Control Engineering*, 219(I5):359–369, August 2005.

P. J. Gawthrop, Jones, R. W., and D. G. Sbarbaro. Emulator-based control and internal model control:complementary approaches to robust control design. *Automatica*, 32(8):1223–1227, August 1996.

P.J. Gawthrop, S.A. Neild, M.I. Wallace, and D.J. Wagg. Robust real-time substructuring techniques for lightly-damped systems. *Journal of Structural Control & Health Monitoring*, 14(4):591–608, 2007.

P.J. Gawthrop, M.I. Wallace, and D.J. Wagg. Bond-graph based substructuring of dynamical systems. *Earthquake Engineering and Structrural Dynamics.*, 34(6):687–703, 2005.

A. Gonzalez-Buelga, D. J. Wagg, and S. A. Neild. Parametric variation of a coupled pendulum-oscillator system using real-time dynamic substructuring. Structural Control and Health Monitoring: In Press, 2007.

G. C. Goodwin, S. F. Graebe, and M. E. Salgado. *Control System Design.* Pearson, 2000.

K. S. Hong, H. C. Sohn, and J. K. Hedrick. Modified skyhook control of semi-active suspensions: A new model, gain scheduling, and hardware-in-the-loop tuning. *Journal of Dynamic Systems Measurement and Control-Transactions of the ASME*, 124(1):158–167, March 2002.

T. Horiuchi, M. Inoue, T. Konno, and Y. Namita. Real-time hybrid experimental system with actuator delay compensation and its application to a piping system with energy absorber. *Earthquake Engineering and Structural Dynamics*, 28:1121–1141, 1999.

W. Hu and N. M. Wereley. Magnetorheological fluid and elastomeric lag damper for helicopter stability augmentation. *International Journal of Modern Physics B*, 19(7-9):1471–1477, April 2005.

S. Jezernik. Hardware-in-the-loop simulation and analysis of magnetic recording of nerve activity. *Journal of Neuroscience Methods*, 142(2): 295–304, March 2005.

R. Y. Jung and P. B. Shing. Performance evaluation of a real-time pseudo-dynamic test system. *Earthquake Engineering & Structural Dynamics*, 35(7):789–810, June 2006.

D. L. Kunz. Elastomer modelling for use in predicting helicopter lag damper behavior. *Journal of Sound & Vibration*, 226(3):585–594, September 1999.

Y. N. Kyrychko, K. B. Blyuss, A. Gonzalez Buelga, S. J. Hogan, and D.J. Wagg. Real-time dynamic substructuring in a coupled oscillator-pendulum system. *Proceedings of the Royal Society A*, 426(2068):1271–1294, 2005.

P. Lambrechts, M. Boerlage, and M. Steinbuch. Trajectory planning and feedforward design for electromechanical motion systems. *Control Engineering Practice*, 13(2):145–157, February 2005.

C.M. Lim, S.A. Neild, D.P. Stoten, C.A. Taylor, and D. Drury. Using adaptive control for dynamic substructuring tests. *Proc. of the 3rd European Conf. on Structural Control*, 2004.

Yu-Yuan Lin, Kuo-Chun Chang, and Yuan-Li Wang. Comparison of displacement coefficient method and capacity spectrum method with experimental results of rc columns. *Earthquake Engineering & Structural Dynamics*, 33(1):35–48, 2004.

D. Maclay. Simulation gets into the loop. *IEE Review*, 43(3):109–112, May 1997.

S. P. Mansoor, D. I. Jones, D. A. Bradley, F. C. Aris, and G. R. Jones. Hardware-in-the-loop simulation of a pumped storage hydro station. *International Jornal of Power and Energy Systems*, 23(2):127–133, 2003.

S. McGreevy, T. T. Soong, and A. M. Reinhorn. An experimental study of time delay compensation in active structural control. In *Proceedings of the 6th International Modal Analysis Conference-IMAC*, volume 1, pages 733–739, 1988.

W. E. Misselhorn, N. J. Theron, and P. S. Els. Investigation of hardware-in-the-loop for use in suspension development. *Vehicle System Dynamics*, 44(1):65–81, January 2006.

M. Nakashima. Development, potential, and limitations of real-time on-line (pseudo dynamic) testing. *Philosophical Transactions of the Royal Society A*, 359(1786):1851–1867, 2001.

M. Nakashima, H. Kato, and E. Takaoka. Development of real-time pseudo dynamic testing. *Earthquake Engineering and Structural Dynamics*, 21: 779–92, 1992.

M. Nakashima and N. Masaoka. Real-time on-line test for mdof systems. *Earthquake Engineering and Structural Dynamics*, 28:393–420, 1999.

S. A. Neild, D. P. Stoten, D. Drury, and D. J. Wagg. Control issues relating to real-time substructuring experiments using a shaking table. *Earthquake Engineering and Structural Dynamics*, 34:1171–1192, 2005.

P. Pan, M. Nakashima, and H. Tomofuji. Online test using displacement-force mixed control. *Earthquake Engineering & Structural Dynamics*, 34 (8):869–888, July 2005.

B. Panda, E. Mychalowycz, and F. J. Tarzanin. Application of passive dampers to modern helicopters. *Smart Materials & Structures*, 5(5): 509–516, October 1996.

A.V. Pinto, P. Pegon, G. Magonette, and G. Tsionis. Pseudo-dynamic testing of bridges using non-linear substructuring. *Earthquake Engineering and Structural Dynamics*, 33:1125–1146, 2004.

A. R. Plummer. Model-in-the-loop testing. *Proc IMechE Part I-Journal of Systems and Control Engineering*, 220(I3):183–199, May 2006.

A. Preumont. *Vibration control of active structures*. Kluwer: Dordrecht, 2002.

J. S. Prezemieniecki. *Theory of matrix structural analysis*. McGraw-Hill, 1968.

W. Rulka and E. Pankiewicz. Mbs approach to generate equations of motions for hil-simulations in vehicle dynamics. *Multibody system dynamics*, 14(3-4):367–386, November 2005.

R. T. Severn. Ecoest—european consortium of earthquake shaking tables—overview. In *10th European Conference on Earthquake Engineering*, pages 2987–2992. Balkema:Rotterdam, 1995.

P-S. Shing and S. A. Mahin. Cumulative experimental errors in pseudodynamic tests. *Earthquake Engineering and Structural Dynamics*, 15: 409–424, 1987.

P-S. Shing, M. T. Vannan, and E. Cater. Implicit time intergration for pseudodynamic tests. *Earthquake Engineering and Structural Dynamics*, 20:551–576, 1991.

M.V. Sivaselvan, A. Reinhorn, Z. Liang, and X. Shao. Real-time dynamic hybrid testing of structural systems. *Proc. 13th World Conf. Earthquake Engineering*, 2004.

E. C. Smith, K. Govindswamy, M. R. Beale, and G. A. Lesieutre. Formulation, validation, and application of a finite element model for elastomeric lag dampers. *Journal of the American Helicopter Society*, 41(3):247–256, July 1996.

T. T. Soong. *Active Structural Control: Theory and Practice.* Longman, London and Wiley, New York, 1990.

G. Stépan. *Retarded Dynamical Systems: Stability and Characteristic Functions.* Longman Scientific & Technical, 1989.

C. Thewalt and M. Roman. Performance parameters for pseudodynamic tests. *Journal of Structural Engineering*, 120(9):2768–2781, 1994.

D. J. Wagg and D. P. Stoten. Substructuring of dynamical systems via the adaptive minimal control synthesis algorithm. *Earthquake Engineering and Structural Dynamics*, 30:865–877, 2001.

M. Wallace, D.J. Wagg, and S.A. Neild. A stability analysis of substructured systems for real-time dynamic substructuring. In *Proceedings of the 1st International Conference on Advances in Experimental Structural Engineering*, volume 1, pages 385–392, 2004a.

M. I. Wallace. *Real-time dynamic substructuring for mechanical and aerospace applications: control techniques and experimental methods.* PhD thesis, University of Bristol, 2006.

M. I. Wallace, D. J. Wagg, and S. A. Neild. Use of control techniques for error analysis of real time dynamic substructure testing. In *Proceedings of the Third European Conference on Structural Control*, volume 1, pages 59–62, 2004b.

M. I. Wallace, D. J. Wagg, S. A. Neild, P. C. Bunnis, N. A. Lieven, and A. J. Crewe. Testing coupled rotor blade-lag damper vibration dynamics using real-time dynamic substructuring. Journal of Sound and Vibration: In Press, 2007.

M.I. Wallace, J. Sieber, S. A. Neild, D.J. Wagg, and B. Krauskopf. Stability analysis of real-time dynamic substructuring using delay differential equation models. *International Journal of Earthquake Engineering and Structural Dynamics*, 34(15):1817–1832, 2005a.

M.I. Wallace, D.J. Wagg, and S. A. Neild. An adaptive polynomial based forward prediction algorithm for multi-actuator real-time dynamic substructuring. *Proceedings of the Royal Society of London A.*, 461(2064): 3807–3826, 2005b.

T. Wang, M. Nakashima, and P. Pan. On-line hybrid test combining with general-purpose finite element software. *Earthquake Engineering & Structural Dynamics*, 35(12):1471–1488, October 2006.

B. Wu, G. Xu, Q. Wang, and M. S. Williams. Operator-splitting method for real-time substructure testing. *Earthquake Engineering and Structural Dynamics*, 35:293–314, 2006.

R. Zhang and A. G. Alleyne. Dynamic emulation using an indirect control input. *Journal of Dynamic Systems Measurement and Control - Transactions of the ASME*, 127(1):114–124, March 2005.

W. D. Zhu, S. Pekarek, J. Jatskevich, O. Wasynczuk, and D. Delisle. A model-in-the-loop interface to emulate source dynamics in a zonal dc distribution system. *IEEE Transactions on Power Electronics*, 20(2): 438–445, March 2005.